Galbraith's Construction and Land Management Law for Students

Eighth Edition

Edited by Carrie de Silva

With contributions from

Mark Allison

Jennifer Charlson

Michael Fay

Stephen Hardy

Laura Healy

Emma Pierce-Jenkins

Mark Simcock

Routledge
Taylor & Francis Group
LONDON AND NEW YORK

Eighth edition published 2026

by Routledge
4 Park Square, Milton Park, Abingdon, Oxon OX14 4RN

and by Routledge
605 Third Avenue, New York, NY 10158

Routledge is an imprint of the Taylor & Francis Group, an informa business

First edition published by Butterworth-Heinemann 1989
Seventh edition published by Routledge 2021

British Library Cataloguing-in-Publication Data
A catalogue record for this book is available from the British Library

Library of Congress Cataloging-in-Publication Data
Names: De Silva, Carrie editor | Allison, Mark D. (Lawyer)
Title: Galbraith's construction and land management law for students / edited by Carrie de Silva ; with contributions from Mark Allison, Jennifer Charlson, Michael Fay, Stephen Hardy, Laura Healy, Emma Pierce-Jenkins, Mark Simcock.
Other titles: Construction and land management law for students
Description: Eighth Edition. | Abingdon, Oxon [UK] ; New York, NY : Routledge, 2026. | Includes bibliographical references and index.
Identifiers: LCCN 2025040177 (print) | LCCN 2025040178 (ebook) | ISBN 9781041095965 paperback | ISBN 9781041108498 hardback | ISBN 9781003657101 ebook
Subjects: LCSH: Building laws—Great Britain | Construction industry—Law and legislation—Great Britain | LCGFT: Textbooks
Classification: LCC KD1140 .G355 2026 (print) | LCC KD1140 (ebook) | DDC 343.4107/869—dc23/eng/20250828
LC record available at https://lccn.loc.gov/2025040177
LC ebook record available at https://lccn.loc.gov/2025040178

ISBN: 978-1-041-10849-8 (hbk)
ISBN: 978-1-041-09596-5 (pbk)
ISBN: 978-1-003-65710-1 (ebk)

DOI: 10.1201/9781003657101

Typeset in Optima
by Apex CoVantage, LLC

Contents

Statutes

Regulations, Rules and Orders

Treaties

Cases

Courts and other forums are given in abbreviated form after the date. Abbreviations are as follows:

Admin	Administrative Court (within Kings Bench Division)
CA	Court of Appeal
CACh	Court of Appeal in Chancery*
CCP	Court of Common Pleas*
CEx	Court of Exchequer*
CExC	Court of Exchequer Chamber*
Ch	Chancery Division, High Court
CC	County Court (specific court given to aid access to report)
CrC	Crown Court (specific court given to aid access to report)
ECHR	European Court of Human Rights
ECHR GC	European Court of Human Rights, Grand Chamber
EAT	Employment Appeal Tribunal
ET	Employment Tribunal
ExCh	Exchequer Chamber
ExD	Exchequer Division, High Court**
HL	House of Lords (function transferred to Supreme Court on 1 October 2009)
IT	Industrial Tribunal
KB	Court of King's Bench*
KBD	King's Bench Division, High Court
NYCA	New York Court of Appeals
PD	Privy Council
QB	Queen's Bench*
QBD	Queen's Bench Division, High Court
SC	Supreme Court
TCC	Technology and Construction Court (within King's Bench Division)
UTLC	Upper Tribunal Lands Chamber

* These courts were abolished with the significant courts reorganisation under the Judicature Acts 1873–1875.
** The Exchequer Division was operative from the 1873–1875 changes to 1881.

A v UK (1998) ECHR 15
A and Others v UK (2009) ECHR GC 18
Adams v Lindsell (1817) KB 83
AG Securities v Vaughan and Others (1990) HL 276
AG v PYA Quarries (1957) CA 204–205
Alcock v Chief Constable of South Yorkshire Police (1991) HL 221
Aldred's Case (1610) KB 268
Devine Children's Cancer Trust v Housing Solutions Ltd (2020) SC 273
Allen v Gulf Oil Refining Ltd (1981) HL 209
Aluminium Industrie Vaassen BV v Romalpa Aluminium (1976) CA 140
Amalgamated Investment Co. Ltd v John Walker & Sons Ltd (1976) CA 116
AMF International Ltd v Magnet Bowling Ltd (1968) QBD 230
Ammah v Kuehne & Nagel Logistics Ltd (2009) CA 253
Anchor Brewhouse Developments Ltd v Berkley House Ltd (1987) Ch 204
Andreae v Selfridge and Co. Ltd (1938) CA 205
Andrews v Schooling (1991) CA 235
Annamunthudo v Oilfields Workers Union (1961) PC 62
Anns v Merton London Borough Council (1977) HL 222, 234
Anton Piller KG v Manufacturing Processes Limited (1975) CA 127
Applegate v Moss (1971) CA 126
Appleton v Garrett (1996) QBD 203
Arcadis Consulting (UK) Ltd v AMEC BCS Ltd (2018) CA 160
Armstrong v Armstrong (2024) Ch 286
Arnold v Britton & others (2015) SC 97
Ashburn Anstalt v WJ Arnold & Co (1989) CA 289
Arthur J S Hall & Co. v Simons (2000) HL 219
ASLEF v UK (2007) ECHR 193
Associated Dairies v Hartley (1979) ET 245
Atlantic Baron, The (1979) QBD 88
Autoclenz Ltd v Belcher (2011) SC 166
Ayres v Odedra (2013) QBD 229

Bailey v Stephens (1862) CCP 268
Balfour v Balfour (1919) CA 92
Balfour v Barty-King (1957) CA 240
Bank of New York Mellon (International) Ltd v Cine-UK Ltd (2021) QBD 117
Bank of Scotland plc v Burnley Borough Council (2024) UKUT 273
Barber v RJB Mining (UK) Ltd (1999) QBD 171
Barnard v National Dock Labour Board (1953) CA 63
Barnett v Chelsea and Kensington Hospital Management Committee (1968) QBD 226
Batchelor v Marlow (2001) CA 268
BBC v Hearn (1977) CA 197
Beard v London General Omnibus Company (1900) CA 239
Bellman v Northampton Recruitment Ltd (2018) CA 239
Benedetti v Sawiris (2013) SC 126
Benn v Hardinge (1992) CA 270
Bernstein v Skyviews and General Ltd (1978) QBD 203, 262
Beswick v Beswick (1968) QBD 121
Bissett v Wilkinson (1927) PC 102
Blyth v Birmingham Waterworks (1856) Exch 224

Contributors

Mark Allison, Coppergate Training

Jennifer Charlson, National Infrastructure and Service Transformation Authority

Michael Fay, Harper Adams University

Stephen Hardy, University of Hull

Laura Healy, University of Law

Emma Pierce-Jenkins, Harper Adams University

Mark Simcock, Harper Adams University

Preface to the eighth edition

The eighth edition of *Galbraith's Construction and Land Management Law for Students* acknowledges the work of Anne Galbraith and the earlier authors and editors and has been updated and extended by a team of subject specialists to reflect the broad scope of the topics covered.

The new edition has provided the editors with an opportunity to revise the existing chapters of the book and to add further material. Examples of significant new developments that are dealt with include reference to the Building Safety Act 2022, the considerable changes in employment law, particularly with regard to equality and diversity, and provisions that are striking, the ongoing developments in residential tenancy reform and developments in planning. There is, of course, much else covered in each chapter.

The law is stated as the editors consider it to be as of 31 October 2025.

Carrie de Silva
Mark Allison
Jennifer Charlson
Michael Fay
Stephen Hardy
Laura Healy
Emma Pierce-Jenkins
Mark Simcock

Note: Unless otherwise stated, the law in this textbook applies to England. Those in Scotland, Wales and Northern Ireland should check for differences which may be applicable in their regions. The contents are not to be interpreted as legal advice.

Case citation: Note that, for simplicity, all case dates have been shown with round brackets (). You will see in other sources that square brackets [] are sometimes used. This is where the date is required to find the correct law report, i.e. where reports are by year rather than by sequential volume. With the ease of internet searching, no citations have been given, but when citing cases formally, for example, in academic or professional writing, you will need to ascertain the full legal citation referencing the precise law report for any particular case. Full details of the citation of cases, and all other legal materials, may be found in OSCOLA, *The Oxford University Standard for the Citation of Legal Authorities* (2012).

1 The nature of law

Carrie de Silva

Introduction

Any society quickly finds that it needs rules or laws to enable it to function smoothly. Consider the results if you had decided to drive to work today on the right instead of the left. People need to be able to live and go about their business with certainty, knowing that they can expect others to abide by the same rules. It is probably true that most people obey most of the rules most of the time. They may not have any precise knowledge of the actual rules which they are obeying, but that precise knowledge only becomes important when disputes arise or problems have to be solved.

As society evolves and becomes more complex, the number of laws increases and the types of laws become more sophisticated. Initially people need laws to protect their person or their property. Some rules accord with basic ideas of morality, e.g. not killing or not stealing, but it is not essential for all the rules to have a moral base. For instance, when we create a rule for motorists that they must drive on the left, the rule commands obedience simply because it is necessary and certain, not morally superior to driving on the right.

Although certainty is an important principle in any legal system, the rules of law should also aim to achieve justice. This is sometimes accomplished by giving a discretion. Some areas of English law are subject to discretionary principles known as equity. It should be remembered that laws are equally binding even if they are thought to be unjust. The remedy then is to seek to reform the law, not to violate it.

Sanctions

If a law is disobeyed, there must be some effective means by which it can be enforced. This may be described as a sanction and could take the form of a fine, imprisonment, an award of damages or an injunction, which is an order of the court forbidding certain actions or behaviour. Although such sanctions may deter people from breaking the law, in fact, the reasons for obeying the law may be far more complex. A person may obey the law because he or she fears the disapproval or hostility of others, because he or she believes in the 'rightness' of the rule or because he or she sees obedience to the law as a duty owed as a citizen.

Divisions of the law

It should be noted that all the contents of this book relate to English law. Although it is appropriate to talk about the British Constitution, it should be remembered that Scotland

DOI: 10.1201/9781003657101-1

has a different system of law from England. Whereas some of the rules are common to both countries, there are still significant differences. These are most apparent in areas like the system of land holding and in the structures and procedures of the courts. Although lacking the underlying and fundamental difference in the development of certain Scottish law, there is also a growing body of law applicable only in Wales.

English law may be subdivided in a number of ways. A simple division would be into:

- Criminal law
- Civil law

Criminal law

It is quite common for the layperson to believe that the bulk of our law consists of criminal law. This inevitably results from the publicity given in the media to major criminal cases. In this sense the law is like an iceberg, the criminal law is the part seen above the water and the civil law is the mighty bulk that lies below. The state has an interest in preserving our society and upholding law and order, and it makes criminal laws to secure those objectives. A person who infringes those laws commits a criminal offence for which he or she can be prosecuted by the state. The principal objectives of the criminal law may, therefore, be seen to be to deter and to punish, although there are additional aims of criminal sanctions, such as the rehabilitation of offenders to help ensure they do not re-offend.

Civil law

The civil law protects rights and creates obligations between individuals, although sometimes those individuals will be large public corporations, government departments, local authorities, etc.

There is no simple way to classify civil law, but it may be subdivided into:

- the law of contract,
- the law of tort,
- the law relating to property and
- the law relating to persons.

The rules of civil law must deal with matters as diverse as contracts for the sale of goods, actions for damages for negligence, planning applications and compulsory purchase, divorce, tenancy, creation of companies, contracts of insurance, making a will, money lending, recovery of debts, defective workmanship and unfair dismissal.

Language of the law

Every profession has its language or jargon, which is often a convenient shorthand method of communication between people engaged in that profession. With its long history, the language of the law is especially rich as it frequently uses Latin expressions such as *res ipsa loquitur* (the facts speak for themselves) or *caveat emptor* (let the buyer beware). Different terminology applies to criminal and civil law. In criminal proceedings, a person is arrested, charged with an offence and prosecuted in summary or indictable proceedings by a prosecutor. If the accused person is found guilty, he or she will be sentenced and punished. If found not guilty, he or she will be acquitted. In a civil action a claimant (formerly known as a plaintiff) sues a defendant. If the defendant is found to be liable, he or she may be ordered to pay damages or an order of the court, such as an injunction, may be made against him or her. Where the terminology is correctly used, it is possible to tell whether a case concerns criminal or civil law.

Although it is convenient to break down the mass of English law into subdivisions for the purposes of study, it should be realised that these various divisions of the law are not mutually exclusive. It is possible for one set of facts to give rise to both civil and criminal liability. The simplest example is the motorist who drives carelessly and injures a pedestrian. A motorist may both be prosecuted in criminal proceedings for an offence under the Road Traffic Acts and be sued in civil proceedings by the injured pedestrian hoping to recover compensation. Similarly, an employee injured at work may wish to bring civil proceedings against his or her employer to recover damages where the employer has been negligent in caring for the employee's safety. Arising out of the same set of facts, the Health and Safety Executive may prosecute the employer for breaches of the Health and Safety at Work etc. Act 1974.

Making the law

Unless some catastrophic event like a war or a revolution occurs, which may cause a country to adopt a completely new system of law, the law-making processes will have developed over a period of centuries. Many of the rules of English law which are still in force are of considerable antiquity. Laws do not become ineffective merely because they are very old. Indeed, some old rules have been used very imaginatively by the judges in the courts to create principles relevant to modern life. The oldest provisions still on the statute books at the time of writing are parts of the Statute of Marlborough, sealed by Henry III in 1267.

In the early stages of law-making, custom will usually play a large part. As the needs of a society become more sophisticated, custom as a source of law tends to be superseded by more formal sources. Local custom may still be upheld as a valid part of the law where it can be shown that the custom is reasonable and certain and has been continued without interruption since time immemorial. For practical purposes, this normally means that the custom must be shown to have existed during living memory. Occasional examples still come before the courts, for example, *Egerton v Harding* (1975) CA.

Sources of law

The two main sources of law in our system today are **legislation** (laws made or approved by Parliament) and **judicial precedent** (binding decisions of judges). Between 1 January 1973 and 31 January 2020, both of these sources of law had to account for the fact that the United Kingdom (UK) was a member of the European Union. Note, also, the impact of the Human Rights Act 1998 (HRA) (see later in this chapter).

New legislation

Law-making authority is vested in the Crown and Parliament, although the role of the Crown is now almost entirely formal. Our complex modern society requires rule-making techniques which are capable of coping with economic, social and welfare problems.

Parliament, consisting of the House of Lords and the House of Commons, is said to be supreme, but in practice its ability to make rules must be viewed subject to a number of limitations. By virtue of the European Communities Act 1972, a new element of European Community law was introduced into our system. European Community law took precedence over the national law of any member state. This, therefore, imposed a limit on the ability of Parliament to make whatever laws it wished. If English law was found to conflict with European Community law, the Community law prevailed. The final transition period ended with the Retained EU Law (Revocation and Reform) Act 2023, operative from 1 January 2024, although for commercial and policy reasons, European legislation is likely to inform and impact some elements of national legislation in future, although legal sovereignty is within the UK.

The UK is a signatory to the European Convention on Human Rights and is obliged by international law to ensure that English law does not contravene the Convention. This is separate from European Union membership. Under the HRA, the main provisions of the European Convention on Human Rights became part of UK law and thus enforceable through the English courts. Other factors that would limit the power of Parliament are the existence of an opposition party, or parties, whose duty is to seek to curb or limit government proposals; Parliamentary question time which is held daily, when ministers, including the Prime Minister, must justify their activities; the two-chamber system whereby the House of Lords can at least delay the passing of legislation and public opinion, pressure groups, freedom of speech and publicity through the media. Once laws are enacted by Parliament, the interpretation of those laws is carried out by the judges in the courts. If Parliament has passed a law which is seen to be too harsh or too extensive in its application, it may be possible for the judges, by a restrictive interpretation of the words of the statute, to limit the scope of the new rule.

Acts of Parliament, which can also be called statutes, contain the main laws made by Parliament, acting in its legislative role. Until the statute or Act has passed through all its stages in both Houses and received the Royal Assent, it is referred to as a bill.

Development of new law

The inspiration for new legislation may come from a number of sources. A new government will have made manifesto commitments and will have outlined its own policies. During its first year or so in office, it will be keen to push through those changes which were outlined in its election manifesto. Inevitably, this source of new legislation becomes comparatively less significant as the government's term of office, a maximum of five years (under the Fixed-term Parliaments Act 2011), progresses. Each of the major government departments will have a programme of legislation, which it would like to introduce. Parliamentary time is at a premium, and this may well produce competition between departments. Where a small measure is needed, a department may seek to persuade a private member who has won a high place in the private members' ballot to introduce a bill on its behalf.

New laws may also be needed to implement recommendations of the Law Commission, a body established in 1965 to review the law with a view to its systematic development and reform. The Law Commission is an active body, and their website should be regularly reviewed for activity.

Occasionally the government of the day sets up Royal Commissions or other special committees of enquiry. These are usually established to investigate one specific topic area, and they will be disbanded once their reports are published. Many of these reports gather dust and do not result in their recommendations being implemented in legislation.

In other cases, the recommendations may be implemented. Examples include the Robens Report on Health and Safety, which very quickly became the Health and Safety at Work etc. Act 1974 and the Royal Commission on Criminal Justice, appointed in the wake of miscarriages of justice such as the 'Birmingham Six', which resulted in changes to the law (see the Criminal Appeal Act 1995). Royal Commissions are now rare, with only three since 1980, the last being the Royal Commission on the Reform of the House of Lords in 1999, led by Lord Wakeham.

There is a limited scope for individual politicians to introduce private members' bills, by which means changes may be introduced into our law. Although sometimes, matters with wider government support are introduced through the mechanisms of a private members' bill, notably where the issue may be controversial, for example, the Abortion Act 1967 introduced by Liberal MP David Steel.

Occasionally new laws will be required to meet a sudden emergency or to meet unforeseen events and, where necessary, Parliament can act with considerable speed. This was seen with the Banking (Special Provisions) Act 2008, which nationalised the Northern Rock bank in the teeth of the financial crisis. It is possible for an Act of Parliament to pass through all its stages in both Houses and receive the Royal Assent in a day if necessary.

Once an idea for a piece of legislation has been accepted by the Legislation Committee of the Cabinet, there will usually follow an intensive period of deliberation and consultation. Parliamentary Counsel (the draftsmen of the bills) will then be required to draw up the bill clause by clause. Many bills will be significantly redrafted before they are enacted in their final version. The problems which beset the draftsmen include the lack of precision of our language, trying to reconcile many conflicting demands, attempting to cover situations which can be envisaged but which have not yet arisen in practice and pressure of time. Not surprisingly many Acts of Parliament are subsequently found to create interpretative difficulties which must be resolved in the courts. Occasionally, legislative amendment is required due to unanticipated ambiguity or outright error. In *Fisher v Bell* (1961) QBD, a shopkeeper could not be convicted under the Restriction of Offensive Weapons Act 1959 for selling a flick knife as it was simply in his shop window. The Restriction of Offensive Weapons Act 1961 amended the offence to include 'exposes or has in his possession for the purpose of sale'.

Acts of Parliament begin life as bills, which may be either public, private or private members' bills. Public bills comprise the vast majority passed in each parliamentary session and are general in their operation. An example of a private bill may be one promoted by a local authority to authorise some activity specifically in its own local area. Private members' bills provide the limited opportunity available in each parliamentary session for an MP to introduce some proposals for change in the law on a topic of his or her choice; private members' bills are public bills. Public bills are usually, however, introduced by government departments and may be introduced in either House. The likelihood is that most of these public bills will be passed. A government, particularly one with a large majority, has an effective stranglehold on procedures in Parliament. Non-controversial bills are often introduced in the House of Lords, but all money bills must be introduced in the Commons.

The normal procedure in both Houses is as follows:

First reading. When the bill is introduced, it will have its first reading. Normally the sponsors of the bill present it in dummy form at the table of the House, and one of the clerks reads out the title. The bill is then deemed to have been read for the first time. It is next ordered to be printed and published, and a date fixed for the second reading.

Second reading. This is the occasion for a parliamentary debate on the principles of the bill. The bill is not considered clause by clause at this stage. There will be a vote at the end of the debate, and the bill could be rejected at this stage.

Committee stage. Assuming that the bill is not lost, it is then referred to a committee, to undergo its committee stage. It will usually go to a standing committee of between 16 and 50 members, chosen to reflect the relative strength of the political parties in the House, and having regard to the special qualifications, concerns and interests of the MPs in question. These standing committees will consider the bill in detail, examining it clause by clause, trying to produce a result which is unambiguous. The committee will also deal with proposed amendments to the bill. Where a bill is of major constitutional importance, the committee stage may be taken in a Committee of the Whole House. However, the drawback of this system is that, while it sits as a committee, no other business can be conducted by the House.

Many of the amendments proposed to the bill at the committee stage may be put down by the minister in overall charge of the bill himself. These proposed amendments may reflect afterthoughts by him/her or officials or they may be the result of concessions to outside pressure groups.

Report stage. Once the committee stage is completed, the bill is then reported back to the House, where the changes introduced in committee will be outlined. There is an opportunity for debate and voting on the amendments.

Third reading. There is then a third reading of the bill, often done immediately after the report stage is concluded, and the debate on the third reading may culminate in a vote.

Where the measures proposed are controversial, both the government and the opposition will be anxious to have as many of their supporters voting as possible, and each may have issued a three-line whip, which is an instruction to their supporters to attend and vote in accordance with instructions.

The whip system is the means by which political parties control and organise their members in Parliament. There is a system of whipping in the Lords, but its principal importance is in relation to the activities of the House of Commons. The name comes originally from the world of hunting. The whipper-in is a hunt official charged with the control of the hounds. The word has now transferred itself to the parliamentary context. A whip in Parliament is a person whose job is to give information to his or her party members and to maintain discipline among them.

Royal Assent. Once the bill successfully passes all its stages in both Houses, it receives the Royal Assent and becomes an Act of Parliament. It takes effect from the date on which it receives Royal Assent, from the date of commencement set out in the Act itself or from a date to be fixed in the future. Power to fix that date will usually be given to an appropriate minister. The Act then remains in force until it is repealed. Repeals are effected by exactly the same process. It is usual to refer to the Act by its short title, for example, the Sale of Goods Act 1979 or the Arbitration Act 1996. Unless the Act states to the contrary, it will apply throughout the UK. All legislation since 1988, and much from before, is available at www.legislation.gov.uk and there is a useful feature to show 'geographical extent', i.e. the application of the legislation as across England, Wales, Scotland and Northern Ireland.

Nowadays Acts of Parliament follow a fairly standard pattern. Before setting out any of the main text, there is a long title which establishes the purpose of the Act. For example, the Arbitration Act 1996 states that it is 'an Act to restate and improve the law relating to arbitration pursuant to an arbitration agreement; to make other provision relating to arbitration and arbitration awards; and for connected purposes'. The main text is then divided into sections and subsections and, if appropriate, the whole Act will be set out in parts. Where the Act contains detailed lists, these will often be contained in a schedule to the Act. Most modern Acts have sections which deal with definitions, repeals, date of commencement and area of operation. The interpretation section in an Act of Parliament is important because, within it, the draftsman of the Act can set out the precise meaning of a word for the purpose of that Act only. This can be of considerable help to judges when they subsequently need to interpret the exact meaning of the legislation. The repeals section will indicate which earlier Acts or regulations or parts of Acts and regulations have been repealed by the present Act.

Pressure groups

It is undoubtedly true today that in drawing up and seeking to implement its legislative programme, no government can afford to ignore the views of pressure groups. Joining a pressure group is one of the ways in which individuals have an opportunity to influence government decisions and future legislation. There are two main types of pressure groups – interest groups, where the group exists primarily to promote the interests of its own members (for example, the Automobile Association, the National Farmers Union or the various trade unions) and cause groups, which are usually formed for the purpose of lobbying for a specific cause (for example, Age UK, The National Society for the Prevention of Cruelty to Children or the Campaign for Nuclear Disarmament). Professional bodies can cover both areas in supporting their members, but often have wider policy remits (for example, the Royal Institution of Chartered Surveyors and the Chartered Institute of Building).

The word 'lobbying' means influencing members of the legislature. The word itself derives from the practice of people meeting their MPs in the lobby of the House of Commons. The process of lobbying can be extended over a very long period and requires considerable energy, organisation and resources to carry it out effectively. A large organisation may employ professional lobbyists to pursue its cause.

The government of the day is likely to be influenced by pressure groups for a number of reasons – the pressure group may well be expert within its particular field or the government may need active cooperation from the members of the pressure group in order to implement its policies. Some pressure groups are immensely wealthy and can promote their cause by advertising and other forms of publicity, while other pressure groups are powerful in terms of the significant proportion of the community which they represent.

Pressure groups can seek to exercise influence in a number of ways and at a number of points in the legislative process. Before any legislation is introduced, members of pressure groups may be represented on advisory bodies connected with appropriate government departments. If legislation is proposed, consultation with appropriate groups will frequently take place before a bill is introduced. A pressure group will always be working to advertise and promote its cause, possibly by liaising with sympathetic MPs, even when no immediate legislation is proposed. Once a bill has been introduced, MPs sympathetic to particular causes may participate at the detailed committee stage, when a number of amendments to a bill may be proposed. At that point, the pressure group will wish to stay in touch with its 'tame' MPs. Even after an Act has been passed, a watching and monitoring role can be played by pressure groups who may seek changes by regulations if an Act is seen to have loopholes. Much of the success of a pressure group may depend on making and maintaining good personal contact with MPs, peers who sit regularly in the Lords, civil servants and ministers.

Delegated legislation

It can easily be seen that pressure on parliamentary time is so great that all the rules necessary in a complex modern society cannot be made in the labourious and time-consuming way outlined previously. It is therefore common for Parliament to lay down a framework in an Act (an 'enabling Act') and then grant power to some other person or organisation to make the detailed regulations. For example, by power given under the Building Act 1984, the Secretary of State has made the Building Regulations, the latest being the Building Regulations 2010.

Delegated legislation can take one of the three forms:

1 *Orders in Council* – where the rule-making power is vested in the Privy Council. This is normally confined to emergency situations and was much used during the Second World War.
2 *Statutory instruments* – this is the form commonly used when ministers make regulations by virtue of delegated powers. For example, the Building Regulations are in this form.
3 *Bye-laws* – made by local authorities or other public bodies under powers granted to them by Parliament.

Although delegated legislation can be seen to be necessary and to have some advantages as a rule-making technique, there must be adequate control of the process. The advantages include:

- *speed* – where there is an emergency or a temporary or rapidly changing situation;
- *flexibility* – outdated rules can be more easily changed and
- *expertise* – consultation with outside bodies may harness the experience of technical and other experts.

Despite these advantages, it will be realised that rule-making by these methods gives great power to persons who have not been elected. Critics of delegated legislation would argue that it vests too much power in the hands of the executive, and that scrutiny and control of the process are vital. That control may come through the procedures themselves, for example, by delegating powers on strict terms to prevent abuse, by subjecting any regulations proposed to an affirmative resolution procedure of one or both Houses of Parliament or by requiring prior consultation with an advisory body.

If the power which had been granted is exceeded, control can be exercised by the application of the *ultra vires* (literally, beyond the powers) principle. This would require a person aggrieved by the abuse of power to claim in court that the rule maker had acted outside the scope of his or her authority, i.e. that the action has been *ultra vires*. If so found by the court, the *ultra vires* action will be declared to be void, i.e. of no legal effect.

Parliament itself is aware of the danger of abuse of power and has established committees to scrutinise statutory instruments. Parliament might pass only 50 Acts of Parliament in any one year, often fewer, yet in the same period, more than 3,000 Statutory Instruments might be made, so the potential size of the problem can be seen.

Statutory interpretation

When an Act of Parliament comes into force, it may cause disputes as to its true meaning. The written word is not an exact form of communication, and it will ultimately be the duty of the judges, when cases arise on disputed points in an Act, to determine the 'intention of Parliament'.

Faced with such a problem, the judge starts by looking at the Act itself. Most Acts contain a definition section, in which words will be given a specific meaning for the purposes of that Act only. For example, in s.14 of the Unfair Contract Terms Act 1977, definitions are given of the words 'business', 'goods' and 'notice', thereby giving those words a specific and restricted meaning confined to that Act. 'Business' is defined there as including a profession and the activities of any government department or local or public authority.

Where no definition is given within the Act itself, judges may take the meaning of a word from a dictionary, but if they do so, they must bear in mind the principle expressed in Latin, *noscitur a sociis*, which means 'a word is known by the company it keeps'. The meaning of a word may well be governed by the context in which it is used, e.g. in the phrase 'interest, annuities or other annual payments', the word 'interest' means annual interest. The judge may also turn to the Interpretation Act 1978, which gives meanings for standard phrases and expressions like day, week and month. And again, the word or phrase in question may have been interpreted and applied in an earlier case heard by the court, so that the present judge would be obliged by the rules of judicial precedent to accept that meaning in the case before him or her.

Assuming that these aids do not solve the problem, the judge may then adopt one of the following approaches to the question:

1 The literal approach
This approach dictates that words should be given their ordinary, plain and natural meaning, even if this produces an absurd result. Judges favouring this approach take the view that it is then Parliament's responsibility to put the situation right by amending the law. In using this approach, two general rules of construction should be remembered:

2 *Ejusdem generis* (sometimes known as Lord Tenterden's rule)
When specific words are followed by general words, the general words must be read in the light of the specific words, for example, 'on a pond, lake, river, canal or other water'. By applying this rule, the words 'other water' would be read to include a reservoir but not the open sea. It is clearly intended by the specific words to confine the stretches of water to inland water.

3 *Expressio unius est exclusio alterius*
If one thing is specifically mentioned in an Act, the Act is presumed not to apply to another similar but unmentioned thing, e.g. some sections of an Act may refer to the seller and the buyer. A subsequent section mentioning only the seller is presumed not to apply to the buyer.

The literal approach to interpretation is not favoured by some judges. Lord Denning, then Justice Denning, famously said: 'When a defect appears a judge cannot simply fold his hands and blame the draftsman' (*Seaford Court Estates Ltd v Asher* (1949) CA).

1 The golden rule
  This allows the judge to interpret so as to avoid the absurdity or to avoid a result which is inconsistent with the Act itself or common law principles. This approach was used in *Re. Sigsworth* (1935) Ch, in a case where a man claimed his mother's property, despite having murdered her. Under the Act which governed the distribution of her property, this ought to have been shared amongst her children. The son was the only child. The judge held that the common law rule that a murderer cannot benefit from the person he has murdered prevailed over the clear words of the Act.
2 The mischief rule
  Another approach adopted is where the judge looks to see which defect of the law is intended to be corrected by the Act and then interprets to correct the defect.

An important development in the interpretation of statutes was seen in *Pepper v Hart* (1993) HL. In this case, the House of Lords changed the existing judicial practice of not allowing reference to be made to parliamentary materials in the interpretation of statutes. The use of Hansard (a record of debates in Parliament) is now allowed within the following limits:

a The legislation must be ambiguous or unclear or a literal interpretation would lead to absurdity.
b The material relied upon consists of statements made by the minister or other promoter of a bill, together with such other parliamentary materials as are necessary to understand such statements.
c The statements relied upon are clear.

This allows the courts to ascertain the mischief from another source outside the words of the legislation. The judges also sometimes refer to a purposive approach to interpretation, meaning that the courts' task is to give effect to Parliament's purpose, i.e. the object of the legislation.

A simple example from a case may illustrate some of the problems of interpretation faced by judges. In a case concerning single payments under the old Supplementary Benefit scheme, a claimant requested a single payment to help towards the purchase of wallpaper to redecorate her lounge and kitchen. The wording of the appropriate regulation required that payments could only be made in respect of expenses of *essential* internal redecoration. The court had to consider the meaning of the word 'essential'. It was not defined anywhere in the Supplementary Benefits Act 1976 or the relevant Regulations, and it fell, therefore, to be given its ordinary meaning in everyday use. When a dictionary was consulted, the *Shorter Oxford English Dictionary* gave two rather differing meanings of the word 'essential'. It can mean 'material' or 'important' in the sense of 'your work is essential to the success of this project' or it can mean 'indispensably requisite', as, for example, 'water is an essential ingredient of a cup of tea'. Relying on the stricter meaning of the word, the benefit officer had refused a single payment. On appeal to the tribunal, however, the judge favoured the less strict interpretation of the word. He was guided in doing so by the fact that the same word 'essential' was used elsewhere in the Regulations, where its meaning tended to be 'material' and 'important' rather than 'indispensably requisite'.

Following the introduction of the HRA, the courts are, so far as it is possible to do so, to interpret Acts of Parliament and delegated legislation to ensure compatibility with Conventions Rights. If it is considered that national legislation is incompatible with the Convention, a declaration of incompatibility can be issued, but the legislation must still be enforced as is until the national Parliament removes the incompatibility. It is not possible for the courts to declare legislation invalid on the ground of incompatibility.

No one instructs a judge as to which of these approaches should be used. It is clear that the process of interpreting and applying Acts of Parliament gives considerable power to the judges. But their role in the law-making process is not limited to the interpretation of statutes because the other principal source of law is judicial precedent (binding decisions of the judges).

Judicial precedent

When a judge gives his decision in a case before him, this has two elements:

1 The actual decision affecting the parties, e.g. X is found to be liable and must pay damages to Y of £1,000.
2 The principles of law which have caused the judge to arrive at that decision.

Where the case concerns facts or situations which are comparable with earlier cases heard by the courts, the judge will normally apply the principles from the earlier cases. In one well-known case, *Donoghue v Stevenson* (1932) HL, a soft-drinks manufacturer, David Stevenson, was held to be liable to the ultimate consumer who had suffered physical harm by drinking a bottle of the manufacturer's ginger beer which was contaminated by the remains of a decomposed snail. If, on a future occasion, a court hears a case brought by a person made ill by eating a pie containing a dead mouse, the same principles can be applied. By analogy these principles may be extended further. The principle in the 'ginger beer' case was actually applied to a later case where a person contracted dermatitis because of wearing underpants where chemicals had been left in the material during their manufacture (*Grant v Australian Knitting Mills* (1936) PC).

Although many systems of law in practice apply this principle of deciding cases in the same way as previous cases of a similar nature, the English system goes further. The judge is *bound* by the earlier decisions, and hence the system is often referred to as the system of binding precedent. In the English legal system, there is a hierarchy of courts:

The **Supreme Court of the United Kingdom** (the Judicial Committee of the House of Lords prior to October 2009) binds all the courts beneath it by its decisions.

The **Court of Appeal**, both Civil and Criminal Divisions, is bound by the Supreme Court and binds all the courts below.

The **High Court** is bound by the Supreme Court and the Court of Appeal and binds all the courts beneath it.

Finally, the **Crown Court**, **magistrates' courts** and the **county courts** are bound by all the courts previously mentioned but do not bind any courts.

Also note that some courts are bound by their own previous decisions and some are not. For example, the Supreme Court is not strictly bound by its own previous decisions, although it usually follows them, whereas the Court of Appeal is bound by its own previous decisions, subject to certain exceptions.

One difficulty presented by the doctrine of precedent lies in finding the actual part of the case which is binding. When a judge gives a decision, his legal reasoning, when set out in the law reports, can be several pages long. Much of what is said may relate to why he or she rejected other arguments put by counsel or what would have influenced him or her if the facts had been slightly different. These parts of the judgment are not binding in later cases. They are matters 'said by the way' or in legal terms, *obiter dicta*. The judge in a later case must search to find the rule of law upon which the decision is based, properly called the *ratio decidendi*.

The process of identifying the ratio is a vital part of the training of a lawyer. In each case the first step is to establish the material facts, meaning facts relevant to the decision. For example, say Mrs Smith, a mother of four aged 45, knocked down and injured Mr Brown last Friday. The accident was caused because Mrs Smith had gone to sleep at the wheel. The only material facts here are that a motorist injured a pedestrian by falling asleep at the wheel of a car. Once the material facts have been established, if they correspond with the material facts of an earlier decision, the judge will usually be bound by the principle of law in the earlier case. However, judges are normally only bound by decisions given in a court superior to their own. This allows the system to be flexible (as it is open to a higher court to overrule decisions of lower courts) but certain (as it obliges lower courts to follow the decisions of higher courts). The structure of the courts is discussed in Chapter 2.

Advantages and disadvantages of the system of precedent

One of the greatest advantages of the system of binding precedent is that the rules have evolved from real-life cases and are, therefore, essentially practical. Again, the 'binding' feature of the system makes it reasonably certain. However, two major criticisms are often levelled against the system. The first is that it creates a bulky system of very detailed rules which requires the production, at considerable expense, of large numbers of law reports to enable the previous cases to be checked and quoted. A more significant problem is that the system can be very uncertain, due to the powerful role played by the judges. Consider the case where a judge is faced with a precedent which, in theory, is binding on him, but which he prefers not to apply. There is a technique called 'distinguishing' one case from another, which may allow him to avoid applying the earlier decision. The judge is allowed to distinguish (and therefore not apply) any case where the material facts are not the same as in the case before him. When this technique is properly used, it allows the law to be flexible. When improperly used, it results in hair-splitting distinctions, which can leave two apparently conflicting decisions reported in the casebooks.

If a person is dissatisfied with the outcome of a case, he or she may have grounds for appeal. The final court of appeal in the English system is the Supreme Court of the UK. In relation to European Community law, the decisions of the Court of Justice of the European Communities (the European Court) bound the Supreme Court and all other English courts prior to the UK's departure from the European Union. Until 1966, even the House of Lords was bound by its own previous decisions, but it was then announced that: 'Their Lordships propose to modify their present practice, and while treating former decisions of this House as normally binding, to depart from a previous decision when it appears right to do so'. An example of such a departure can be seen in *Murphy v Brentwood DC* (1990) HL.

Applying a precedent

Examples taken from actual cases demonstrate how the case law system works. To see how the normal processes of applying a precedent operate, look first at a case from 1932,

Donoghue v Stevenson, introduced earlier. A manufacturer of ginger beer which was marketed in an opaque bottle was held to be liable to the ultimate consumer of the drink in the tort of negligence. (A tort is a civil wrong.) The consumer had become ill on discovering the remains of a decomposed snail in the bottle. If this precedent had subsequently been confined by judges to sets of facts involving drinks sold in opaque bottles, its value as a precedent would have been very limited. However, in 1936, another court seized the opportunity to extend the scope of the precedent. In the case of *Grant v Australian Knitting Mills*, the claimant, Dr Grant, had purchased a pair of long woollen underpants, manufactured by the defendants. He contracted a severe form of dermatitis after wearing them, which was shown to have been caused by excessive amounts of sulphur which had not been successfully washed out during the manufacturing process. Were the defendants liable? The judges examined the Donoghue case and extracted from it the following statement of principle:

> A manufacturer of products which he sells in such form that he intends them to reach the ultimate consumer with no reasonable possibility of intermediate examination, and where the absence of reasonable care in the preparation of the products will result in injury to the consumer, owes a duty to the consumer to take reasonable care.

Undoubtedly a manufacturer of underpants must realise that potential customers cannot carry out scientific tests to measure sulphur levels before purchasing, and the courts had no difficulty in applying the Donoghue principle. If the court in the Grant case had thought the principle was unsound, it could have sought to distinguish the two cases on the ground that the material facts were not similar. For example, the snail case concerned goods to be consumed internally, whereas the underpants case concerned goods to be worn externally. Such fine distinctions could have been justified if the earlier precedent had been unpopular for any reason, but the law develops best when the scope of a sound principle is gradually extended.

Common law and equity

The expression 'common law' is frequently used and can have a number of meanings. It may mean the law which is common throughout England. Indeed, the phrase was originally used in that sense to distinguish it from local rules and customs. The expression may also mean rules developed through precedents, rather than created by Acts of Parliament. It may further be used to mean rules which are not derived from equity. An unusual system emerged in England between the twelfth century and 1875, whereby a completely separate system of courts developed, administering quite separate rules – the principles of equity. These rules were usually developed to meet situations where the common law had no remedy to offer or where the common law was in some way deficient or unjust. The equitable principles are discretionary in nature but are part of a body of case law which operates according to the doctrine of precedent. Since 1875, the principles of equity have been administered side by side with the rules of common law (Judicature Acts 1873–75). If the two sets of rules should ever conflict, the principles of equity will prevail. Examples of equitable principles are promissory estoppel (whereby the strict requirements of land law, such as written transfers, can be overridden where it is deemed just to do so) and undue influence (whereby transactions will be set aside in the interests of fairness).

In the civil courts, the distinction between common law and equity can be seen in remedies awarded. Compensation is a common law remedy available as of right (subject to judicial discretion as to amount). But fairness and true resolution might require something other. Equitable remedies are available to cover this, such as an injunction to prevent trespass or specific performance to enforce the completion of a contract.

The European Convention on Human Rights

The UK has been a signatory to the European Convention on Human Rights since 1950 and UK citizens have, since 1966, been able to take cases to the European Court of Human Rights in order to obtain a remedy therefrom. Moreover, human rights arguments based upon the Articles of the Convention have long been deployed before UK courts. It was only in October 2000, however, when the majority of the provisions of the HRA came into force, that the Convention became part of domestic UK law. This is a separate regime to the European Union, so the UK's withdrawal from the EU does not impact engagement with the European Convention on Human Rights.

Before considering the significance of the HRA, it is first necessary to briefly consider both the nature of the rights that the Convention's Articles protect and that of the European Court of Human Rights itself.

The rights were set out in Articles from 1953, with amendment and expansion through Protocols.

The Convention rights

Article 1 Obligation to respect human rights

Article 1 of the Convention requires the contracting parties (including the UK) to secure the rights that the Convention protects for those within their jurisdiction. Article 1 has not been incorporated into UK law, but, given the obligations imposed upon public authorities by the HRA, this does not appear to be problematic.

Article 2 Right to life

Article 2 requires that the right to life must be protected by law. Thus, as a general rule, the state (e.g. the police) should not intentionally deprive the citizen of his life. Further, as the decision of the European Court of Human Rights in *Osman v UK* (1998) ECHR made clear, the state must take appropriate steps to protect life (e.g. where the life of a person is at risk from the unlawful acts of a third party). Moreover, the decision of the European Court of Human Rights in *McCann v UK* (1995) ECHR demonstrates that where death results from the use of force by agents of the State, Article 2 implicitly requires an effective official investigation into the killing.

Whilst Article 2 does not prevent the execution of a lawfully imposed death penalty following a criminal trial, the UK is a party to the Convention's Sixth Protocol, which requires the abolition of the death penalty, and this has been incorporated into the HRA.

Article 3 Prohibition of torture

Article 3 prohibits torture or inhuman or degrading treatment or punishment. Thus, for example, the state (e.g. the police) must not torture a person suspected of a crime when they are interviewing him. Equally, the state must protect persons from torture, e.g. by ensuring that the law prevents parents from violating the Article 3 rights of their children (*A v UK* (1998) ECHR). The Article 3 right is an absolute right, from which derogation is not permitted (see Article 15 later on). Thus, for example, even in the context of the fight against terrorism, the Article 3 rights are absolute (*A and Others v UK* (2009) ECHR).

Article 4 Prohibition of slavery or servitude

Article 4 prohibits the holding of persons in slavery or servitude and also provides that persons shall not be required to perform forced or compulsory labour. Article 4 does not, however, prohibit what would otherwise be termed forced or compulsory labour where it is required in the ordinary course of detention or conditional release from detention, where it amounts to service of a military character or is required of a conscientious objector as an alternative to compulsory military service, where it is required in the context of an emergency or a calamity which threatens the life or well-being of the community or where it forms part of normal civic obligations. Like the Article 3 right, the Article 4 right is an absolute right, from which derogation is not permitted (see Article 15 later on).

Article 5 Right to liberty

Article 5 protects the right to liberty and security of a person, providing that a person shall only be deprived of his or her liberty in any of a number of cases specified by Article 5and only in accordance with a procedure prescribed by law. The domestic law must permit the arrest or detention, must be sufficiently precise to enable persons to understand when they might be affected by it and must be accessible (e.g. must not be secret) (*Hashman and Harrup v UK* (1999) ECHR). Because the Article 5 right is subject to specified exceptions, however, the right is not an absolute right but, rather, is a limited right. The situations in which Article 5 does not prevent a person's liberty from being taken away comprise lawful detention after conviction by a criminal court; lawful arrest or detention for non-compliance with a court order or to secure the fulfilment of legally prescribed obligations; lawful arrest or detention on reasonable suspicion of committing an offence to bring a person before the competent legal authority or where reasonably necessary either to prevent the commission of an offence or to prevent flight after its commission; lawful detention of a minor for educational supervision or to bring him before the competent legal authority; lawful detention to prevent the spreading of disease or of persons of unsound mind, alcoholics, drug addicts or vagrants or lawful arrest or detention to prevent unauthorised entry into the country or in the context of action to deport or extradite a person. Article 5 also, essentially, gives a person who is arrested or detained the right to be informed of the reasons for his arrest and of the charges against him; the right to be brought promptly before a judge or a person authorised to exercise judicial authority and to be tried within a reasonable time and released on bail (*Wemhof v Germany* (1968) ECHR), the right to take proceedings in court to speedily determine the lawfulness of the detention and the right to compensation where arrested or detention is in contravention of Article 5.

Article 6 Right to a fair trial

Article 6 guarantees the right to a fair trial. Thus, in the context of the determination of his civil or criminal rights, a person is entitled to a fair and public hearing which takes place within a reasonable time and is held by an independent and impartial tribunal that has been established by law. Article 6requires that the court must pronounce its verdict in public but permits the exclusion of the press or public from the trial for a variety of specified reasons, e.g. in the interests of national security, in the interests of juveniles or to protect the private lives of the parties. Article 6 expressly provides that a person charged with a criminal offence shall be presumed innocent until proven guilty in accordance with law. Article 6also expressly gives a person charged with a criminal offence a number of minimum rights, namely to be informed promptly of the nature and cause of the charge; to have adequate time and facilities to prepare a defence; to defend him- or herself in person or through chosen legal advisers or, if he or she cannot afford legal assistance, to be provided with free legal assistance where this is required by the interests of justice; to examine the witnesses against him or her or to have them examined and to secure their attendance and examination under the same conditions as these witnesses are required to attend and are examined and to have an interpreter free of charge if he or she cannot understand the court or cannot speak the language used.

In determining whether a trial was fair within the meaning of Article 6, it is necessary to consider the issue of 'equality of arms'. The significance of 'equality of arms' is, essentially, that the conditions under which a party is permitted to present their case in court should not place him or her at a substantial disadvantage when compared to the conditions under which the other party is permitted to present their case (*Neumeister v Austria* (1968) ECHR).

Article 7 Retroactivity of criminal offences

Article 7 provides both that a person shall not be found guilty of a criminal offence in consequence of an act or omission which did not amount to a criminal offence at the time when the person acted or omitted to act and that a person who is convicted of an offence shall not receive a heavier sentence than could have been imposed at the time the offence was committed. Article 7 does not prevent the normal development of the law via case law, provided that such development (e.g. a change in the common law to the effect that a man can rape his wife) is reasonably foreseeable (*SW v UK* (1996) ECHR).

Article 8 Right to privacy

Article 8 concerns the right of persons to respect for their private and family life, for their home and for their correspondence. Article 8 does, however, permit interference with the exercise of the Article 8 right by a public authority if such interference is in accordance with the law and is necessary in a democratic society for any of a number of specified reasons. These reasons comprise the interests of national security, the interests of public safety, the country's economic well-being, the prevention of disorder or crime, the protection of health or morals and the protection of the rights and freedoms of others. Article 8 right was violated where, for example, personal data, in the form of fingerprint and DNA information relating to persons who had not been convicted of offences, was retained by the police under powers which the European Court of Human Rights regarded as having a 'blanket and indiscriminate nature' (*S and Marper v UK* (2008) ECHR GC).

Article 9 Freedom of thought, conscience and religion

Article 9 protects a person's right to freedom of thought, freedom of conscience and freedom of religion and encompasses both a person's right to change religion or belief and the freedom, whether alone or with others and whether in private or in public, to manifest this religion or belief via worship, teaching, practice or observance. Article 9 provides, however, that a person's freedom to manifest religion or beliefs may be subject to limitations which are prescribed by law and are necessary in a democratic society for any of a number of specified reasons. These reasons comprise the interests of public safety, the protection of public order, health or morals and the protection of the rights and freedoms of others. The fact that a Christian employee was required by her contract of employment to work on Sundays and was dismissed when she refused to do so did not violate her Article 9 right because the matter was contractual, she had been free to resign and she had not been dismissed from her employment because of her Christian beliefs (*Stedman v UK* (1997) ECHR). The meaning of 'prescribed by law' is the same as that of 'in accordance with the law', which was considered in the context of Article 8 (mentioned previously).

Article 10 Freedom of expression

Article 10 concerns the right to freedom of expression, including the freedom to hold opinions and receive or impart information and ideas in the absence of interference by public authorities and regardless of frontiers. Article 10 does not prevent states from requiring broadcasting, TV or cinema enterprises to be licensed. The exercise of the Article 10 freedoms may be subject to formalities, conditions, restrictions or penalties which are prescribed by law, provided that they are necessary in a democratic society for one or more of a number of specified reasons. These reasons comprise the interests of national security, territorial integrity or public safety; the prevention of disorder or crime; the protection of health or morals; the protection of the reputation or rights of others; the prevention of the disclosure of information received in confidence and maintaining the authority and impartiality of the judiciary.

Article 11 Freedom of assembly

Article 11 protects a person's right to freedom of assembly and their right to freedom of association, including the right to form or join a trade union. Restrictions may be placed on the exercise of the Article 10rights provided they are prescribed by law and are necessary in a democratic society for one or more of a number of specified reasons. These reasons comprise the interests of national security or public safety, the prevention of disorder or crime, the protection of health or morals and the protection of the rights or freedoms of others. Article 11 does not prevent members of the police, the armed forces or the administration of the state from imposing lawful restrictions on the exercise of Article 11 rights. The Article 11 rights appear to encompass the right not to join an association, such as a trade union (*Young v UK* (1981) ECHR) and the right of peaceful protestors to state protection (*Plattform 'Ärzte Für Das Leben' v Austria* (1988) ECHR). The meaning of 'prescribed by law' is the same as that of 'in accordance with the law', which was considered in the context of Article 8 (mentioned previously).

Article 12 Right to marry

Article 12 concerns the right of men and women of marriageable age to marry and found a family according to domestic laws which govern the exercise of the Article 12 right. It appears that the Article 12 right does not apply to marriages between persons of the same sex or to transsexuals (*Cossey v UK* (1990) ECHR), though the European Court of Human Rights has recognised the right of a transsexual to marry in her new gender (*Goodwin v UK* (1996) ECHR).

Article 13 Effective remedy

Article 13, which concerns the right to an effective remedy when a person's Convention rights and freedoms have been violated, has not been incorporated into UK law, though ss.7 and 8 of the HRA (which are considered later on) were intended to give effect to the obligations imposed by Article 13.

Article 14 Discrimination

Article 14 provides that persons are entitled to enjoy the Convention rights regardless of discrimination on grounds such as, but not restricted to, sex, race, colour, language, political or other opinion, national or social origin, association with national minorities, birth or status of some other sort. Article 14 may be violated where people in similar situations are treated differently, but differences in treatment do not violate Article 14 if they pursue a legitimate aim and the method used to achieve the legitimate aim is a proportionate response to the legitimate aim (*Burden v UK* (2006) ECHR GC). In assessing whether differences are justified, the national authorities are allowed a 'margin of appreciation' which is usually wide in the context of matters of economic or social strategy (*Burden v UK*).

Article 15 Derogations in war or public emergency

Article 15 gives a contracting state (e.g. the UK) limited power to derogate from Convention obligations during a war or public emergency (e.g. in the context of terrorism) which threatens the life of the nation. Derogation from Articles 3, 4(1) and 7 is not permitted, however, and derogation from Article 2 is only permitted in relation to deaths resulting from lawful acts of war. The European Court of Rights has recognised with regard to the operation of Article 15 that the Contracting States have a wide 'margin of appreciation' when determining whether derogation is necessary, but has also made clear that they do not have an unlimited discretion in this regard (*A and Others v UK* (2009) ECHR GC).

Article 16 Restrictions on foreign nationals

Article 16 permits a contracting state (e.g. the UK) to impose restrictions on the freedom of expression, freedom of assembly and freedom of association of foreign nationals.

Article 17 Abuse of rights

Article 17 essentially provides that the Convention rights and freedoms do not give any state, group or person the right to perform acts which are either aimed at the destruction of the Convention rights and freedoms or which are aimed at limiting those rights and freedoms to a greater extent than the Convention itself makes provision for.

Article 18 Permitted restrictions

Article 18 provides that the restrictions permitted under the Convention rights and freedoms must not be used for any purpose other than those which the Convention prescribes.

First Protocol

Article 1of the Convention's First Protocol provides, essentially, that persons (including non-natural persons such as companies) are entitled to the **peaceful enjoyment of their possessions**, shall not be deprived of their possessions other than where this is in the public interest and shall only be so deprived subject to the conditions for which domestic law and the general principles of international law provide. Article 1of the First Protocol does not, however, restrict the right of a state such as the UK to enforce the use of those domestic laws that it believes necessary either to control the use of property in the general interest or to secure the payment of taxes, etc. Deprivation of possessions, interference with peaceful enjoyment of possessions or control of possessions by the state can only be justified where there is a fair balance between the interest of the community and the interests of individuals (*Sporrong v Sweden* (1982) ECHR), and the domestic law which provides for such deprivation, interference or control must be accessible, must be sufficiently certain and must provide protection against arbitrary decisions. In determining whether such a fair balance has been achieved in the context of a deprivation of property, one matter which will be of relevance will be the adequacy of the compensation, if any, payable to the property owner, though it appears that the public interest is capable of justifying a level of compensation below market value (*Lithgow v UK* (1986) ECHR). Examples of matters which are potentially capable of giving rise to issues under Article 1 of Protocol 1 in appropriate circumstances include planning decisions, compulsory purchase or nationalisation of property and statutory controls restricting the ability of landlords to recover possession of rented property.

A landowner, who lost a substantial holding through the provisions of adverse possession (see Chapter 11), unsuccessfully challenged in the European Court of Human Rights, and on appeal to the ECHR Grand Chamber, that the law of adverse possession was contrary to Article 1 (*J A Pye (Oxford) Ltd v UK* (2007) ECHR GC).

Article 2 of the First Protocol concerns the **right to education**, providing that no one shall be deprived of that right and that the state must respect the parental right to ensure that education and teaching conform with their religious and philosophical convictions. The right to education does not require the state to provide education, but rather, imposes conditions where the state does provide education. A UK reservation concerning Article 2 of Protocol 1 (which has never been tested by the European Court of Human Rights) means that the right to education has only been accepted by the UK to the extent to which it is compatible with efficient instruction and training and avoiding unreasonable public expenditure.

Article 3 of the First Protocol requires contracting states (e.g. the UK) to hold **free elections by secret ballot at reasonable intervals** under conditions which ensure the free expression of the opinion of the people. This has led, for example, to prisoners being able to successfully challenge the blanket ban on their right to vote in elections (*Hirst v UK* (2006) ECHR).

The Convention's other Protocols are not further considered in this book.

The European Court of Human Rights

The European Court of Human Rights has one judge from each of the contracting states (one of which is the UK) who are elected from a list of three candidates nominated by the contracting state, and sit independently, i.e. not as a representative of their country. They must be persons of high moral character who are either qualified for high judicial office or are juriconsults of recognised competence. They sit for six years and may be re-elected, but their terms of office expire at the age of 70. When Protocol No. 14 has been ratified by all the contracting states (see later on), judges will be elected for a nine-year period, but when the nine-year period has ended, re-election will not be permitted.

For the purpose of considering cases, the European Court sits in Committees of three judges, Chambers of seven judges and a Grand Chamber of 17 judges. The full Court elects its President, Vice President, Presidents of Chambers, etc. Committees may, upon a unanimous vote, finally declare applications from individuals, groups of individuals or non-governmental organisations inadmissible or strike them out, where they can make such a decision without further examination. Unless a party objects, a Chamber may relinquish jurisdiction to the Grand Chamber where a serious question concerning the interpretation of the Convention or protocols arises or where the result of the case might be inconsistent with an earlier decision of the Court. A party to the judgment of a Chamber may, within three months, request the referral of the case to the Grand Chamber. The request will be considered by a panel of five judges of the Grand Chamber and will be accepted if it raises a serious question concerning the Convention's interpretation or application or concerning the interpretation of application of a protocol or if it raises a serious issue of general importance. If the request is accepted, the judgment of the Grand Chamber will determine the case. The Grand Chamber may also give advisory opinions on legal questions concerning the interpretation of the Convention and protocols, where requested to do so by the Committee of Ministers.

The European Court of Human Rights can only receive applications from a person, group of persons or non-governmental organisation claiming to be the victim of a violation of Convention rights by a High Contracting Party (e.g. the UK). The applicant may be a person who has been affected by the violation or is under a real risk of being so affected. Thus, the family of a deceased person may be a victim (*McCann v UK* (1995) ECHR GC), but it appears that a pressure group is not a victim if not itself a victim.

The jurisdiction of the Court encompasses all matters concerning the interpretation and application of the Convention and its protocols. The Court can only deal with a matter if all domestic remedies have been exhausted and within six months of the making of the final decision. It will not deal with anonymous applications, with applications that are substantially the same as matters that the Court has already examined or with applications that have been submitted to other forms of international investigation or settlement and do not contain new relevant information. It will declare an application inadmissible if it is incompatible with the Convention, is manifestly ill-founded or is an abuse of the right of application. It may reject an application which it considers inadmissible at any stage of the proceedings. It may strike an application out if the applicant does not intend to pursue the application or the matter has been resolved or if, for any other reason, the examination is no longer justified.

Where an application is admissible, the Court will pursue the examination of the case together with the parties' representatives and undertake an investigation if necessary. It will also place itself at the disposal of the parties in an attempt to secure a friendly settlement. If a friendly settlement is effected, the Court will strike the case out. A hearing will normally be in public, but proceedings to secure a friendly settlement will be confidential.

Where no friendly settlement is reached, if the Court finds that the Convention or the Protocols have been violated and the relevant domestic law (e.g. UK law) only allows for partial reparation, the Court will, if necessary, afford 'just satisfaction' (i.e. compensation). A judgment of a Chamber becomes final either when the parties declare that they will not request referral to a Grand Chamber or if no such request has been made within three months of the judgment or when the request is rejected by the panel of the Grand Chamber.

A principle which the European Court of Human Rights applies in appropriate circumstances when determining whether an Article of the Convention has been violated is that of the 'margin of appreciation'. This essentially means that when determining matters such as whether interference with a Convention right is 'necessary in a democratic society', the European Court of Human Rights recognises that the authorities of the individual state concerned are better placed than the European Court to determine whether this is so in the circumstances of their own nation (*Handyside v UK* (1976) ECHR). Whilst the European Court recognises that this is so, it also recognises, however, that the final decision is one for the Court, not for the state, and that the principle of the margin of appreciation does not give the state unlimited power to determine what is best for its people. The principle of the margin of appreciation is not a principle to be applied by the UK courts when deciding whether there has been a violation of Convention rights. The UK courts will, however, be required to apply the concept of proportionality (examples of the application of which were seen previously) and, in so doing, in a democracy, may, at times, find it appropriate to defer to the opinion of an elected body or individual (*R v DPP, ex parte Kebeline* (1999) HL). Examples of the application of this principle were identified in the context of specific Articles of the Convention, which were examined earlier.

It should also be noted that the Convention is a 'living instrument' which must be interpreted in the light of current values and of social change (*Tyrer v UK* (1978) ECHR; *Cossey v UK* (1978) ECHR). Consequently, there is no strict doctrine of precedent in Convention law, and a previous decision of the European Court of Human Rights will not be followed by the European Court on a subsequent occasion if it does not reflect current values. This is a matter which the UK courts are required to take into account when interpreting the Convention and considering earlier decisions of the European Court of Human Rights.

The Human Rights Act 1998

The majority of the provisions of the HRA came into force in October 2000. Essentially, its effect was to make the Convention part of UK law, though, as was seen previously, not all of the Convention rights embodied in the Convention's Articles and Protocols became part of UK law. Thus, for example, s.1 of the HRA provides that, for the purposes of the HRA, the 'Convention rights' do not include the rights set out in Article 13.

When the UK courts are required to interpret Convention rights, the effect of s.2 of the HRA is essentially that they must *take into account* relevant decisions of the European Court of Human Rights, decisions of the European Commission of Human Rights (which no longer exists) and decisions of the Committee of Ministers. The Act does not require the UK courts to *follow* decisions of the European Court of Human Rights, though where they do not do so, an aggrieved party will still have the option of taking the matter to the European Court of Human Rights itself.

Section 3(1) of the HRA requires the UK courts, so far as this is possible, to read both primary and subordinate legislation in a way which is compatible with the Convention rights. At times, the courts may find that legislation is irredeemably incompatible with a Convention right. In the case of primary legislation (i.e. Acts of Parliament plus certain other types of legislation), the court, under HRA s.3(2), must give effect to the legislation even though it is incompatible with the Convention right. In the case of subordinate legislation (i.e. many but not all forms of delegated legislation), the courts, again under HRA s.3(2), must give effect to the legislation if primary legislation prevents the removal of the incompatibility. If this is not the case, however, then the validity, continuing operation or enforcement of subordinate legislation may be affected by its incompatibility with a Convention right.

Where a court (Supreme Court, the Judicial Committee of the Privy Council, the Courts-Martial Appeal Court, the Court of Appeal and the High Court) determines that primary legislation is incompatible with a Convention right, the court may, under HRA s.4, make a declaration of incompatibility. The court may also make a declaration of incompatibility where subordinate legislation is incompatible with a Convention right, if primary legislation prevents the removal of the incompatibility. A declaration of incompatibility does not affect the validity, continuing operation or enforcement of the relevant legislation and does not bind the parties to the relevant proceedings. Rather, the significance of a declaration of incompatibility is that, under HRA s.10, if no appeal is being made from the relevant decision or after any appeal has been determined, the government may, by making a 'remedial order', amend the relevant primary legislation so as to remove the incompatibility (or so as to allow the incompatibility between subordinate legislation and a Convention right to be removed). A remedial order may also be made where, following a decision of the European Court of Human Rights in proceedings brought against the UK, it appears that UK legislation is incompatible with UK obligations under the Convention. A remedial order may be made with retrospective effect, i.e. it may take effect from a date earlier than the date on which it was made. The use of a remedial order may be appropriate where legislation requires minor amendment but may not be appropriate if substantial amendment is required to primary legislation or the issue is controversial. This is so because whilst remedial orders are subject to Parliamentary scrutiny, they cannot be amended in Parliament. Thus, by 2009b, the government was disappointed that it had not been able to make greater use of remedial orders (*Responding to Human Rights Judgments,* Government Response to the Joint Committee on Human Rights, Great Britain, 2009b).

HRA s.6 makes it unlawful for a 'public authority' to act in a way which is incompatible with a Convention right. Section 6 also makes a failure to act in a particular way unlawful where this failure is incompatible with a Convention right. A failure to introduce or lay proposed legislation before Parliament or to make primary legislation or a remedial order will, however, never be unlawful under s.6. Moreover, it is not unlawful under s.6 for a public authority to act in a way which is incompatible with a Convention right if in consequence of primary legislation the public authority could not have acted differently or if the public authority was acting so as to give effect to or enforce legislative provisions which could not be read or given effect to in a way which was compatible with the Convention rights. A court or a tribunal is a public authority for the purposes of s.6. Even where a person or body performs some functions of a private nature and some of a public nature, the person or body will still be a public authority for the purposes of s.6 so far as public acts which he/she/it perform are concerned. The Houses of Parliament are not public authorities for the purposes of s.6, however, and neither are persons who exercise functions in connection with Parliamentary proceedings.

Where a public authority fails to comply with its s.6 duty (i.e. when it acts unlawfully in the s.6 sense), it does not commit a criminal offence (unless its conduct is criminal under the general criminal law). Rather, the potential effects of an act which is (or would be if it took place) unlawful under s.6 are that, under s.7, a person who is or would be a victim of the relevant act (in the Article 34 sense, seen earlier) may either bring proceedings against the public authority under the HRA (e.g. via an application for judicial review) or may rely on the relevant Convention rights in the course of other legal proceedings, whether civil or criminal. When a court (including a tribunal) finds that an act or a proposed act of a public authority is unlawful, the court, under s.8, may grant any remedy, etc., which is within its powers but damages may only be awarded by a court which possesses the power to award damages or order the payment of compensation in civil proceedings (which does not include the Crown Court). Moreover, the court may only award damages if, in all the circumstances and taking into account the principles applied by the European Court of Human Rights in this context, the award of damages is necessary to award 'just satisfaction'. In practice, where human rights have been violated and a civil claim for damages results, it seems that, where this is possible, it would be sensible for the victim to rely on 'traditional' common law or statutory rights (such as a civil claim in tort) as well as relying on s.7 of the 1998 Act, because the level of damages awarded under s.8 may not equate with that which might be awarded in the context of a claim based on such rights.

Where a person's Convention rights are violated not by a public authority but, rather, are violated by a private person or body, the conduct of the person or body will not be unlawful under HRA s.6. Moreover, the victim will not, under s.7, be entitled to bring proceedings under the HRA. This does not mean, however, that the violation of Convention rights will be of no significance. This is so because if the dispute between the parties comes before a court of law or a tribunal, the court or the tribunal will be under the s.3(1) duty to read legislation where possible in a Convention-compatible way and the s.6 duty not to act unlawfully. Thus, it appears that the violation of Convention rights may still be of some relevance even though the rights were not violated by a public authority (*Venables v News Group Newspapers* (2001) QBD).

HRA s.12 concerns those circumstances in which the court is considering granting any relief which might affect the Convention right to freedom of expression (guaranteed by Article 10). Essentially, s.12 requires the court to pay particular regard to the importance of the Convention right as well as to other specified matters. The section also provides that relief should not be granted in the absence of the respondent unless either the applicant has taken all reasonable steps to notify him or there are compelling reasons for not notifying him. Moreover, s.12 also provides that relief must not be granted so as to restrain pre-trial publication unless the applicant is likely to establish that publication should not be allowed. One particular problem which the courts have had to deal with in the context of s.12 is that of balancing the right to freedom of expression against the right to respect for private and family life, etc. (guaranteed by Article 8). Neither right takes precedence over the other, with the court being required to balance the proportionality of interfering with one right against the proportionality of interfering with the other (*Campbell v MGN Ltd* (2005) HL).

Finally, the HRA s.13 provides that when a judicial determination of a question under the HRA might affect the exercise of the Convention right to freedom of thought, conscience and religion by a religious organisation or its members (guaranteed by Article 9), the court must have particular regard to the importance of the Convention right.

Discussion questions

1 To what extent should business people be able to have the freedom to develop their businesses without legal constraints?
2 Why does the law impose constraints on many aspects of business and personal life?
3 Is Parliament (through legislation) or the courts (with judicial precedent) best placed to develop and change the law in this country?
4 Is a breach of the law ever justified? If you think 'yes', on what basis?
5 In a property context, what is the role of the Human Rights Act 1998?

2 Dispute resolution

Mark Allison

In any situation where things go wrong, causing a dispute or conflict between people or organisations, the parties involved may think of turning to the 'law' for a remedy or a solution. So, if you have bought faulty goods, or want to claim against a motorist who has damaged your car or are owed money and want to recover the debt, you may wish to take proceedings in the courts against the relevant party.

It would be wrong to think only of the courts when considering the settlement of disputes. Some measures of self-help can be very effective. Alternative forms of dispute resolution, such as mediation, are now well-established. In other instances, the parties may prefer the privacy and convenience of arbitration. Sometimes legislation has provided that a forum other than the courts is more appropriate for the settlement of disputes, for example, those cases where any rights must be pursued through specialist tribunals. There are also occasions where a dispute reveals no cause of action capable of being pursued through the courts, and in those cases, it may be that intervention by an ombudsman may be more appropriate. Ombudsmen cover many sectors, including Property, Financial Services and Local Government and Social Care. The scope of these various procedures is considered separately.

Self-help

Where a dispute arises between parties, it may be possible to settle matters by negotiation without the need to go to court. For example, in the law of trespass, the person in possession of land can ask the trespasser to leave, and if he does not do so, may then use a reasonable amount of force to eject him. What is reasonable will vary with the circumstances of each case.

More specialised examples of self-help can be found in the rules relating to set-off and lien. Set-off can occur where one party owes money to another and is, in turn, owed money by that other. So, if A owes B £10, and B owes A £5, then A may set off the £5 he is owed by B and thereby reduce the amount he must pay to B.

A lien is the right of a person in possession of goods which belong to someone else to retain them until some demand, usually a demand for payment, has been met. This remedy can be particularly useful to people such as garage owners who have carried out expensive repairs on a vehicle. They can hold on to the vehicle, i.e. exercise their right of lien, until their bill is paid.

These are legally recognised forms of self-help, but less formal methods may be equally effective. Large numbers of pressure groups now exist which may support individuals who have grievances or who are involved in disputes; for example, *Which?* in consumer rights

DOI: 10.1201/9781003657101-2

or *Shelter*, which provides advice and advocacy on housing problems. The power of the press and the media can sometimes achieve more than the use of formal legal procedures; for instance, media coverage has led to rogue landlords being prosecuted or prompted public bodies to reverse unfair decisions. Moreover, the intervention of an advice agency may be sufficient to prompt the other party to a dispute to resolve matters without resorting to court action.

Court procedures

If proceedings in court become necessary, there is an elaborate system of civil and criminal courts. The vast majority of civil and criminal matters are disposed of by the inferior courts, respectively, the county courts and magistrates' courts (though it should be noted that magistrates' courts do also have limited civil jurisdiction). The more valuable or complex civil cases may be heard in the High Court, and the more serious criminal offences may (some of them must) be tried with a jury in the Crown Court. Following the trial of a civil or criminal matter, an appeal or appeals may be possible. The most important appellate courts are the Court of Appeal and, most important of all, the Supreme Court (which replaced the Judicial Committee of the House of Lords in October 2009).

Civil cases

The county courts and the High Court

Civil proceedings may concern disputes between individuals, disputes between individuals and companies, disputes between companies and companies or even disputes between individuals or companies and public bodies or government departments. Essentially, the party who brings the claim is known as the claimant (he or she was formerly known as the plaintiff) and the party against whom it is brought is known as the defendant. Civil proceedings may have a variety of purposes which may, for example, include obtaining damages as compensation for some civil wrong (for example, where the claimant suffered personal injuries in consequence of the defendant's negligence) or in respect of a breach of contract, enforcing performance of a contract, preventing the continuation of a civil wrong, such as a nuisance, by obtaining an injunction or recovering land from a squatter. Often a civil dispute can be resolved by agreement between the parties prior to commencing proceedings in the civil court, and it is only if such negotiations between the parties fail to result in a settlement that the expense and uncertainty of proceedings in the civil courts seem necessary.

Civil proceedings may take place in a county court or in the High Court. County courts are essentially local courts which deal with the less valuable or less complex civil matters, whereas the High Court deals with the more valuable or more complex civil matters, but since the introduction of the Civil Procedure Rules 1998 (CPR 1998), the distinction between the role of county courts and that of the High Court has become more blurred.

Most of the judges who sit in county courts are district judges, but the most senior judges who sit in the county courts are circuit judges, most of whom also sit in the Crown Court (see criminal proceedings later on). Whilst most of the work of the county courts is done by district judges, certain types of claim must be heard by a circuit judge, not by a district judge, and some types of claim must be heard by a circuit judge unless the parties and the judge agree otherwise.

The High Court is based at the Royal Courts of Justice in London, but also has district registries around the country, and High Court judges also sit outside London in the various county court districts. Many applications to the High Court, for example, applications for injunctions, are dealt with not by High Court judges but, at the Royal Courts of Justice, by Masters or otherwise by district judges. The High Court is divided into three divisions: the Chancery Division, the Family Division and the King's Bench Division (formerly the Queen's Bench Division). The King's Bench Division deals, for example, with actions in contract and tort, actions for the recovery of land, applications for judicial review and encompasses the Divisional Court, the Commercial Court, the Admiralty Court, the Technology and Construction Court and the Administrative Court. The matters for which the Chancery Division is responsible include, for example, matters concerning dealings in land, mortgages, bankruptcy, administration of estates and intellectual property and the Chancery Division encompasses the Bankruptcy and Companies Court and the Patents Court. The Family Division is responsible for matters such as divorce, adoption and Children Act 1989 proceedings.

For debts of up to £100,000, the action can be processed without legal assistance through the government's online civil money claims system, provided the claim is for money only and otherwise straightforward.

In general, a claimant can choose whether to commence proceedings in the High Court or in a county court, though some claims (e.g. a money claim for £100,000 or less or a personal injuries claim for less than £50,000) should not be commenced in the High Court. Moreover, legislation or rules of court may specify that a county court does not have jurisdiction in respect of certain claims. Certain types of claim (such as applications for judicial review) must always be commenced in the High Court. Where proceedings could be commenced either in the High Court or in a county court, the factors that should govern the claimant's choice are the financial value of the claim, its complexity and whether it is of general public importance.

Pre-action protocols

A Pre-action Protocol (annexed to the Civil Procedure Rules) sets out what must be done, in relation to a claim to which it applies, before court proceedings are issued. The types of claims which have Pre-action Protocols include the following:

- Clinical Negligence
- Personal Injury Claims
- Construction and Engineering Disputes
- Debt Claims
- Professional Negligence
- Possession Claims based on Mortgage or Home Purchase Plan Arrears in Respect of Residential Property
- Possession Claims by Social Landlords
- Housing Conditions Claims (England)
- Housing Disrepair Cases (Wales)
- Claims for Damages in Relation to the Physical State of Commercial Property at Termination of a Tenancy (the 'Dilapidations Protocol')

The aims of Pre-action Protocols include ensuring both parties have access to documents and information at an early stage and to allow appropriate offers to resolve the issues to be made. Failure to comply with a Pre-action Protocol will be taken into account in any court proceedings which follow. The defaulting party may be ordered to pay additional costs resulting from his or her failure.

A full list of current Pre-action Protocols is set out in the Practice Direction – Pre-Action Conduct and Protocols, which was most recently updated in 2025.

Claim form

Essentially, civil proceedings are commenced by the court issuing a claim form (which, amongst other matters, identifies the parties, provides basic details of the claim and specifies the remedy which the claimant seeks), which the claimant or his legal advisor has prepared. The claim form must then be served on the defendant, normally within four months of being issued, and will normally contain or be served with particulars of claim (which, amongst other matters, contains a concise statement of the facts as alleged by the claimant), though the particulars can follow within 14 days if this is within the four-month period following the issuing of the claim form. The particulars of claim will normally be accompanied by the defendant's 'response pack' (containing various forms that the defendant may need), though the response pack will not be required in relation to certain claims that do not raise substantial factual issues (known as 'Part 8 claims').

Defence

When the defendant receives the particulars of claim he or she may, within 14 days, either admit the claim by filing (i.e. delivering to the court) a form of admission or contest the claim by filing a defence. Alternatively, if the defendant needs more than 14 days before filing a defence or if they dispute the court's jurisdiction, they may, again within 14 days, file an acknowledgement of service. If an acknowledgement is filed, the time for filing a defence is extended to 28 days from the date of service of the particulars of the claim. If the defendant does not admit the claim but fails to file either a defence or an acknowledgement of service (other than in the context of a Part 8 claim), the result may be that the claimant can obtain a default judgment, which, depending upon the nature of the claim, may be obtained with or without a hearing, though the court may set aside a default judgment upon application by the defendant.

If the defendant files a defence, the defence should, amongst other matters, indicate which of the claimant's allegations the defendant admits, which they deny and which they can neither admit nor deny but require the claimant to prove. The defendant may also make a 'Part 20 claim' against the claimant or against some other party (such as a co-defendant or a third party), for example, where he or she claims a remedy against the claimant (i.e. a 'counterclaim') or where he or she claims 'contribution' or 'indemnity' from the co-defendant or third party in respect of potential liability to the claimant.

After the defendant has filed a defence or an acknowledgement of service, the claimant may apply for summary judgment against the defendant. Equally, the defendant may apply for summary judgment against the claimant. The court may give summary judgment if, respectively, the claimant or the defendant has no real prospect of success, and there is no other compelling reason for disposing of the case at trial. Summary judgment may not be given in certain types of proceedings (e.g. proceedings for the possession of residential premises brought against a mortgagor or a tenant who is protected by the Rent Act 1977 or the Housing Act 1988). The most extreme consequences of a summary judgment hearing are that the court may give judgment on the claim or, conversely, may strike out or dismiss the claim. Another possibility is that the court might make a conditional order, which requires a party to take certain steps, and if he fails to do so, his statement of case will be dismissed or his claim struck out.

Directions and interim remedies

Prior to the trial itself, one or more of the parties to civil proceedings may find it necessary to apply to the court for directions or for interim remedies. The former may relate to matters such as the use of expert evidence (the civil courts now possess broad powers to limit the admissibility of expert evidence and to control the form in which such evidence will be given). The latter may concern remedies, such as an interim injunction, which can either prohibit a party from acting in a certain way or, less commonly, require him or her to perform a particular act. In some circumstances, it may be necessary to apply for an interim remedy, such as an injunction, before proceedings have commenced. Normally, notice of the making of an interim application must be given to the other parties, but in some circumstances, this will not be possible or (where secrecy is required) desirable. Some applications (such as those for interim injunctions) must be supported by evidence. It may be necessary for the court to hold a hearing to deal with an interim application, but some interim applications may be dealt with or without a hearing. At the end of an interim application, the court will normally make a costs order, which will normally, but not always, be made in the successful party's favour.

Apart from the injunction, another important interim remedy is the court's power to strike out part or all of the statement of case (i.e. claim form, particulars of claim, defence or Part 20 claim) of the claimant or defendant. The court may do so where the statement of claim does not disclose reasonable grounds for the bringing or defending of the claim, where the statement of case abuses the court's process or otherwise obstructs the just disposal of the proceedings (for example, where the claimant could reasonably have brought the claim in the context of earlier proceedings against the defendant) or where the relevant party has failed to comply with a rule of court, a practice direction or a court order.

A further important interim remedy is the court's power to order an interim payment. Essentially, the court may so order if the defendant admits his or her liability to pay damages or money or the claimant has obtained judgment for damages or a sum of money to be assessed against the defendant or where the court is satisfied that if the claim went to trial the claimant would obtain judgment for a substantial sum or the claimant seeks possession of land and the court is satisfied that if the case went to trial the defendant would be held liable to pay the claimant a sum of money. The interim payment must not exceed a reasonable proportion of the likely final amount, and in reaching its decision, the court must take contributory negligence, set-off and counterclaim into account. Clearly, if the interim payment exceeds the amount finally awarded at the trial, an order for repayment (and for the payment of interest) will be required.

Another matter which may be the subject of interim applications is that of the disclosure and inspection of documents. Where a case is allocated to the fast track or the multi-track, the court will normally order disclosure of documents between the parties.

In most multi-track cases proceeding in the Business and Property Courts, disclosure is governed by Practice Direction 57AD, which sets out a scheme of *extended disclosure* in place of standard disclosure. In other multi-track and fast track cases, the court will usually order standard disclosure, which essentially requires the parties to disclose documents on which they rely, which adversely affect their case or another party's case or which support another party's case. This means that the party must state in a list of documents that the document does or did exist. The other party then has the right to inspect (and to request a copy of) the relevant document unless the party who disclosed it no longer has it in their control or has a right or duty to withhold inspection of it (e.g. because it is privileged) or because he or she consider that it would be disproportionate to permit inspection of it. A party will find it necessary to apply to the court for an order where he or she seeks specific disclosure or specific inspection (i.e. the disclosure or inspection of specific documents or classes of document), where he or she seeks disclosure before the proceedings have started, where he or she seeks disclosure against a person who is not a party or where he or she seeks an order permitting him/her to withhold disclosure or inspection of a document.

Case management

Under the CPR 1998, the civil courts possess extensive case management powers and are required to actively manage cases. The courts should exercise these powers so as to give effect to the overriding objective of the CPR 1998 – namely, dealing with cases justly. This essentially requires the court, to the extent to which this is practicable, to ensure that the parties are on an equal footing; to save expense; to deal with cases in a way which is proportionate to the money involved, their importance, their complexity and the respective financial positions of the parties; to ensure that cases are dealt with speedily and fairly and to allot an appropriate share of the court's resources to each case.

Active case management includes encouraging the parties to cooperate, identifying the issues as early as possible, deciding quickly which issues need to be resolved at trial, deciding in which order issues should be resolved, encouraging the parties to use alternative dispute resolution where appropriate, helping the parties to settle the case, fixing timetables and controlling the case's progress, considering whether the benefits of a step justify the expense of taking it, dealing with as many issues as possible on the same occasion, dealing with a case without requiring the parties to attend court, using technology and giving directions so as to ensure that trial is fast and efficient. The use of remote hearings and electronic communication has become increasingly common in furthering these aims.

The court possesses the power to give directions concerning the issues in relation to which evidence is required, the nature of the requisite evidence and the method by which it is to be put before the court, and in exercising its powers may both exclude admissible evidence and limit cross-examination. Where a party fails to comply with a rule, a practice direction or a court order, the court can impose a variety of sanctions. These may vary at one end of the spectrum from sanctions concerning costs to, at the other end, the draconian sanction of striking out a party's statement of case. The court may deprive a party of the ability to rely upon certain evidence if, for example, a party failed to disclose a witness statement or an expert's report as directed.

Track allocation

Important case management decisions which the court (often a district judge or a master) will normally be required to make include both whether the case should be transferred to a different court (e.g. to the county court for the district where the defendant lives or works) and which 'track' the case should be allocated to. A case may be allocated to one of the three tracks:

- the small claims track,
- the fast track or
- the multi-track.

Essentially, the small claims track is the track for the least valuable or important cases, e.g. for claims the value of which does not exceed £10,000. In personal injury cases, the position is more nuanced: claims arising from road traffic accidents will usually be allocated to the small claims track unless the general damages exceed £5,000, while for other personal injury claims, the threshold remains £1,000. The small claims track also applies to claims by residential tenants against landlords, requiring the landlord to carry out repairs where the cost of the repairs does not exceed £1,000.

Where the small claims track is not the appropriate track, the fast track will be the normal track if the financial value of the claim does not exceed £25,000, the length of the trial is not likely to exceed one day, and oral expert evidence at trial will be limited to no more than two expert fields and no more than one expert witness per party per field.

In deciding which track a case should be allocated to, the court will normally take into account Directions Questionnaires (formerly allocation questionnaires) which have been filed by the parties and may find it necessary to require further information or hold an allocation hearing. The matters that the court will take into account when allocating a claim to a track will include its financial value, the remedy sought, the complexity of the case, the number of parties, the value or complexity of a Part 20 claim, the amount of oral evidence that will be required, the importance of the claim to non-parties, the views of the parties and their circumstances. Essentially, the consequences of allocation to the small claims track are that the pre-trial procedure and the trial itself will be simpler and cheaper, the rules concerning costs are different and the parties' rights of appeal are more limited. Conversely, the more valuable and/or complex cases will be allocated to the multi-track. Cases which are allocated to the multi-track will normally take longer to come to trial (and to try) than fast track cases and will often raise more complex case management issues than fast track cases. Cases allocated to the small claims track will be tried in a county court. It is normally only multi-track cases that will be tried in the High Court, but the vast majority of multi-track cases are tried in county courts and only the most valuable, complex or important multi-track cases will be tried at the Royal Courts of Justice in London.

Trial

Civil trials differ from criminal trials in a number of respects. One obvious one is that a civil trial, unlike a criminal trial in the Crown Court, is normally heard by a judge sitting alone. Jury trial in civil proceedings is now extremely rare. Defamation claims (i.e. libel and slander) were historically the best-known exception, but even in those cases, the default is now trial by a judge alone.

Another significant respect in which civil trials now differ from criminal trials is that the evidence-in-chief of a witness who has been called (i.e. the evidence that the witness gives for the party who calls him or her) now normally takes the form of the witness's written witness statement as opposed to oral testimony, though cross-examination (i.e. examination of the witness on behalf of the party who did not call him or her) will be oral. Further, expert evidence is normally given in civil proceedings by written report, experts now rarely being called to give oral evidence at a civil trial. Moreover, a civil court may, in appropriate circumstances, limit expert evidence in relation to a particular issue to the evidence of a single joint expert who is jointly instructed by claimant and defendant.

The legal burden of proof in civil proceedings is borne by the party who raises an issue. Thus, in the context of a negligence claim, the claimant would bear the legal burden of proving the defendant's negligence, but the defendant, if he or she relied upon the defence, would bear the legal burden of proving the claimant's contributory negligence. The standard of proof is that of proof upon the balance of probabilities, i.e. the party who bears the legal burden of proving an issue must prove that it is more probable than not that the issue is as he or she asserts.

Civil trials normally take place in public, but the court may order that a hearing take place in private for a variety of reasons and, for example, the hearing concerns confidential matters, such as personal financial matters, or where it would be in the interests of justice to do so. In particular, claims by mortgagees against individuals for possession of land and claims by residential landlords against tenants for repossession of dwelling houses in consequence of non-payment of rent will normally be heard in private.

At the end of the trial, when the court gives its judgment, the judge may find it necessary to make a variety of orders. Thus, for example, the court may find it necessary to order the sale, mortgage, exchange or partition of land, and consequent upon such an order, may find it necessary to order a person to deliver up either or both the possession of land or the rents or profits thereof. In particular, the court will normally make a costs order, normally, though not necessarily, in the successful party's favour. Matters that the court will take into account in the context of making a costs order will include whether a party made an offer to settle (including a formal offer to settle under Part 36) or a payment into court, whether part of a party's case was successful and the conduct of the parties in the course of the proceedings.

Civil appeals

Essentially, in the context of civil proceedings, the appellate structure is that a decision of a county court district judge may be appealed to a circuit judge, a decision of a master, a High Court district judge or a circuit judge may be appealed to a High Court judge, and a decision of a High Court judge may be appealed to the Court of Appeal. Where the county court or the High Court would normally hear an appeal but the appeal either raises an important point of principle or practice or there is some other reason why the Court of Appeal should hear it, the court from or to which the appeal is made or the court from which permission to appeal is sought may order the appeal to be transferred to the Court of Appeal.

Permission to appeal is normally required either from the court which made the decision at the time of the hearing or from the appellate court. Where an appeal to the Court of Appeal from the High Court or the county court itself concerns an appeal to the High Court or the county court, permission to appeal to the Court of Appeal will be required from the Court of Appeal. Permission to appeal will only be given either where the appeal would have a real prospect of success or where there is some other compelling reason why the appeal should be heard.

An appeal will not normally take the form of a rehearing of the case (but, rather, will take the form of a review of the lower court's decision) unless it is in the interests of justice to hold a rehearing. The appeal court will not normally receive either oral evidence or evidence that was not before the lower court, though it does possess the power to receive either or both of these types of evidence. An appeal will be allowed either where the decision of the lower court was wrong or where it was unjust in consequence of a serious procedural or other irregularity. The appeal court has a variety of powers (e.g. it could affirm, set aside or vary the lower court's judgment or order a new trial).

The Court of Appeal normally sits at the Royal Courts of Justice in London, and the Court normally consists of three Lords Justices. Exceptionally, an appeal may be made from the Court of Appeal to the Supreme Court (which has replaced the Judicial Committee of the House of Lords). A case coming before the Supreme Court will involve points of law of general public importance, and the permission of either the Court of Appeal or the Supreme Court will be required in order to bring the appeal. The Supreme Court must comprise at least three Justices of the Supreme Court (formerly, Lords of Appeal in Ordinary), though, in practice, usually five will sit to hear an appeal and at times more Justices will sit, e.g. seven or nine (there must always be an odd number of Justices).

Occasionally, an appeal may go directly from a High Court judge to the Supreme Court using what is known as the 'leapfrog procedure', but this will only be possible if the trial judge, upon application by a party, certifies that he is satisfied that the case involves a point of law of general public importance, that the point of law concerns the construction of a statute (and has been fully argued in the proceedings and fully considered in the judge's judgment) or a matter in relation to which the judge is bound by the doctrine of precedent, that a sufficient case to justify a leapfrog appeal has been made out and that the parties consent to the grant of a certificate by the judge. Even where the trial judge does grant a certificate, the leapfrog appeal can only be brought with the permission of the Supreme Court.

Criminal cases

The scope of the criminal law is much wider than merely dealing with cases, like murder and theft, which are obviously criminal. Rather, the modern criminal law regulates a variety of matters which do not at first glance appear to concern the criminal law (everything from health and safety at work, to unlawfully felling trees protected by a preservation order, where one is required, to corporate manslaughter). Thus, today, no business or commercial organisation can afford to ignore the criminal law. Criminal trials are conducted in accordance with Criminal Procedure Rules and Criminal Practice Directions.

Crown Court

Most criminal trials take place in magistrates' courts, such trials being known as 'summary trials' (e.g. driving without insurance). Some, however, including all of those for the most serious offences, take place in the Crown Court, such trials being known as 'trials on indictment' (e.g. murder or rape). (Note: the document containing the charges against the accused to which he or she pleads guilty or not guilty in the Crown Court is called the 'indictment' and, consequently, trial in the Crown Court is known as 'trial on indictment'.) Some criminal offences (summary offences, such as assault and battery) must be tried in a magistrates' court. Other criminal offences (offences triable only on indictment, such as rape) must be tried in the Crown Court. A third category of case (offences triable either way, such as health and safety offences or theft) may be tried either in a magistrates' court or in the Crown Court.

The Crown Court sits at numerous centres around England and Wales, the most famous being the central Criminal Court or 'Old Bailey' in London. In the Crown Court, a case is tried by a judge sitting with a jury. Depending upon the nature of the offence with which the accused is charged, the judge may be a High Court judge, a circuit judge or a recorder (a part-time Crown Court judge). The jury consists of 12 persons who are between 18 and 75 years old, are registered as electors, have been ordinarily resident in the UK for a period of at least five years since they were 13 years old, are not mentally disordered and are not disqualified (in consequence of their criminal record or because they are on bail). Persons eligible to serve on a jury who believe that there is a good reason why they should be excused from serving may apply to be excused or to defer their service, but this is at the discretion of the court.

The judge governs the trial process, determines questions of law, such as the admissibility of evidence, sums up the case for the jury (both directing them as to the law and summarising the evidence for them) and determines the accused's sentence, if he or she is convicted. The jury determines questions of fact and, fundamentally, determines whether the accused is guilty or not guilty of the offence(s) with which he or she is charged.

Magistrates' courts

Magistrates' courts also sit in areas all around England and Wales. In a magistrates' court, the case will be tried (and a sentence imposed if the accused is found to be guilty) either by three unpaid justices of the peace (i.e. magistrates) who are advised by a legally qualified clerk or by a district judge. Justices of the peace need not be (and most will not be) legally qualified.

Criminal proceedings always commence in a magistrates' court, and this is so even where the accused is charged with an offence which is triable only on indictment. The accused will normally have ended up before the magistrates either because he or she has been arrested by the police, interviewed by them and charged with a criminal offence or because he or she has been summoned to attend court. Where the accused has been arrested and charged, the decision to charge will normally have been made by the Crown Prosecution Service, although in some less serious cases, the police may charge without referring to the CPS.

In relation to those offences in respect of which there is not normally a power of arrest (e.g. speeding), the magistrates will issue a summons requiring the accused to attend court. Even where the accused has been charged with a criminal offence, the Crown

Prosecution Service may decide not to prosecute, perhaps because there is not a realistic prospect that the accused will be convicted given the weakness of the evidence against him or perhaps because it would not be in the public interest to prosecute him, given his age, poor state of health, the trivial nature of the offence, etc.

Bail

Whether the accused is eventually to be tried in a magistrates' court or in the Crown Court, a decision which the magistrates will be required to make when the accused first appears before them (unless the accused is attending in response to a summons and has not been remanded in custody or on bail) is whether the accused should be remanded in custody or remanded on bail.

Essentially, subject to certain exceptions, there is a general presumption in favour of bail, but bail can be refused on a variety of grounds. For example, where the offence is imprisonable, bail may be refused: if there are substantial grounds for believing that the accused might fail to surrender to custody, might commit an offence whilst on bail or might interfere with witnesses; if the defendant should be kept in custody for his own protection; if the defendant is already serving a custodial sentence; if there has not been sufficient time to gather sufficient information to make a decision regarding bail or if the defendant has previously failed to surrender in connection with the current offence.

If the accused is granted bail, this may be subject to conditions, such as a curfew, a requirement that he resides in a bail hostel or the taking of sureties (persons who guarantee that the accused will attend court and who will potentially lose a specified sum of money if the accused fails to attend) or electronic monitoring, for example, through the use of an electronic tag to enforce a curfew or monitor compliance with location restrictions.

Mode of trial hearing

In relation to 'either way' offences, the magistrates will be required (at a 'mode of trial hearing', formally known as the allocation decision) to determine whether to hear the case themselves or whether to commit the accused for trial in the Crown Court. If the accused indicates that he will plead guilty, the magistrates must sentence the accused unless they decide that their sentencing powers are not sufficient, in which case they must commit the accused to the Crown Court for sentence.

Trial preparation

Prior to a criminal trial, the prosecution is required to disclose to the defence both the material that they intend to rely on at the trial and 'unused material' (i.e. material they are not relying on which might undermine their case). The defence then may (*must* in the case of Crown Court trial) serve a defence statement which, amongst other matters, sets out the nature of the accused's defence, indicates the matters in relation to which the accused takes issue with the prosecution and indicates why he does so. This may then draw the attention of the prosecution to other unused material, which they will be under a duty to disclose. The prosecution has a continuing duty to keep under review whether there is any unused material which might reasonably be considered capable of undermining the prosecution's case or of assisting the defence case.

In the context of a Crown Court trial, there will be a Plea and Trial Preparation Hearing (PTPH) (formerly known as a Plea and Case Management Hearing or PCMH) before the trial for the purpose of determining the accused's plea and resolving a variety of procedural, legal and evidential issues pre-trial. There is no PPTH in the context of a summary trial, though there may be a 'pre-trial hearing' to deal with the admissibility of evidence or questions of law. Moreover, it should be noted that in relation to some summary offences it is possible for the accused to plead guilty by post.

Trial

Following a not guilty plea, the trial process itself is similar whether the trial takes place in the Crown Court or in a magistrates' court. It should be noted, however, both that there is provision in the context of summary trial for the trial to take place in the accused's absence in certain circumstances, particularly where the offence is not punishable by imprisonment or where it is in the interests of justice to proceed, and that whilst in the context of trial on indictment prosecution and defence counsel normally both make opening and closing speeches, the usual practice in a magistrates' court is that the prosecution make an opening speech and the defence make a closing speech.

Essentially, the trial process is that the prosecution calls their witnesses first, the witnesses being examined in chief by counsel for the prosecution, cross-examined by defence counsel and then may be re-examined by prosecution counsel. The defence may then, if appropriate, submit that the accused has no case to answer. If the judge or magistrates accept this submission, the accused will be acquitted. If the submission fails, then the defence will call their witnesses whom they will examine in chief, the prosecution will cross-examine, and the defence may re-examine.

Essentially, the purpose of examination-in-chief is, if possible, to obtain the evidence from the witness upon which the party calling the witness intends to rely in support of his case. The purposes of cross-examination may include obtaining evidence from the witness which supports the cross-examining party's case, contradicting the witness's evidence-in-chief or discrediting the witness (for example, by showing that he has previous convictions or is biased against the accused). The purpose of re-examination is to deal with matters arising from cross-examination, not to permit the re-examining party to adduce new evidence that he failed to adduce during examination-in-chief.

When all of the prosecution and defence witnesses have been examined and following any closing speeches (and, in the case of a trial on indictment, the judge's summing up to the jury), the magistrates, district judge or jury determine the guilt or innocence of the accused.

Sentence

If the accused, having pleaded not guilty, is convicted, or if the accused pleads guilty, the judge or the magistrates will determine the sentence. They may do so immediately or may adjourn to await the production of a pre-sentence report (which will normally be prepared by a probation officer). Before imposing a sentence, the judge or the magistrates will have the facts of the case summarised for them by the prosecution and will be informed by the prosecution of the accused's antecedents. The accused will be entitled to a defence mitigation speech, in which defence counsel has an opportunity to put before the court factors which may potentially persuade the court to impose a lesser sentence.

The maximum sentence that the court is entitled to impose in relation to an offence may vary from a small fine at one end of the spectrum to life imprisonment at the other. Examples of other possible sentences which the court may be entitled to impose, apart from fines and custodial sentences, include an absolute or conditional discharge, a suspended sentence or a community sentence. A community sentence will include one or more of the following requirements: an unpaid work requirement, an activity requirement, a programme requirement, a prohibited activity requirement, a curfew requirement, an exclusion requirement, a residence requirement, a mental health treatment requirement, a drug rehabilitation requirement, an alcohol treatment requirement or an attendance centre requirement

The maximum sentence that a magistrates' court is entitled to impose for a single offence is 12 months' imprisonment (s.13 Judicial Review and Courts Act 2022).

Tribunals

A significant feature of dispute-solving since 1945 has been the rise in the number of tribunals created by Act of Parliament to deal with specific questions. The reasons for the growth of the tribunal system were that:

- The volume of work could not be given to the ordinary courts as the system would have become overloaded.
- The questions to be resolved by tribunals were frequently specialised, and the expertise of specialists could be used in the decision-making.
- Tribunals could dispose of cases quickly, cheaply and informally.

Each type of tribunal had its constitution fixed by the statute creating it. For example, the Lands Tribunal was established by the Lands Tribunal Act 1949. Many of the earlier tribunals have since been replaced or restructured and a new tribunal structure was created, under the Tribunals, Courts and Enforcement Act 2007, consisting of the First-tier Tribunal and the Upper Tribunal along with the Employment Tribunal and others, such as the Valuation Tribunal (dealing with council tax and business rates disputes), and those dealing with specialist matters, for example, mental health, school exclusions, plant varieties and forestry.

The First-tier Tribunal is currently divided into seven chambers, including the Property Chamber, which handles applications, appeals and references relating to disputes over property and land. Residential property disputes handled include the following:

- Rent increases for 'fair' or 'market' rates
- Leasehold disputes
- Leasehold enfranchisement
- Improvement notices and prohibition orders where notices are given under the Housing Act 2004

Land registration matters handled include the following:

- Disputes over a change to the land register
- Applications to correct or cancel certain documents relating to registered land

Agricultural land and drainage matters handled include the following:

- Some disputes between agricultural tenants and landlords in relation to Agricultural Holdings Act 1986 tenancies (note that Agricultural Tenancies Act 1995 disputes and some 1986 Act matters are dealt with in arbitration)
- Applications in respect of certain drainage disputes between neighbours

The Upper Tribunal (Lands Chamber) handles appeals against decisions made by the First-tier Tribunal (Property Chamber) and handles applications about cases, including the following:

- Compensation for the compulsory purchase of land
- Discharge or modification of land affected by a 'restrictive covenant'
- 'Right to light' disputes
- Compensation for blighted land
- Electronic Communication Code – disputes involving masts and other telecommunications equipment on land

The other three Upper Tribunal Chambers are the Administrative Appeals Chamber, the Immigration and Asylum Chamber and the Tax and Chancery Chamber.

Arbitration

Going to court can be an expensive, public and time-consuming activity for the parties to a dispute. It may be better to agree to arbitrate when the parties themselves can exercise some control over the choice of arbitrator. This process is particularly useful where a dispute involves points of a technical nature. In some types of contract (e.g. construction contracts), it is common for the parties to agree not to refer disputes arising from those contracts to the court. Then, if a dispute does arise, it will be a breach of contract if one of the parties tries to take the case to court instead of arbitration. Although the parties must finance the arbitration themselves, the process has appeal because of its speed and privacy. Once the arbitrator gives their decision (known as 'the award'), it can be enforced like a court judgment.

The law governing arbitration is now to be found in the Arbitration Act 1996 and the general law of contract. The 1996 Act is intended to restate and improve the law relating to arbitration and is founded on the following principles:

a the object of arbitration is to obtain the fair resolution of disputes by an impartial tribunal without unnecessary delay or expense;
b the parties should be free to agree how their disputes are resolved, subject only to such safeguards as are necessary in the public interest and
c in matters governed by Part I of the Act, the court should not intervene except as provided by Part I.

For the Act to apply to an agreement to submit to arbitration in present or future disputes, it must be in writing, although this is defined very widely. The written agreement need not name the arbitrator or indeed specify the number of arbitrators. In the latter situation, the Act states that in the absence of agreement as to numbers, there shall be one arbitrator.

On normal contractual principles, parties are, in general, free to fix the terms of their own agreement and can decide which disputes the arbitration agreement will cover. The House of Lords (in *Fiona Trust and Holding Corp. v Privalov* (2007)) held that where the issue of whether a dispute is covered by an arbitration clause arises, unless the arbitration clause makes a clear provision to the contrary, the assumption is that the parties intended it to be determined by arbitration.

Whilst parties cannot oust the jurisdiction of the courts, it is possible for them to exclude by agreement an appeal against the arbitrator's award. In any event, it is only possible to appeal on a point of law, lack of jurisdiction or procedural irregularity, and not on findings of fact. The courts have power to supervise arbitration, and may remove an arbitrator, for example, for lack of impartiality, or for a failure 'to use all reasonable dispatch in conducting the proceedings or making an award and substantial injustice has been or will be caused to the applicant' (s.24(1)(d)(ii) Arbitration Act 1996).

A number of terms are implied into an arbitration agreement by the Arbitration Act 1996, unless the parties have agreed otherwise. These include that the parties to the agreement must be prepared to be examined on oath by the arbitrator; that they must produce all documents required by the arbitrator; that the arbitration will be final and binding on the parties and that the arbitrator can award costs. Once the arbitrator makes his decision, 'the award', it can be enforced just like a court order or judgment.

The distinction between arbitration and valuation has always been regarded as problematic. In arbitration the aim is to settle an existing dispute. A valuation seeks to prevent a dispute arising by allowing a third party to fix the value or price. The distinction is not always clear, and the third party may be anxious to be sure of his status, as a valuer can be sued for negligence, but an arbitrator cannot be sued for acts or omissions unless done in bad faith (see s.29 of the 1996 Act in this respect). This was an important point in issue in the case of *Sutcliffe v Thackrah* (1974) (HL), where an architect issued an interim certificate on the basis of which his employer paid the contractor. The employer subsequently sacked the contractor and wanted to recover damages from him for proven defective work. The contractor went into liquidation, so the employer then sued his architect for his negligence in certifying the poor work. In issuing the certificate, does the architect act as a valuer or an arbitrator? Once the architect issues a certificate, the employer is then obliged to pay. But the architect was not determining a dispute between the parties, only certifying, and his duty was to act impartially. The House of Lords held that the architect could be liable in negligence for his certification.

Whilst arbitration does have some advantages over litigation (for example, it is private and the arbitrator may have technical expertise that a judge may not possess), arbitration can sometimes take just as much time and be just as expensive as litigation. Indeed, since the Civil Procedure Rules were introduced in 1998, judges have been given significant powers to reduce the length and expense of civil proceedings (for example, by directing that expert evidence be given by a single joint expert). It is interesting to note that in the JCT Design and Build Contract 2016, if the parties wish arbitration to be the method of dispute resolution, this must be expressly stated in the contract. If this is not expressly stated, the default method of dispute resolution will be litigation.

Alternative Dispute Resolution (ADR)

Recent decades have seen the increased use of alternative forms of dispute resolution (ADR) as an alternative to litigation in the civil courts. Indeed, under the Civil Procedure

Rules 1998, a civil court must encourage the parties to use ADR, where this is appropriate. Alternative dispute resolution refers to alternatives to going to court and, therefore, strictly speaking, includes arbitration. However, as arbitration, at least prior to the Arbitration Act 1996, was widely seen as very court-like, ADR may also be viewed as an alternative to arbitration.

ADR has arisen because of concerns over the time and expense associated with litigation and, indeed, arbitration. Additionally, litigation and arbitration are adversarial processes that tend to destroy existing relationships. Hence the cost of litigating a dispute will often include the breakdown of long-term commercial dealings. The purpose behind ADR is to 'blunt adversarial attitudes' and to promote settlement of disputes, speedily and with the minimum of expense. In order for ADR to succeed, it depends upon the parties exhibiting a spirit of cooperation and ultimately compromise.

There are several forms of ADR, including mediation, expert adjudication, non-binding arbitration, the mini-trial and early neutral evaluation. However, although recognised as a form of ADR, note that adjudication is also an adversarial process. It is most commonly used in construction disputes and has statutory status under the Housing Grants, Construction and Regeneration Act 1996.

While parties are encouraged to engage with ADR voluntarily, recent case law has confirmed that the courts may, in appropriate cases, order parties to engage in non-court dispute resolution mechanisms without violating their right of access to the courts. This principle was confirmed in the Court of Appeal decision in *Churchill v Merthyr Tydfil County Borough Council* (2023) CA, which held that courts may stay proceedings to require participation in ADR, such as mediation (in this case, the Council's in-house dispute resolution process), provided this does not unjustly prevent access to justice.

Mediation

Mediation involves the use of a third party, a mediator, to promote communication between disputing parties. After a joint meeting, the mediator will then listen to each party's side of the dispute in private, but must not divulge such information, without permission, to the other party. The process will end with another joint meeting.

Usually, no view as to the dispute will be given by the mediator. However, by exploring each party's case, it is hoped that common ground between the parties will emerge, which may form the basis for settlement of the dispute by the parties themselves. Some mediators may adopt a more evaluative approach if this is agreed between the parties, but most mediations remain facilitative in nature.

A drawback in this process is that, unlike with litigation or arbitration, there is no guarantee that the dispute will be resolved. If the parties cannot settle the dispute after mediation, then the cost of mediation, which is borne by the parties, is in addition to the cost of going to court or arbitration. This drawback applies to many forms of ADR, though it is less of a concern in the case of adjudication, which results in a decision that is binding unless and until overturned.

Adjudication

Adjudication under the Housing Grants, Construction and Regeneration Act (HGCRA) 1996 (commonly known as the Construction Act) has been put on a statutory footing in relation to 'construction contracts'. Sections 104–106 provide a definition of this term.

Clearly construction work is covered, but the term 'construction contracts' also includes agreements to do architectural, design or surveying work and to give advice on building, engineering, interior or exterior decoration or on the laying out of landscape, in relation to construction operations. Certain types of contracts are excluded from the definition, such as contracts with residential occupiers.

Under the Act, any party to a construction contract has the statutory right to refer a dispute arising under the contract to adjudication, at any time, in accordance with the HGCRA. Note there is no obligation upon the parties to submit a dispute to adjudication, but if a party wishes to do so then such dispute must go to adjudication; another party cannot prevent this.

An adjudicator's decision is binding until the dispute is finally determined by legal proceedings, arbitration or agreement. The decision is intended to provide a swift, interim resolution so that construction work can continue. However, the parties may agree to treat the adjudicator's decision as final. Adjudication is now a widely used and essential feature of dispute resolution in the construction industry.

Should a construction contract not contain an adjudication provision or if such a provision fails to comply with the terms of the HGCRA, a default adjudication mechanism will be imposed by law through delegated legislation. This is done via the Scheme for Construction Contracts (England and Wales) Regulations 1998, as amended by the 2011 Regulations, which set out the default procedure to be followed.

Other forms of ADR

Non-binding arbitration is similar to mediation, but if the parties fail to settle the dispute, the arbitrator will give a view as to what will be the likely outcome of a trial of the dispute. This may have the effect of encouraging the parties to settle.

A mini-trial is a form of structured settlement. Senior representatives of the disputing parties sit with a neutral chairman and listen to arguments for each side presented by advocates and experts. The panel is then able to assess the strengths and weaknesses of each side's case. Again, the object is to encourage the senior representatives to settle the dispute.

A further option is early neutral evaluation. In this process, a neutral third party (often a judge or senior legal figure) provides a non-binding assessment of the likely outcome of the case. The aim is to give the parties a realistic picture of their position and to encourage early settlement. Early neutral evaluation has been increasingly supported by the courts, and recent case law confirms that it may be ordered even where one party does not consent.

Ombudsmen

Where a dispute involves allegations of maladministration by a local or central government department or agency, it may be possible to refer the dispute to the appropriate ombudsman. For example, complaints concerning government departments or the National Health Service in England may be referred to the Parliamentary and Health Service Ombudsman, while complaints concerning local authorities may be made to the Local Government and Social Care Ombudsman.

Where the Parliamentary and Health Service Ombudsman accepts a complaint, the ombudsman investigates the complaint and sends a report both to the government department or health authority and to the person who made the complaint. If the ombudsman finds that the complaint is justified, the report will make recommendations, but the ombudsman does not possess the power to require the government department or health authority to comply with the recommendations. Similarly, a local authority is not obliged to act in compliance with recommendations made by the Local Government and Social Care Ombudsman, though, in practice, local authorities normally do comply with such recommendations.

Other examples of ombudsmen include the Legal Ombudsman, who deals with complaints about legal professionals in England and Wales; the Housing Ombudsman, who deals with complaints against certain landlords, such as housing associations and other registered providers; and the Property Ombudsman, who handles complaints about estate agents and letting agents who are members of the Property Ombudsman Scheme.

While most ombudsmen do not have the power to enforce their decisions, their findings carry significant weight and organisations usually comply in order to maintain public confidence and avoid reputational damage.

Affording the law

Potential sources of funding civil litigation might include a conditional fee agreement (sometimes referred to as a 'no win, no fee' arrangement, under which, for example, the client may be required to pay no legal fee if the case is lost, but may have to pay an additional success fee if the case is won).

Another option is legal expenses insurance. This may be in place before the event (e.g. under a home or motor insurance policy), or the client may take out after-the-event insurance once a dispute has arisen, to protect against the risk of paying the other side's costs.

A party may also fund the litigation privately. However, civil litigation is expensive, and legal aid is now only available in limited categories of civil cases, such as some housing, family or discrimination claims. For most litigants, the decision to pursue legal action remains a significant financial commitment.

The legal profession

Practising lawyers in England and Wales have traditionally been divided into solicitors and barristers, each with distinct roles and training routes, though it is possible to convert from one profession to the other. In recent years, the legal services sector has expanded to include a wider range of regulated professionals, such as Chartered Legal Executives, Licensed Conveyancers, Licensed Probate Practitioners, Patent Attorneys, Trade Mark Attorneys, Notaries Public and paralegals. Each of these has its own route to qualification and regulatory body, and all play an important role in the delivery of legal services.

Solicitors

Solicitors are sometimes compared to general medical practitioners, with barristers being seen as the equivalent of consultants. This can be rather misleading, however, as a solicitor, more usually in a large firm, may have an extremely specialised practice, while a young barrister seeking to make his name as an advocate may have to be prepared to be a 'jack of all trades'.

Most solicitors work in private firms. These vary from very small high street firms of a few partners to large city firms. Some solicitors work for other organisations, for example, for companies, for local authorities or for the Crown Prosecution Service. Some firms specialise in very narrow areas of legal work, whereas others offer a broad range of legal services.

Solicitors possess rights of audience in county courts and magistrates' courts, but most solicitors do not possess rights of audience in the Crown Court, the High Court, the Court of Appeal or the Supreme Court. Solicitors can, however, obtain 'higher rights' and, thus, some solicitors do appear as advocates in the higher courts. Many solicitors, however, never appear in court at all but, rather, specialise in non-contentious work, such as conveyancing.

When litigation arises, the work may be passed on to their firm's litigation department and/or a barrister may be instructed to represent the client in court.

Solicitors' governing body is The Law Society, and they are regulated by the Solicitors Regulation Authority.

Barristers

Barristers in private practice are self-employed. Groups of self-employed barristers operate in sets of chambers, which provide the administrative and clerking support needed to manage their practices. In most cases, they are instructed through an intermediary such as a solicitor, patent attorney or licensed conveyancer. However, under the Public Access Scheme, it is now possible for members of the public and certain professionals (such as surveyors) to instruct authorised barristers directly. While this route remains less common than traditional referral, its use has increased in recent years.

Like solicitors, some barristers specialise in narrow specialist areas of law, whereas others have a much more general practice. Unlike most solicitors, barristers have higher rights of audience. This does not mean, however, that every barrister will spend the majority of his or her time in court. Many barristers will spend much of their time writing opinions for solicitors (i.e. advising solicitors in relation to the application of specialist areas of law to the facts of specific cases), engaged in negotiations or conferences on behalf of their clients or drafting documents on their clients' behalf.

Moreover, a considerable number of barristers are employed by other organisations (e.g. companies, local authorities or the Crown Prosecution Service). Indeed, some barristers are now employed by firms of solicitors. Some employed barristers (e.g. those employed by the Crown Prosecution Service) will spend much of their time in court, whereas others (e.g. those employed by companies) may never go near a courtroom.

Barristers are represented by The Bar Council and are regulated by the Bar Standards Board (BSB).

Other legal professionals

In addition to solicitors and barristers, legal work is carried out by a range of other qualified professionals. These include Chartered Legal Executives (qualified members of the Chartered Institute of Legal Executives (CILEx)), Licensed Conveyancers and Licensed Probate Practitioners (regulated by the Council for Licensed Conveyancers), Notaries Public and Patent and Trade Mark Attorneys (regulated by the Intellectual Property Regulation Board).

Legal work is also frequently carried out by paralegals, who may be law graduates, part-qualified solicitors or barristers, qualified legal executives or members of recognised bodies such as the National Association of Licensed Paralegals (NALP). While many paralegals work in unregulated roles, there is an increasing move towards professionalisation and voluntary regulation within this sector.

Each of these professionals plays a valuable role in the delivery of legal services and may, in some cases, have rights to conduct litigation, represent clients or carry out reserved legal activities.

Alternative Business Structures

In consequence of reforms introduced by the Legal Services Act 2007, lawyers and non-lawyers can now form Alternative Business Structures (ABSs); organisations in which legal services can be provided alongside other professional or commercial services. For example, solicitors and estate agents may operate together under a single business structure, enabling them to offer integrated services to clients.

Firms wishing to operate as ABSs must apply for authorisation from a recognised licensing authority, such as the Solicitors Regulation Authority or the Council for Licensed Conveyancers. Since their introduction, ABSs have become a common feature of the legal services market, particularly among larger firms and organisations looking to combine legal services with other areas of expertise.

Discussion questions

1 Why might parties choose to resolve a dispute through alternative dispute resolution (ADR) rather than through court proceedings? Can you think of any disadvantages to ADR in a construction or land management context?
2 How do the principles of case management and the Civil Procedure Rules 1998 help to ensure that civil cases are dealt with justly? Are there situations where these rules might create challenges for claimants or defendants?
3 What are the main differences between the roles of solicitors and barristers? In what types of legal work might their functions overlap, particularly in construction disputes?
4 Construction contracts often include arbitration or adjudication clauses. Why might one be preferred over the other, and how do these mechanisms differ in terms of finality and speed?
5 What are the implications of the public now having direct access to certain barristers? Do you think this is a positive development for access to justice?
6 Do you think the growing role of professionals such as legal executives, conveyancers and paralegals enhances or complicates the delivery of legal services? Why might it be important for clients to understand these different roles?

3 Central and local government

Carrie de Silva

The structure of central government

When society develops to the point where the activities of its members need to be directed and controlled, it will require a government. The tasks of government are to formulate and carry out policies, which is its **executive** function; to frame laws, which is its **legislative** function and to enforce those laws, which is its **judicial** function. At the same time, as it grants power to a government to fulfil these functions, society wants to see control exercised over the government so that it does not become too powerful or dictatorial. Such control is exercised through the rules of the constitution, rather as a club or association is controlled by means of its rules. A striking feature of the British Constitution is that it is unwritten. Unlike countries such as France or the USA, we cannot point to one document embodying our guaranteed rights. Nevertheless, important constitutional laws are contained in a number of historic Acts of Parliament. These Acts are not invested with any special protection and could be revoked or altered by the same processes as any other Act of Parliament. Thus, whilst the European Convention on Human Rights is now part of UK Law and the UK was formerly a member of the European Community, the Acts of Parliament which give effect to Human Rights Law in the UK could be repealed by the UK Parliament (as was the case with the withdrawal from the European Community). By contrast, in certain foreign constitutions, there are entrenched provisions which can only be changed by special procedures.

The leader of the political party which commands a majority in the House of Commons will be invited by the monarch to form a government. Parliament consists of the monarch, the House of Lords and the House of Commons. The King is a constitutional monarch who exercises his powers only on the advice of his ministers. The House of Lords, which is the upper chamber of Parliament, is a non-elected assembly. Its members include hereditary peers, life peers and the Lords Spiritual. The Lords Spiritual are not peers. They comprise the Archbishops of Canterbury and York; the Bishops of London, Durham and Winchester and the next 21 most senior bishops of the Church of England. Inevitably, the major criticism of this element of representation in the Lords is that it is exclusive to the Church of England. Indeed, over the years, the make-up of the House of Lords has been subjected to regular criticism. This mainly centred on the fact that the House was largely aristocratic and non-elected, with an inevitable inbuilt permanent Conservative majority. In 1999, however, the number of hereditary peers entitled to sit and to vote in the House of Lords was reduced from 759 to 92. Thus, the vast majority of the members of the House of Lords (over 800 are currently qualified to sit) are now life peers. It is proposed to remove the

DOI: 10.1201/9781003657101-3

last hereditary peers in the near future. Members of the House of Lords do not receive a salary, but are entitled to claim expenses.

House of Lords

Until October 2009, the House of Lords also possessed a judicial function, via the Lords of Appeal in Ordinary (a maximum of 12 judges who were specifically appointed to the House of Lords to hear appeal cases from the civil and criminal courts but who were also entitled to take part in all business of the House of Lords). In October 2009, however, the judicial function of the House of Lords was transferred to the new Supreme Court, the former Law Lords becoming Justices of the Supreme Court, Lord Phillips of Worth Matravers becoming the first President of the Supreme Court, with Baroness Hale of Richmond becoming the first female President in 2017.

It is often thought that the House of Lords has little power today since the Parliament Acts of 1911 and 1949 restricted its role to the delaying of legislation rather than its out-and-out rejection. In answer to the criticisms of the House of Lords, however, it is worth remembering the quality of membership of the Lords overall, reflecting a wide range of backgrounds and experience, with life peers being selected from trade unions, commerce and industry, public life and the armed forces. Standards of debate are generally accepted to be high. The part which the House of Lords can play in improving and refining legislation sent from the Commons is regarded as an important and significant part of its functions.

House of Commons

The House of Commons is the elected chamber of Parliament with 650 Members of Parliament. Each MP represents a constituency, and elections must be held at least once every five years, at a time chosen by the Prime Minister when he or she believes it will be politically opportune.

Once the result of the general election is known, the leader of the political party which commands a majority of the House of Commons will be invited to form a government. Many of the MPs will be hoping to hold office in the government. The new Prime Minister will then select the Cabinet team and ministers for the various government departments.

The Cabinet represents the most important departments but is of no fixed size, and the selection of ministers to be in the Cabinet may reflect the government's policies and priorities. Traditionally, the Cabinet will always include the Chancellor of the Exchequer, the Home Secretary, the Foreign Secretary and the Lord Chancellor (the Lord Chancellor formerly acted as the Speaker of the House of Lords and sat as one of the Lords of Appeal in Ordinary but the House of Lords now has a Lord Speaker and has lost its judicial function and the current Lord Chancellor, who is also the Secretary of State for Justice, sits in the House of Commons). To be of a reasonable working size, the Cabinet is likely to consist of about 24 ministers.

Examples of other key cabinet posts include the Secretaries of State for Defence; Health and Social Care; Environment, Food and Rural Affairs; Business, Energy and Industrial Strategy; Work and Pensions; Transport; International Development and Housing, Communities and Local Government.

Outside the Cabinet, however, there are other Ministers of State and Parliamentary Under Secretaries of State (for example, the Minister of State for Housing is in the

Department for Communities and Local Government), and in total the government is likely to have over 100 members. Some ministers will be members of the House of Lords, but it would be an unpopular move to appoint too many from the Lords because they cannot be questioned in the House of Commons, a corrective and control feature much prized by MPs. The government, once formed, is His Majesty's Government, and the ministers are His Majesty's Ministers. The new session of Parliament will be formally opened by the King, reading the King's Speech (which outlines the government's legislative programme) from the Throne in the House of Lords. Originally, all executive power was vested in the monarch, but a series of historical events reduced the power and produced the modern figurehead monarchy of today. All acts of government are done in the name of the monarch, but not necessarily with the monarch's personal participation. The King does, however, preside at meetings of the Privy Council. All bills going through the two Houses of Parliament require the Royal Assent to become Acts of Parliament. In theory, the monarch could refuse to Assent, but in this, as in most of the King's activities, he is bound by constitutional conventions. The Assent has not been refused since 1708 in the reign of Queen Anne. The principal convention is that the monarch exercises his formal legal powers only on the advice of ministers. The role was set out by Walter Bagehot in an often-quoted phrase: 'the sovereign has, in a constitutional monarchy such as ours, three rights: the right to be consulted, the right to encourage and the right to warn' (1867).

There is very little formal law governing our constitution and, therefore, many of the 'rules' under which central government and the Crown operate are simply conventions. A constitutional convention is a rule of political conduct, not a strict law in the sense that there is no sanction to enforce such a convention. Conventions can change imperceptibly over a period of time, and it can therefore be argued that they keep the constitution flexible. Conventions have been described as 'the flesh which clothes the dry bones of the law, they make the legal constitution work, they keep in touch with the growth of ideas' (Jennings, 1959).

It has been seen that the government of the day will emerge largely from members of the House of Commons. The House of Commons itself has three main functions:

- to make laws,
- to control national expenditure and taxation and
- to criticise policy.

Its law-making function has already been considered. In order to carry on its business, the first task of the newly elected members of the Commons following a general election is to elect their Speaker, who is an impartial chairman of the proceedings in the House. Usually this will mean the re-election of the previous Speaker if he or she is prepared to stand. Proceedings in the House of Commons take place in the Chamber with the government side sitting to the right of the Speaker's chair and the opposition to the left. The Chamber is not large enough to accommodate all the members at once, but usually they only want to be present in force for a limited number of great debates or when there is going to be a division (i.e. a vote). MPs are paid salaries and receive allowances regarding the expenses of running an office, provision of accommodation in London and in their constituencies and travelling between their constituencies and Parliament. MPs who are Government Ministers, or who hold certain other jobs, receive additional salaries.

There is no formally recognised career path towards becoming an MP. The 650 MPs in the House of Commons represent a wide range of backgrounds. After the 2024 general election, most MPs were in their 50s but there were 29 under 30 years and 20 over 70 years, 263 (40%) MPs were female, 90% were graduates with 20% having been to Oxford or Cambridge as undergraduates, 24% had attended a fee-paying secondary school, 14% came from an ethnic minority background and 64% were openly LGBTQ+ (lesbian, gay, bi-sexual, trans or queer).

Procedure of the House of Commons

There is a 'bible' of procedure in the House of Commons: *Parliamentary Practice* by Erskine May (Natzler et al, 2019). It started off in 1844 as a small handbook, but as the procedure of the House of Commons has become more complex, the book has grown and grown. The procedure is complex because of the range of activities and interests with which the House must deal. Inevitably, rules are important where the Speaker needs to control the activities of 650 volatile and articulate MPs of strong conviction, many of whom hold violently opposing views. If there are disputed points of procedure, the final word lies with the Speaker.

After numerous assorted items of business, including question time, private notice questions, ministerial statements, introductions of new members and requests for emergency debates, the House eventually gets down to the main business of the day, which might be a major debate or the second reading of an important bill. When an issue is put to the vote in the House of Commons, the Speaker puts the question and he must then weigh up whether the 'ayes' or the 'nos' have succeeded. If his assessment is challenged, he orders the lobbies to be cleared and the division bells will be rung; tellers are appointed and members must file through the appropriate lobby to record their votes. The tellers give the results to the Speaker, who announces the outcome to the House.

Parliamentary questions

Question time takes place in the House of Commons on Mondays to Thursdays and lasts for one hour. On Wednesdays the Prime Minister answers questions. Other ministers take it in turn to be questioned on a rota basis.

There are a number of reasons why an MP may wish to ask a question. He or she may wish to elicit factual information or may, indeed, be asked by a minister to table a question to give the minister an opportunity to make a public statement in the House. The question may seek redress of a grievance for a constituent. Very often in such a case the question will only be asked if initial approaches to the minister or department concerned have drawn a blank. Questions in Parliament are not the only option open to an MP acting for a constituent. He or she can also refer the matter to the Parliamentary Commissioner for Administration. The question may be designed to embarrass a minister. This can be achieved particularly successfully in the follow-up to the oral question in what are called supplementaries. Although the minister will have notice of the question tabled, a cunning supplementary question might find him or her unprepared. It is likely, however, that he or she will be well briefed by his/her civil servants with information to anticipate any supplementaries. Again, the question may be designed to enhance the reputation of the MP concerned, as opportunities to speak in the House are extremely limited.

The value of the parliamentary question can be considerable, especially if it ensures that ministers give their personal attention and consideration to issues or problems which would otherwise be handled by officials in their department. Often decisions will have been taken at quite a low level within the department. On a closer look at the file, ministers may have occasion to reconsider their decision or redefine the policy within their department.

Question time can be regarded as an important time for the ordinary Member of Parliament, the backbencher, particularly when it is remembered that this is part of the day when the parliamentary whips are not in control. Moreover, starred questions for oral reply are not usually asked by leading frontbenchers (members of the Government or Shadow Cabinet), although supplementaries may be put by them. The publicity potential is very great as question time takes place at a fixed time each day at the start of the day's business when the chamber is likely to be at its fullest and the press are in attendance, as there is always the prospect that a minister may be caught out or shown up in some way. A leading writer on constitutional law, Alexander de Smith, has said of question time:

> Parliamentary reputations have been made and ruined in the rapid cut and thrust of question time. For a few minutes the House comes to life audibly and visibly; wit, feigned or genuine outrage, cheers and jeers intrude upon the solemnity of the proceedings; government and opposition are briefly locked in verbal combat; the Prime Minister and the Leader of the Opposition may gain or lose a point or two in the public opinion polls; a backbencher shows his ministerial potential, and the House wonders how much longer the Minister of Cosmology can last.
>
> (1971)

Private members' bills

If the opportunities to ask questions are restricted, then the opportunities to introduce a private member's bill are even more limited. The number of private members' bills passed in each session is always small, often not running into double figures. As a general rule, private members will seek to introduce bills which do not require public expenditure, because if this were necessary, the bill would face further hurdles. The government would then need to be persuaded to put forward a financial resolution. If a member wishes to introduce a private member's bill, there are three possible ways of doing so:

- by winning a high place in the annual ballot
- under the ten-minute rule or
- under Standing Order No. 37.

The safest of these three courses is winning a high place in the ballot for promoting a private member's bill. Successful MPs may have some personal preference for the measure which they wish to introduce, or an MP may be solicited by various pressure groups anxious to promote some particular measure or even an individual constituent. Occasionally, the government itself has small measures that it wishes to promote. Where it has no time in the government programme, it may approach an MP who has been successful in the ballot, asking him or her to take on a particular bill. Once a member knows what the subject matter of a bill is to be, he or she must then see to its drafting. This can be a costly process, especially if the bill is lengthy.

The committee system of the House of Commons

Whenever there is a very large organisation with significant amounts of business to transact, it is common to find the delegation of some of that business to specialised committees. In the House of Commons, delegation can be to a Committee of the Whole House, when the committee consists of all of the MPs. But, usually, two main types of committees are used – the general committee (formerly known as the standing committee) and the select committee. The most common form of general committee is the public bill committee. A public bill committee is specifically set up to consider the committee stage of a particular bill, after which the committee will be suspended. Select committees, on the other hand, will usually operate for the duration of a Parliament.

Public bill committees

These consist of 16 to 50 members who are chosen to reflect the political strength of each party in the House. A government with a large majority will, therefore, enjoy a large majority in these committees. The actual members for each committee are chosen by the Committee on Selection, which is a select committee. The chairman of the standing committee will be chosen by the Speaker from a panel of members which is selected at the start of the parliamentary session. The main task of the standing committee is to consider amendments to individual clauses of the bill before it. This is the first time that the specific terms of the bill will be given detailed clause-by-clause consideration. Because of the number of votes on amendments, many of them of critical importance to the government if the substance of its policy is to be actively carried through, the whip system operates fully over the activities of public bill committees.

Select committees

Some of the select committees are permanent features of the life of the House of Commons, but with extremely varying functions. One of the most important select committees is the Committee on Standards, which, amongst other matters, oversees the work of the Parliamentary Commissioner for Standards and considers complaints relating to the declaring of interests or breaches of the Code of Conduct by MPs. There are a number of departmental select committees, such as the Science and Technology Committee and the Home Affairs Committee, which scrutinise the work of the government departments. Select Committees have a permanent membership for the duration of a Parliament. The committees are small, many only having 11 members, and the chairmanships are divided between the political parties, although the government would tend to retain the chairmanship of the more important committees like Home Affairs. They are free to determine for themselves which areas they will investigate.

Once a committee decides upon an area of investigation, it will usually appoint its own expert advisers. It can call before it witnesses from industry and commerce, from professional bodies (such as the Royal Institution of Chartered Surveyors), from the government and from the civil service. Persons who are called to give evidence take an oath and, if they lie, may be prosecuted for perjury. It publishes its findings in a report to the House, to which the government must respond. Because of a lack of parliamentary time, only a very small proportion of these reports is ever debated in the House, but there is, nonetheless, considerable value in the work of these committees. Their work is a means by which scrutiny and control of the executive is provided. They produce a considerable amount of information with which MPs may arm themselves for debate and questions. That information can be useful to add to the armoury of material used by pressure groups. Because issues are considered in some detail, MPs can develop considerable expertise. Membership of these committees is highly prized amongst MPs, and they foster a cooperative sense of cross-party unity.

Financial procedures of the House of Commons

The Treasury, under the Chancellor of the Exchequer, is the government department with responsibility for the management of the economy. Inevitably, as Treasury approval is required for all large-scale expenditure, it is a significant department, but its influence may wax and wane depending on the forcefulness of individual Chancellors.

In order to run the essential services of the country, money must be raised and collected and some priorities have to be established about how it should be paid out. There must also be mechanisms for controlling how that money is used. Whilst the government does obtain revenue from other sources (e.g. from Crown lands and loans), significant amounts have to be raised by taxes, which may be annual taxes like Income Tax or transaction taxes like Stamp Duty Land Tax.

The major financial statement of the year in the House of Commons is the Budget speech delivered by the Chancellor of the Exchequer. The proposals contained in the Chancellor's speech will ultimately be implemented in the Finance Act, but it may be several months before the Act receives Royal Assent. In the meantime, authority to collect annual taxes like income tax needs to be renewed, and the gap between the Budget and the Finance Act will be bridged using the authority contained in the Provisional Collection of Taxes Act 1968 (as amended). The appropriate resolutions will be passed as soon as the Chancellor has finished delivering his Budget speech.

In former times, it was common to raise taxes specifically for particular items, but now all moneys raised are lumped together into one fund, the Consolidated Fund, which is effectively just an account at the Bank of England.

There is a Treasury Committee, a select committee which monitors the Treasury, His Majesty's Revenue and Customs, the Bank of England and the Financial Conduct Authority as well as other bodies. The banking sector is regulated by the Prudential Regulation Authority. There is also a Public Accounts Committee, a select committee which examines reports concerning the economy, efficiency and effectiveness demonstrated by government departments and other bodies in using the resources allocated to them so as to further their objectives.

There are many ancient conventions with regard to financial matters in Parliament. Most significant amongst these are that the granting of public money and the imposing of taxes is the function of Parliament, not the government; that the granting of public money and imposing of taxes must begin in the House of Commons and be finally determined by the Commons and that the redress of grievances must precede any grant of public money. There must be no taxation without representation.

The Civil Service

Initially, when a minister takes over a new department or ministry, it might seem a daunting task to have to quickly learn all that is necessary for him or her to be able to make appropriate decisions. However, the role played by the Civil Service should not be forgotten. Civil servants are the permanent officials employed in the ministries. Unlike the American system, where all officials change when the President changes, our Civil Service is permanent. The civil servants work for the government of the day, whatever its political complexion. The civil servants in most senior positions are in close touch with their ministers and play a part in the formation of policy, watching over bills on their way through Parliament and keeping their ministers supplied with information to enable them to answer questions in Parliament. They will frequently make decisions by applying rules and policy to individual cases, and when this happens, such a decision is regarded as having been made on behalf of the minister, who, according to a constitutional convention, is responsible for what has been done.

The structure of local government

There has been a long tradition of local administration in England, dating back to 1066 and the Norman Conquest. With poor networks of roads and the dangers of travel, the advantages of control and decision-making resting with local organisations were obvious. But as those justifications have gradually disappeared, it may be questioned whether it is necessary in the twenty-first century to divide responsibilities between central and local government. Two of the main advantages are:

1 The variations necessary for different local conditions can be introduced.
2 Local people will be involved, and those living in the area will see some benefit from the effort which they contribute.

Local government may be through:

- county council,
- district council or
- unitary authority.

Local government in England is based upon the division of the country into counties, which in turn are divided into districts, the two basic types of local authority being county councils and district councils. The type of council will determine the range of functions, although some functions may be shared. County councils have responsibility for matters such as highways, fire, police and social services. District councils have responsibility for housing, public health, rates and refuse collection. In some parts of England, there is a third tier of local government, in the form of parish councils and town councils, which are responsible for matters such as footpaths, allotments, burial grounds and bus shelters. Planning is an example of a responsibility shared between county and district councils. Equally, some local authorities share responsibility for services such as police and fire services.

In certain parts of the country, the county/district council dichotomy is replaced by unitary authorities, which exercise the functions of county and district councils. Moreover, London, whilst divided into a considerable number of boroughs, each a unitary authority governed by its own council, now also has its own Assembly and elected mayor, with responsibility for or involvement in matters such as transport policy, development, policing, fire and emergency planning, town planning, environmental issues, culture, media and sport. Local authorities are examples of bodies enjoying corporate personality, i.e. the local authority is a 'legal entity' recognised by the law in the same way that a person or a company is. It is run by a council of elected members within the powers granted to the local authority by the Act of Parliament which created it. The members of the council may change at each election, but the council has permanent officials, called its officers, who fulfil a role rather like that of civil servants. The business of a local authority is traditionally transacted at council meetings, sometimes meetings of the whole council, but more often meetings of smaller committees set up for specific purposes and with powers delegated to them, e.g. the planning committee. These days many decisions are made by the local authority's executive, the decisions of the executive being subject to scrutiny by groups of elected councillors sitting in scrutiny committees or scrutiny panels. A few local authorities have elected mayors (at the end of 2019, there were 19 directly elected mayors in England), in which case the council's executive will either take the form of the elected mayor plus a cabinet of some of the council's elected members or an elected mayor plus a council manager. In most local authorities, however, the executive takes the form of a council leader and cabinet comprising some of the council's elected members. Some local authorities have ceremonial mayors, and in councils that do not, the functions of the ceremonial mayor (e.g. chairing council meetings) are performed by the council's chairperson. Council officers may exercise powers delegated to them by the council (e.g. the power to determine straightforward planning applications may be delegated to planning officers by the planning committee).

Council meetings are open to the public, unless the council resolves to meet in private, usually because publicity about a particular matter would be prejudicial to the public interest.

Council members may represent a particular political viewpoint, and it is quite often the case at local government elections that the votes of the electorate will swing against the political party in power in central government. This can produce conflicts between local and central government, e.g. on issues such as the sale of council houses or the size of rate increases. This naturally raises the question: to what extent is a local authority truly independent? The answer to this lies in the amount of control which can be exercised over the local authority.

Control of local authorities

Every local authority must act within the scope of the powers granted to it. A local authority is a statutory corporation, which can only do such things as are authorised expressly or impliedly by the Act of Parliament which created it. If the local authority acts in excess of its powers, its actions are said to be *ultra vires* (beyond its powers) and void (of no legal effect). Obviously, it would require immensely detailed legislation to spell out everything a local authority could do. Consequently, by s.111(1) of the Local Government Act 1972, authorities are empowered to do anything which is 'calculated to facilitate, or is conducive or incidental to the discharge of any of their functions'.

Where a local authority has acted in a way which causes someone to be aggrieved, it will be necessary to check whether the authority acted under some express power, some implied power or in some way reasonably incidental to an existing power. Not only must the local authority show that it had power to carry out a particular act, it must also be able to show that, procedurally, it carried out the act in accordance with the prescribed rules. A typical example of how things can go wrong occurred in *Bradbury v Enfield London Borough Council* (1967) CA. This case involved the reorganisation of schools and the introduction of the comprehensive school system. Where the local authority intended to 'cease to maintain' a school, it was obliged to submit its proposals to the minister and give public notice of its intentions, to allow affected persons to make objections. The local authority took the view that what it was proposing to do in its reorganisation plans did not amount to ceasing to maintain any of the schools involved. An aggrieved person applied to the court to stop the authority from going ahead with its plans. An injunction was granted because the behaviour of the local authority was *ultra vires* as it had failed to observe a mandatory procedural requirement.

When the help of the courts has to be sought, it is on the basis of judicial review, by the High Court, of the local authority's decision or actions. The High Court has a number of remedies available, including:

- a quashing order, which is used to quash an *ultra vires* decision; a prohibiting order, which is used to prevent *ultra vires* action which is about to take place;
- a mandatory order, which is used to compel the performance of a public duty and
- a declaration, which simply makes clear what the court determines the legal position to be.

One possible way to give greater scope for action to a local authority, without rendering it constantly liable to claims of *ultra vires*, is to give the authority discretion as to whether, or how, it will act. Although this may appear at first sight to be a welcome prospect, the exercise of discretion is fraught with problems. Complaints may arise from the way in which the discretion has been exercised and, in making its decisions, the local authority will usually be obliged to exercise its discretion reasonably, to exercise it proportionately when breaches of European Human Rights law are potentially involved and to observe the rules of natural justice. These rules are usually stated to be *audi alterem partem* (hear both sides) and *nemo judex in causa sua* (no one should be a judge in their own case). In practice, these rules often require a local authority to give prior notice of its intentions to persons who may be affected by their decisions, so that such persons have an opportunity to put forward their own case, and a local authority should exclude from the decision-making process anyone who has some pecuniary interest in the decision, or who has some other bias. Where it is suggested that an application for judicial review to the High Court may be needed to control the activities of local authorities, many individuals would be deterred by the expense involved or by the formality of the procedures. As was seen in Chapter 2, however, an alternative to litigation may be to have recourse to a Local Government and Social Care Ombudsman. A Local Government and Social Care Ombudsman can investigate complaints of maladministration. The word 'maladministration' is not defined, but has been said to cover instances of bias, neglect, inattention, delay, incompetence, ineptitude, perversity and arbitrariness. If the Commissioner finds that there has been maladministration resulting in injustice, the local authority concerned is then under a duty to consider the Commissioner's report and notify the Commissioner of the action it proposes to take. If the Commissioner is dissatisfied with the response to this report by the authority, he or she may issue a second report.

Control by central government

The main type of control which central government can exercise is financial. The control can consist of withdrawing grants, refusing loans or capping expenditure. Without financial support from central government, and without the ability to raise loans, no local authority could function for long. Admittedly, there are other forms of income available to local authorities, especially the rates levied on business property within the area (the sums raised by business rates are pooled nationally and are then redistributed to local authorities) and the sums raised by council tax, which local authorities levy on domestic properties. There are also less significant amounts, such as rents from council houses, profits from undertakings run by the local authority (e.g. swimming pools, markets, buses) and fees from the granting of various licences. However, the inadequacy of their income is apparent when it is realised that their funds have to be supplemented annually by the central government Revenue Support Grant.

When considering the role of central government in controlling local authorities, it is important to emphasise that much of the control is informal, exercised in the process of consultation, which takes place with various government departments. Circulars, codes of practice and memoranda issued by central government departments form the basis of much decision-making at the local authority level. Central government also possesses more formal powers to control the activities of local authorities, however. For example, the Secretary of State (under s.21 of the Planning and Compulsory Purchase Act 2004) possesses the power to call in all or part of a development plan document at any time before it is adopted and to direct that modifications are made to the document by the local planning authority. These powers were further enhanced under s.145 of the Housing and Planning Act 2016.

The range of services and amenities provided by local authorities is vast. Not surprisingly, there will be some people living in the area covered by a council who will be dissatisfied with the council's efforts. It should be remembered that ultimate control lies with the electorate, whether at the time of the elections, or by the strength of opinion they express to their councillors or through the media. It is clear from the low level of votes cast at local elections that some of these controls are not fully used.

Bye-laws

Within its own area, a local authority will have statutory power to regulate activities by means of bye-laws, for example, the power to make bye-laws for 'the good rule and government of' the district, under s.235 of the Local Government Act 1972. When making bye-laws, a local authority is usually governed by the procedures of s.236 of the Local Government Act 1972. A bye-law must be made under the seal of the council and must be confirmed by the appropriate authority, which will usually be the relevant Secretary of State. Before confirmation, the local authority must publish notice of its intention to submit the bye-law. Bye-laws are only valid if they are *intra vires* (i.e. with the local authority's powers). Moreover, in order to be valid, they must be reasonable, certain and consistent with the general law. If valid and effective, a bye-law may impose penalties for breach (e.g. a fine) which may be imposed upon conviction by a magistrates' court.

Non-departmental public bodies

A non-departmental public body (NDPB) is not a government department and does not form part of a government department but, rather, operates at arm's length from the government. The two main types of NDPBs are Executive NDPBs and Advisory NDPBs.

Executive NDPBs (such as the Health and Safety Executive, the Environment Agency, English Heritage and the Judicial Appointments Commission) perform executive, regulatory, administrative or commercial functions. Advisory NDPBs (such as the Low Pay Commission) provide ministers with independent, expert advice.

The minister who is responsible for an NDPB is accountable to Parliament for the NDPB.

Challenging decisions made by public bodies in the courts

A person who is unhappy with a decision made by a public body, such as a local authority or a government department, may be able to challenge the decision in court. There are three ways in which it may be possible to raise an issue of public law before the courts. These are:

- by appeal,
- by application for judicial review or
- an ordinary 'private law' action.

Appeal

There is no automatic right to appeal to the courts against a decision made by a public body. Such an appeal is only possible where an Act of Parliament has created the right to appeal against decisions made by a particular type of public body. For example, the Town and Country Planning Act 1990 permits persons aggrieved by planning decisions made by local planning authorities to appeal to the Secretary of State. If the aggrieved person is not then satisfied with the Secretary of State's decision, the Act allows him or her to make an application to the High Court within six weeks of the making of the Secretary of State's decision.

Where an Act of Parliament creates a right of appeal, it will also lay down the grounds of appeal, i.e. it states the reasons which, if established, justify the courts in substituting their decision for that of the original decision-maker. An appeal against a decision will not succeed simply because the appellant can show that he or she has good reason to be unhappy with the decision. In order to bring a successful appeal, it is necessary to establish one or more of the specified grounds of appeal. An Act of Parliament which creates a right of appeal also specifies the nature of those persons who are entitled to appeal.

Application for judicial review

Where there is no right to appeal against a decision made by a public body, it may still be possible to challenge the decision by making an application for judicial review to the Administrative Court. The Administrative Court is part of the King's Bench Division of the High Court.

Remedies of judicial review

An applicant for judicial review may apply for one or more of the following remedies. All of the remedies of judicial review are discretionary remedies. Thus, even where the claimant has established the grounds on which the application is based, the court may still refuse to award the remedy if, for example, the claimant failed to pursue a right of appeal or if his or her behaviour does not justify the award of a remedy.

a *Quashing orders.* A quashing order (formerly known as a writ of *certiorari*) can be used to quash an unlawful decision made by a public body. It may be awarded if the decision can be shown to be illegal or unauthorised. Following the making of the order, the decision ceases to have legal effect though, if appropriate, the court can require the public body to reconsider the relevant matter and make a new decision in accordance with the law.
An example of the award of a quashing order is provided by *R v Medical Appeal Tribunal, ex parte Gilmore* (1957) CA. The tribunal misinterpreted regulations made under an Act of Parliament with the result that the award of compensation which they made to an injured workman was less than that to which he was legally entitled (an error of law). A quashing order was awarded to quash the tribunal's decision. The court recognised that the workman's case would then go back to the tribunal, where a valid decision would be made.

b *Prohibiting orders.* A prohibiting order (formerly known as a writ of prohibition) can be used to prevent a public body from exceeding or abusing its powers. The effect of the order will be to prohibit the public body from acting or continuing to act in excess or abuse of its powers.

c *Mandatory orders.* A mandatory order (formerly known as a writ of *mandamus*) can be used to force a public body to do something which it is legally required to do. It may be awarded if it can be established that a public body has failed to perform its public duties. A mandatory order cannot be awarded against the Crown, but can be awarded against an officer of the Crown, including a government minister. In *R v Shoreditch Assessment Committee, ex parte Morgan* (1910) CA, the committee refused to initiate a process whereby the rateable value of a house could be reduced to take account of a fall in its value. Since the committee had a duty to initiate this process, a mandatory order was awarded in order to oblige them to do so.

d *Declaration.* This is a court order which declares what the legal rights of the parties to an application for judicial review are.

e *Injunction.* An injunction is a court order which may prohibit an act from taking place or, more unusually, may require an act to take place.

f *Damages.* Where one or more of the five remedies referred to previously is applied for by way of judicial review, the applicant may also claim damages in respect of any matter to which the application relates. The court may award damages if an ordinary civil action for damages (e.g. a contract or tort action) would have succeeded or if the party is entitled to claim damages under the Human Rights Act 1998.

Applications for judicial review and 'private law' matters

The application for a judicial review process is only available in relation to public law matters, not in relation to private law matters. Issues of private law should normally be raised in the form of an ordinary private law action, such as an action for damages in contract or tort, and not by application for judicial review. The distinction between public and private law matters may be difficult to draw in certain cases. The following guidelines may be helpful.

a Where the legal issue involved is one of contract, such as employment law, the issue will normally be one of private law even if the employer is a public body (such as the BBC).
b Where power is given to a body by a contract or by consent, decisions made in relation to the exercise of the power will normally be private law matters.
c Where a power stems from the Royal Prerogative (ancient common law powers of the Sovereign), decisions made in relation to the exercise of the power will normally be public law matters (e.g. refusal or withdrawal of a passport).
d When determining whether the exercise of a power by a body is a public law matter, the courts take into account not just the source of the power (i.e. Act of Parliament, contract or Royal Prerogative) but also consider whether the function performed by the body is of a public or private law nature. For example, decisions of the Panel on Takeovers and Mergers, a body appointed by City of London institutions such as the Stock Exchange, which devises and polices the code on takeovers and mergers, have been held to be subject to the application for judicial review process.

The process of application for judicial review

a *The pre-action protocol.* The steps that parties should follow before making a claim for judicial review are set out in the pre-action protocol. A letter before claim sent by the claimant to the defendant in compliance with the pre-action protocol, to which the defendant responds, may result in the out-of-court settlement of the claim, thus avoiding litigation.
b *Permission to apply.* Before an application for judicial review can be made, permission to apply must be obtained. Normally, an application for permission will be determined by the judge on the papers without a hearing. If permission is refused, it is possible to request a hearing to have the decision reconsidered. Basically, permission will be granted if the applicant has an arguable case. If permission is granted, the application for judicial review will be heard in open court, usually by one judge.
c *Delay.* An application for review must be made promptly and, in any event, within three months from the date when the grounds for the application arose, though the court does possess discretion to extend the time period. Even if made within three months, the court may refuse an application if it has not been made sufficiently promptly. The issue of delay may be considered both by the court when it is deciding whether to grant the claimant permission to apply and, if permission is granted, by the court when it hears the application for judicial review itself and is deciding whether to grant the remedy for which the applicant has applied.
d *Standing.* An applicant for judicial review must have 'standing'. This means that he or she must have a sufficient interest in the matter which the application concerns. Basically, the test is whether the applicant has been or may be affected by the act or decision which he or she wishes to question, as opposed to busybodies who are meddling in matters of no concern to them. Whether a pressure group has standing to bring an application for judicial review will depend upon the facts of the particular case before the court. The issue of standing will be considered by the court both when it is deciding whether to grant the claimant permission to apply and, if permission is granted, at

the hearing of the application for judicial review itself. It should be noted that where it is asserted that a public authority has acted unlawfully within the meaning of s.6 of the Human Rights Act 1998, the effect of s.7 of the 1998 Act is that only the 'victim' of the relevant act may bring proceedings against the public authority under the 1998 Act (see Chapter 1).

Ordinary 'private law' actions

A public law matter should be raised by appeal or application for judicial review and not by bringing an ordinary private law action (e.g. an action in contract or tort). It may, however, be permissible to raise a public law matter in the course of an ordinary private law action as a defence to that action. Further, it may be appropriate to bring an ordinary private law action where a civil claim raises both private law and public law issues.

Privative clauses

Privative clauses (or ouster clauses) are clauses in Acts of Parliament which remove or restrict the courts' power to hear matters. The normal approach of the courts is that such clauses will be interpreted very narrowly, so that if there is any doubt as to their effect, the courts' powers will not be reduced. Where there is no such doubt, however, the effect of the clause will be to restrict or remove the courts' power to hear the matter in question.

An example is provided by *R v Secretary of State for the Environment, ex parte Ostler* (1977) CA. Ostler discovered that a trader had withdrawn his objection to a compulsory purchase order after the Department of Transport had secretly promised him that a road would be widened. The road widening affected Ostler but he had not been told about it so he had not objected to the compulsory purchase order. Ostler claimed that the order was invalid, as it was made in breach of the fair hearings rule of natural justice (see later on). The compulsory purchase order was protected by a statutory clause which provided that it could only be challenged in the courts within six weeks of notice of its making being published. Ostler had not discovered the secret promise and consequently had not challenged the order until more than a year after the order was made. Thus, the court could not interfere with the order.

Application for judicial review: the grounds of challenge

Upon application for judicial review of a decision, the High Court will not award a remedy simply because it disagrees with the decision. It is not for the court to substitute its decision for that of the decision-maker. Rather, a remedy will only be awarded if the decision is unlawful or unauthorised. Traditionally, the High Court may be prepared to award a remedy upon:

- proof of procedural impropriety,
- illegality,
- irrationality or
- proportionality.

More recently, the courts have also demonstrated an increasing willingness to intervene in circumstances in which the conduct of a public body has given rise to a legitimate expectation that it will act in a particular way and the public body subsequently acts in a different way without a lawful reason for failing to give effect to the legitimate expectation.

Procedural impropriety

a *Failure to observe procedural requirements laid down by Parliament.*
Procedural requirements specify the process which a public body must adopt in acting or making decisions. If, in the course of making a decision, a public body fails to observe procedural requirements laid down either by Parliament or under statutory authority, this may render its decision invalid, but will not necessarily do so. The traditional approach of the courts is that breach of a mandatory procedural requirement normally makes a decision invalid. Breach of a merely directory procedural requirement will not usually have this effect.

Mandatory procedural requirements are those that serve an important purpose. Directory procedural requirements are those which are not so important.

An example is provided by *Howard v Secretary of State for the Environment* (1975) CA, which concerned a requirement of the Town and Country Planning Act 1968. Notice of appeal against an enforcement notice had to be served on the Secretary of State within 42 days of service of the enforcement notice. (The nature and purpose of enforcement notices is considered in Chapter 13.) The Court of Appeal classified this requirement as mandatory. They did so because, without a time limit, the enforcement provisions of the Act would not have worked effectively, so the time limit was of great practical importance. The Act also required that the notice of appeal contain certain specified information relating to the grounds of appeal and the facts of the appeal. The Court of Appeal classified this requirement as directory. They did so because once the notice of appeal had been served, the appellate mechanism began to operate. Failure to provide the specified information could easily be remedied at a later date and thus it was not essential that the information be contained in the notice. On the facts of the case, notice of appeal had been served within 42 days, but a letter containing the required information had not arrived until more than 42 days had passed. The Court held that the procedural defect did not render the appeal invalid.

At times the courts may regard the breach of a minor procedural requirement, which appears to be merely directory, as rendering the decision to which it relates invalid. This may be the case, for example, where the court believes that the requirement has been deliberately ignored. Equally, the courts may at times regard the breach of an important procedural requirement, which appears to be mandatory, as not rendering the decision to which it relates invalid. This may be the case where, for example, the court believes that the breach has not caused any harm.

In *R v Soneji* (2005) HL, Lord Steyn suggested that rather than drawing rigid distinctions between mandatory and directory requirements, the courts should place the emphasis on the consequences of not complying with a requirement and should consider whether Parliament could fairly have intended the consequence of non-compliance to be the total invalidity of the decision. Lord Carswell, however, whilst recognising that the distinction between mandatory and directory requirements has gone out of fashion in recent years, suggested that the principles that underlie this distinction are still of value.

b *Breach of the rules of natural justice (or the requirements of fairness).*
The rules of natural justice are common law rules of procedure which the courts impose both on themselves and on other public bodies. The rules do not apply to all public

bodies and sometimes apply differently to different bodies. It is possible, however, to draw some general conclusions as to what the rules are and how they work. In short, there are two rules:

First, the rule that a decision-maker must not be biased.

Second, the rule that a person whose rights, interests or legitimate expectations are affected by a decision has a right to a fair hearing before the decision is made.

It is also important to remember that apart from the common law rules of natural justice, Article 6 of the European Convention on Human Rights (which was considered in Chapter 1) guarantees that when a person's civil rights and obligations or a criminal charge against a person are determined, they will be determined via a fair hearing by an independent and impartial tribunal.

i The bias rule.

The bias rule itself has two elements. The first requires that the decisions of public bodies must not be tainted by pecuniary bias. The second requires that the decisions of public bodies must not be tainted by non-pecuniary bias.

Pecuniary bias

The decision of a public body is vitiated by pecuniary bias if a member of the body possesses a personal financial interest in the subject matter of the decision. This rule is applied strictly in the sense that even if the court is satisfied that the person who had the interest was not influenced by it, the decision will still be quashed. An example is provided by *R v Hendon Rural District Council, ex parte Chorley* (1933) KBD. The case concerned a councillor who was acting as an estate agent in relation to a land transaction. The transaction would only go ahead if planning permission was granted in relation to a development on the land. The councillor took part in the meeting which decided to grant planning permission for the development and voted in favour of granting permission. The court quashed the grant of planning permission.

Non-pecuniary bias

The decision of a public body is vitiated by non-pecuniary bias if a member of the body has a non-financial interest in it. This could occur, for example, where the decision-maker is a close relative or professional colleague of a person who is affected by the decision. Even where a decision-maker does not have a direct personal interest in the outcome of a decision and even though the court accepts that there was no actual bias, the existence of an indirect interest will still infringe the bias rule if, in the circumstances, there was a real possibility of bias. An example of a case in which non-pecuniary bias arose is provided by *R v Barnsley Council, ex parte Hook* (1976) CA. A market trader urinated in a side street. The market manager reported this to the council, who revoked the trader's licence. The market manager was present whilst the relevant council committee made its decision and thus acted both as prosecutor and judge. The decision was held to be vitiated by non-pecuniary bias and was quashed.

ii The requirement of a fair hearing.

Basically, fairness requires that a person whose rights, interests or legitimate expectations are affected by a decision is entitled to a fair hearing before that decision is made. (A legitimate expectation of a fair hearing may arise where, for example, a promise made by or a policy adopted by a decision-maker leads a person to believe that he or she will be given a fair hearing before a decision is made.) The rule has three basic requirements, though exactly what it requires will vary depending upon the facts of the case (e.g. depending upon the nature of the decision-making body which the case concerns).

First, the person so affected by the decision must be made aware of the charges, complaints, grounds, arguments or evidence against him or her. In *Annamunthudo v Oilfields Workers Union* (1961) PC, A was accused of breaking four union rules, none of which could result in expulsion from the union. He did not attend the meeting at which the charges were to be considered. At the meeting, a new and more serious charge was raised and he was expelled. The expulsion was invalid because, since A had not been made aware of the charge, he had not been given a fair hearing.

Second, the person so affected by the decision must be provided with an opportunity to prepare his or her own defence or arguments. In *R v Thames Magistrates' Court, ex parte Polemis, 'The Corinthian'* (1974) QBD, a ship's captain was charged with discharging oil into navigable waters. His summons was served at 10.30, and he was to appear in court at 14.00. This did not provide time for chemical analysis, which might have shown that the oil did not come from his ship. The magistrates refused to allow an adjournment to allow the analysis to be carried out as the ship was sailing that day. They convicted him and fined him £5,000. Thus, the magistrates, in failing to allow the captain time to prepare his defence, did not give him a fair hearing.

Third, the person so affected by the decision must be provided with an opportunity to present the defence or arguments which he or she has prepared. This does not necessarily mean that he or she must be allowed to appear in person before the decision-maker, and it does not necessarily mean that he or she will be entitled to be legally represented. What is required in order for the hearing to be fair will depend upon the circumstances of the case. It may be sufficient to allow him or her to hand in written arguments, provided that both sides to a dispute are treated in the same way.

In *Errington v Minister of Health* (1935) CA, following a public inquiry, the Secretary of State received further information from a local authority without allowing objectors to the local authority's scheme to comment upon it. Since the objectors had not been given an opportunity to put their arguments forward, they had not been given a fair hearing.

The rules of natural justice do not generally require the giving of reasons, but may do so in appropriate circumstances (*R v Ministry of Defence, ex parte Murray* (1998) CA).

Illegality

A public body may only do those things which it possesses the power to do. (Normally, such power will be given by or under an Act of Parliament or, more unusually, by the common law.) If a public body does not possess the power to make a decision, then, in making such a decision, it exceeds its powers: its decision is thus *ultra vires* and invalid.

In *White and Collins v Minister of Health* (1939) CA, a local authority made a compulsory purchase order in respect of land which formed part of a park. The authority possessed the power to make compulsory purchase orders, but did not possess the power to make such orders in respect of land which formed part of a park. Consequently, the order was quashed.

R v Willesden JJ, ex parte Utley (1947) KBD, concerned an offence of failing to stop a motor vehicle when requested to do so by a police officer. The maximum penalty for the offence was a fine of £5. The magistrates imposed a fine of £15, which was quashed.

INCIDENTAL MATTERS

Where a public body is expressly empowered to do something, it may be implied that the public body can also do things incidental to that which it is expressly empowered to do. In *Loweth v Minister of Housing and Local Government* (1971) QBD, a council was empowered to make a compulsory purchase order to acquire land on which to build a civic centre. The empowering Act did not specify whether the order could also include land on which to construct an access road to the civic centre. The court held that it could, as this was a matter incidental to the development of the civic centre.

UNLAWFUL DELEGATION

Power should only be exercised by the body to which it is given. Only where a decision-maker is given the power to delegate can it lawfully delegate its powers to another person or body.

In *Barnard v National Dock Labour Board* (1953) CA, a local dock labour board, without the power so to do, attempted to delegate disciplinary powers to a port manager. The port manager suspended some dock workers. The suspensions were *ultra vires*.

A common example of lawful delegation is encountered in relation to local authorities. Section 101 of the Local Government Act 1972 empowers a local authority to delegate its powers to a committee, a subcommittee or an officer or to another local authority. Delegation to an individual councillor is not authorised, however.

Irrationality

Where a public body exercises a power which it does possess, it does not exceed its powers, but it may abuse them. Power is abused when its exercise is not in accordance with law. Thus, where in the course of making a decision, a decision-maker makes an error of law, he abuses his power.

An error of law is, essentially, an error of legal reasoning. Such an error may take a variety of forms. In particular, it may consist of:

a the making of a decision which is so unreasonable that no reasonable person properly understanding the law would make it;
b a failure to take legally relevant matters into account or the taking into account of legally irrelevant matters and
c the use of a power for a purpose which is legally improper.

UNREASONABLENESS

The courts may be prepared to quash a decision on the basis of unreasonableness if it is so unreasonable that no reasonable decision-maker would ever have made it. The fact that the court does not agree with a decision does not necessarily mean that it is unreasonable because, sometimes, two reasonable people can come to different conclusions on the same set of facts. Consequently, not every mistake is necessarily an unreasonable mistake. Thus, for a decision to be classified as unreasonable, it should be overwhelmingly clear that no reasonable decision-maker would have regarded it as one which he or she was legally entitled to make in the circumstances.

In *Hall v Shoreham-by-Sea Urban District Council* (1964) CA, a council granted conditional planning permission to an applicant. The condition imposed was that the applicant construct a road on his land and dedicate it to the public. No compensation was to be paid by the council to the applicant for the loss of his land. The court held that the condition was invalid. It was clearly unreasonable of the council to seek to force the applicant to give up his property without compensation by making the grant of planning permission dependent upon this.

FAILURE TO TAKE LEGALLY RELEVANT MATTERS INTO ACCOUNT OR TAKING LEGALLY IRRELEVANT MATTERS INTO ACCOUNT

The courts may be prepared to quash a decision if satisfied that, in reaching it, the decision-maker failed to take a legally relevant matter into account. Equally, they may be prepared to quash a decision if satisfied that, in reaching it, the decision-maker considered a legally irrelevant matter.

Legally relevant matters are matters which, according to law, the decision-maker should take into account in reaching his or her decision. Basically, these include all matters which a reasonable decision-maker would take into account. They also include all matters which an Act of Parliament, expressly or by necessary implication, indicates should be taken into account.

Legally irrelevant matters are matters which, according to law, the decision-maker should not take into account in reaching his decision. Basically, these include all matters which a reasonable decision-maker would not take into account. They also include all matters which an Act of Parliament, expressly or by necessary implication, indicates should not be taken into account.

Roberts v Hopwood (1925) HL concerned the validity of the wage rates paid by a council to its employees. The council was empowered to pay its employees 'such wages as it thought fit'. The council fixed wage levels far higher than those paid either in the public sector or by other councils. The council fixed these wage levels because it sought to become a model employer. In so doing, it failed to take into account ordinary economic considerations, such as the market price for labour. Thus, the council based its decision on a legally irrelevant matter, its desire to become a model employer, and failed to take a legally relevant matter into account, namely the market price of labour. The importance of the latter becomes obvious when it is realised that the effect of paying extremely high wages to council workers was that workers doing the same or similar jobs in the private sector, and being paid the lower market rate, would additionally be required to face higher rates (a past form of council tax) bills in order to meet the increased public sector wage bill.

Sometimes, a public body adopts a policy by which to exercise its powers. If such a policy is rigid, in the sense that exceptions to it will not be considered, the public body is said to 'fetter its discretion'. This means that by rigidly applying the policy, it fails to take into account legally relevant considerations, namely the individual facts of each separate case that comes before it. An example is provided by *Lavender v Minister of Housing and Local Government* (1970) QBD, in which the Minister of Housing adopted the policy of refusing to allow gravel working on agricultural land if the Minister of Agriculture objected. Where a policy is adopted, but it is made clear that exceptions will be considered and, consequently, the individual facts of each separate case are examined in order to determine whether they are exceptional, discretion is not fettered, but rigidity and lack of discretion are impermissible.

IMPROPER PURPOSES

The courts may be prepared to quash an exercise of power if they are satisfied that the power has been exercised for a purpose other than that for which it can lawfully be exercised. In order to decide whether power has been exercised for an improper purpose, the courts must first determine the purpose for which Parliament intends the power to be exercised. Having determined this, they must then decide whether power has been exercised for the intended or for a different purpose.

In *Webb v Minister of Housing and Local Government* (1965) CA, a local council possessed a power to acquire land by compulsory purchase for the purpose of building sea walls for coastal protection. They used this power to acquire land to build a promenade along the seafront. Their decision was invalid because the purpose for which the power had been used was improper.

Sometimes, a power may be exercised for more than one purpose. As long as the main purpose for which it is exercised is proper, it does not matter that a subsidiary purpose is also achieved. In *Westminster Corporation v LNWR* (1905) HL, a council was empowered to construct public conveniences on or under any road. It constructed conveniences under a road with access from both sides of the road. It was asserted that the decision to build the conveniences achieved an unlawful purpose, namely the building of a subway. The House of Lords held, however, that the council's primary object was to build conveniences and thus the fact that the development also achieved the secondary objective of constructing a subway did not make the development unauthorised.

Proportionality

Where rights guaranteed by the European Convention on Human Rights are in issue or, indeed, where other fundamental rights are involved, the principle of proportionality (which was referred to in Chapter 1) comes into play. As Lord Steyn recognised in *R v Secretary of State for the Home Department, ex parte Dayley* (2001) HL, in such circumstances, the intensity of judicial review is greater than would be the case under the traditional grounds of judicial review. This, Lord Steyn indicated, may be so both because the court may be required to consider the balance that the decision-maker struck between the aims that it sought to achieve and the means that it adopted in order to achieve those aims, and because the court may be required to consider the relative weight that the decision-maker accorded to various interests and considerations. For example, in *R (Baker) v First Secretary of State* (2003) CA, a decision to compulsorily purchase property which was unfit for human habitation and the condition of which was having an adverse effect on adjoining residential occupiers was held to be necessary and proportionate.

Legitimate expectations

Where the conduct of a public body has given rise to a legitimate expectation that it will act in a particular way, the public body should not act in a different way without properly taking the legitimate expectation into account and should give effect to the legitimate expectation unless there are lawful reasons for not so doing. For example, in *R v The London Borough of Newham, ex parte Bibi* (2003) CA, where the council promised to provide tenants with suitable accommodation with security of tenure but failed to do so, the Court of Appeal made a declaration that the council was under a duty to consider the housing applications on the basis that the applicants had a legitimate expectation that they would be provided with suitable accommodation on secure tenancies.

Discussion questions

1 What purpose does the House of Lords serve? And is it an effective chamber?
2 What is an appropriate balance between central and local government, and has the optimum balance been put in place in your area of the UK?
3 What is 'the government', as opposed to 'Parliament'?
4 Develop one or other of the following debating points: 'It makes for a strong country to have a hereditary head of state' *or* 'A head of state should be elected or appointed by peers'.

4 Business organisations

Carrie de Silva

Sole trader

When persons want to set up in business, there is nothing to prevent them from simply trading on their own account, with only their own money at risk. As such, the law would describe such a person as a sole trader. Trading in this way, however, may involve a number of drawbacks:

- There may be a lack of adequate funds to develop the business.
- There is no one to share responsibility for running the business.
- If the business gets into financial difficulties, the sole trader will have unlimited liability for its debts. Creditors will not be restricted to seizing the assets connected with the business. Even the person's home may have to be sold.

Partnership

The sole trader may be able to overcome the first two of these disadvantages by forming a partnership. This relationship is described by the Partnership Act 1890 as 'existing between persons who carry on business in common with a view to profit'. The arrangements for a partnership can be very informal, but it is advisable to have a properly drawn-up agreement, in case disputes arise between the parties.

Partnership is a suitable form of business association for professional people, e.g. accountants, architects, quantity surveyors and consulting engineers, although some such firms incorporate or, since 2000, have traded as limited liability partnerships (see later on). Under the 1890 Act, all partners are legally entitled to take part in the running of the business, but in practice, the parties can make whatever arrangements they like. All partners are agents of each other for the purposes of partnership business (e.g. making contracts, paying money and engaging staff on behalf of the firm). For this reason, it may be sensible to limit the number of partners. A partnership is often referred to as a firm, e.g. a firm of solicitors, but legally that term has no significance. The law is concerned with knowing: is this a partnership or a company? The word 'firm' might be equally appropriate to a firm of builders' merchants trading as a limited company, but the legal consequences of dealing with a partnership and a company can be entirely different.

The relationship among partners is a very special one, as potentially each could expose his or her fellow partners to unlimited liability. The law determines that it is a relationship of utmost good faith. This means that intending partners should make a full disclosure

DOI: 10.1201/9781003657101-4

of any relevant facts or circumstances. Wherever possible, potential disputes should be avoided by a carefully drafted partnership agreement. Unless such an agreement is drawn up, the rules of the Partnership Act 1890 will apply. The Act provides that partners are entitled to share equally in the capital and profits of the business. That certainly may not accord with the wishes of partners who have contributed capital in differing proportions. A formal agreement allows for detailed rules about capital and profit sharing, and matters like the running of the firm, engaging in other business activities, retirement of partners and dissolution of the partnership, which under the 1890 Act is automatic on the death or bankruptcy of a partner.

Forming a partnership introduces others who will share liability for the debts of the business, but that liability continues to be unlimited. All the partners are fully liable for all the debts, and again, creditors are not restricted to assets connected with the business. Even the advantages of a partnership (e.g. privacy with regard to the conduct of business affairs and accounting) may not outweigh the drawback of unlimited liability. Alternatives which can avoid this are the formation of a company or a limited liability partnership for the running of the business.

Where there is a partnership under the 1890 Act, a further partner might choose to invest in the business under the Partnerships (Limited Liability) Act 1907. This results in an 'ordinary' partnership with one or more 1907 Act partners whose liability is limited. Contrast businesses operating under the Limited Liability Partnerships Act 2000, whether all partners have limited liability.

Taxation of partnerships

Partnerships are not separate entities in their own right; so, although a partnership tax return is submitted, the charge to tax is against each individual partner who is being charged Income Tax on their share.

Company

The process of forming a company is referred to as incorporation. This is a very significant legal step, because the business then becomes known in law as a corporation. This is to distinguish it from groups or bodies which are not incorporated, and which are usually referred to as unincorporated associations. These latter groups have no separate legal existence or personality distinct from their members. So, the Mid-Tyne Angling Club is simply an identifying name for a group of people who have come together with a common interest. There is no legal person called Mid-Tyne Angling Club, i.e. it is not a separate legal entity or personality.

The phrase 'legal personality' means a person or organisation which is recognised by the law as having legal rights and owing legal duties. All human beings in this country have a legal personality, although the law may impose restrictions on their capacity to act in certain ways (e.g. any child under 18 years of age has legal personality but lacks capacity to make a will (s.7 Wills Act 1837, unless on active service)) or own land (s.1(6) Law of Property Act 1925). The law also confers legal personality on an organisation which has become incorporated. Even a 'one man'/'one woman' company is a separate legal entity to the shareholders and directors (*Salomon v Aaron Salomon and Co. Ltd* (1897) HL).

Incorporation can take place by three methods:

1 **Royal Charter**
This was the method used to create the original trading companies (e.g. the East India Trading Company, established in 1600, and the Hudson Bay Trading Company, established in 1670), but was also, importantly, the mechanism of creating many royal boroughs and institutes.
2 **Act of Parliament**
Many of the public corporations and local authorities have been established in this way, e.g. British Airways (under the Civil Aviation Act 1971).
3 **Companies Act 2006**
Under the rules of the Act, a number of different types of companies can be created. For trading and general business purposes, a company limited by shares is by far the most common. Note that companies are also impacted by other legislation, notably the Insolvency Act 1986 and the Economic Crime and Corporate Transparency Act 2023

The fact that a company has a separate legal personality, quite distinct from its members (i.e. its shareholders), confers a number of advantages. These include the power to sue and be sued in its own name, the right to own and transfer property, the possibility of perpetual succession unless the company is wound up and the possibility for its members to enjoy limited liability.

If a company gets into financial difficulties, creditors will be paid out of the assets of the company. If those assets are insufficient, the individual shareholders do not become personally liable (subject to any personal guarantees which lenders such as banks may have imposed). The extent of the liability of shareholders is the amount they owe for their shares. Once they have paid for their shares, they risk losing that amount if the company founders, but no more. Limited liability is the most appealing feature of forming a company. However, a price must be paid for this advantage; limited liability companies are under statutory duties to publish certain financial details in the interests of both their creditors and the general public. And limitation is not absolute, and the courts will look to the individual(s) behind a corporate structure in certain circumstances, of which more will follow.

Creating a company

The actual creation or formation of a company is achieved by drawing up appropriate documents (including a Memorandum of Association and Articles of Association), which are submitted to the Registrar of Companies. The Memorandum was historically the company's most important document as it set out the objects for which it was formed and thereby determined or limited the powers of the company. The Companies Act 2006 has changed the status of the Memorandum of Association to a historical document simply reflecting that the subscribers to it wish to form a company under the 2006 Act and agree to take at least one share in it. Much of the information which used to be required to be included in the Memorandum is now found in form IN01 (application to register a company).

The company will have a registered name, but it may choose to trade under a different name, its business name. Business names are regulated by statute (Companies Act 2006, Part 5, ss.53–85). The name chosen for the company must not be the same as that of an existing registered company, and if it is too similar to that of an existing company, the new company may be directed to change its name. Permission from relevant bodies has to be sought to use certain words in a company's name, such as bank or charity (the Financial Conduct Authority and the Charity Commission, respectively, must be asked for permission). In the case of a limited company, the word 'limited' must usually be the last word of the company's name. Where the company is a public company, it must include the words 'public limited company' or the abbreviation 'plc' in its name. Both 'limited' and 'plc' can be substituted for the Welsh equivalent for companies incorporated in Wales.

The company must have a registered office. Both the situation of this office and its address must be recorded on form IN01. The location of the registered office establishes the 'domicile' of the company, which may be important in choosing which system of law should apply. For example, Scots law applies to companies registered in Scotland. The company must also file a separate note of the address of its registered office.

Historically, all companies had an objects clause. This clause sets out exactly what business the company can engage in. Traditionally, these clauses have been very long, with the draftsman attempting to cover every possible activity which a company may wish to undertake. This was justified because of the *ultra vires* principle, whereby any activity undertaken by the company outside the scope of its objects clause was *ultra vires* (beyond its power) and void. So far as the *ultra vires* rules are concerned, for any third party who deals with a company, successive Companies Acts (currently s.40(1) of the 2006 Act) have set out that the validity of acts done by the company shall not be called into question on grounds of lack of capacity by the company. Of course, the directors of the company who thus acted beyond the scope of its objects clause would be in breach of duty, and a shareholder could seek to bring proceedings to restrain any proposed activity which would be beyond the scope of the company's capacity. New companies formed under the 2006 Act have unrestricted objects, unless restrictions appear in the company's Articles of Association. The contents of the existing company's memoranda, including any restriction on their objects, are now deemed to be included in the company's Articles of Association.

A statement of the company's initially issued (allotted) share capital appears on form IN01 and includes its division into shares of a fixed amount. For private companies there is no legal maximum or minimum initial share capital fixed, but a public company must have allotted shares with a minimum nominal value of £50,000. Each share in a company has a nominal value – this represents the minimum amount that the company must receive in return for allotting the share. So, if a company has shares with a nominal value of £1 each, the company must be paid at least £1 per share by anyone subscribing for shares in the company. Once shares have been issued by the company to individuals subscribing for them, those shares can be transferred (sold or given away) by the shareholder, subject to any restrictions on transfer of shares in the company's Articles.

The Articles of Association set out the internal rules and regulations for the running of the company, covering matters such as meetings, rights of shareholders, voting, appointment of directors, powers and responsibilities of directors and payment of dividends. There is a model set of Articles for private companies provided by the 2006 Act, called the Model Articles for Private Companies Limited by Shares (the predecessor to these articles was called Table A). It is common for companies to adopt model articles with suitable amendments, although an entirely bespoke set of articles can be used. Which form of articles is being adopted is indicated on form IN01.

The Articles are particularly significant for the members, as they are effectively the terms of the contract between the member and the company. Any violation of the rules contained in the Articles is a breach of contract. Members can sue to enforce their rights in ordinary breach of contract claims.

The Small Business, Enterprise and Employment Act 2015 created a new Streamlined Company Registration Service. The process is easy to follow online and links the Companies House and HMRC registration requirements.

Public and private companies

It is common to think of private companies as small family firms and public companies as the large business giants, and in practice, this is often the case. A company which starts as a private company may expand its activities to such an extent that it decides to 'go public' and seek a change of status to a public company. There are a number of simple differences between public and private companies. For example:

1. A private company can be formed with one member and one director (a single-member company), whereas a public company needs two members and at least two directors.
2. A private company need not obtain from the Registrar a certificate entitling it to do business, whereas a public company must have such a certificate and commits offences by trading without it.
3. A public company must satisfy the minimum capital requirement (i.e. its allotted share capital must be at least £50,000). This rule does not apply to private companies.
4. A private company may exercise its borrowing powers immediately upon incorporation. A public company may not do so until it has allotted shares up to its authorised minimum capital of £50,000.
5. Private companies may still restrict the transferability of their shares if they choose. They would usually choose to do so in cases where they want to preserve control of the company within a family. Restrictions normally operate by giving the directors the power to refuse to register transfers. Where such a discretion is given by the Articles, it places immense power in the hands of the directors, but they must act throughout in good faith. The new Model Articles referred to previously contain provisions allowing the directors to refuse to register a transfer of shares, although the Companies Act 2006 does oblige the directors to give reasons for any such refusal. By contrast, public companies may not restrict the transferability of their shares if they want to be quoted on the Stock Exchange (listed shares). There is a model set of articles for public companies which do not give directors the discretion to refuse to register a share transfer of fully paid shares.
6. Private companies may be freed from some of the rigours of the rules about publication of accounts, depending on their size. For these purposes, size is determined by factors like turnover, assets and numbers of employees.
7. Public companies must have a suitably qualified company secretary. Under provisions in the Companies Act 2006, private companies are no longer required to have a company secretary.
8. Public companies are subject to rules prohibiting their giving financial assistance for the purchase of their shares except in very restricted circumstances. Financial assistance would include, for example, a company lending money to a person, which that person then used to buy shares in the company. Private companies are no longer subject to this restriction.
9. Private companies are no longer required to hold annual general meetings.

Management of a company

Companies are run by a board of directors. The company's first directors are appointed on incorporation. Thereafter, directors can be appointed by the board or the shareholders in a general meeting. With private companies, it is a common requirement that directors must have a certain holding of shares in the company. Where it is a small family business, this is usually not difficult as directors will often also be the members of the company. The company, being an artificial person, must act through human agents, and that normally means through the board of directors. The board will often delegate duties to paid employees, who may also be directors. Once appointed, a director is under a duty to show good faith to the company and to take reasonable care. This means, for example, that a director is obliged to reveal any personal interest in a contract made by the company, particularly if he or she stands to profit by the contract. The duties owed by directors to the company that have been established at common law are now codified and extended under ss.171–177 Companies Act 2006. These duties cover the following: acting within powers; promoting the success of the company; exercising independent judgement; exercising reasonable care, skill and diligence; avoiding conflicts of interest; not accepting benefits from third parties; declaring an interest in proposed transactions or arrangements with the company.

The directors, or those acting under delegated powers, must act within the powers granted to them by the documents creating the company.

A board of directors is likely to have a Chairman, who will preside at directors' meetings. The board may also decide to appoint a Managing Director, who will often have a contract of service with the company and to whom much of the day-to-day management of the company is delegated.

In addition, there is a requirement to register Persons with Significant Control. These are likely to be people who hold more than 25% of shares in the company, more than 25% of voting rights in the company and the right to appoint or remove the majority of the board of directors, but there could be other criteria.

How a company divides its management between directors and shareholders is largely a question to be determined by the Articles. In practice, the right of members to participate in management, particularly in large public companies, is extremely limited. A member does have some say by virtue of voting power at meetings, but he or she may be one of a very large number of shareholders, holding very few shares, and his or her opinion may carry no weight.

Appointment and removal of directors

The Articles should provide mechanisms for the appointment and retirement of directors. If shareholders are dissatisfied with one of the directors, they may seek his or her removal from office by passing an ordinary resolution, requiring a simple majority in favour of the decision. However, so dramatic a step is surrounded by numerous safeguards. The member proposing the resolution must give 28 days' special notice to the company before the relevant meeting; the company must then give other shareholders 21 clear days' notice of it; the company must notify the director concerned and he or she has the right to circulate a written statement setting out his or her point of view prior to the meeting and speak at the meeting personally. It should also be remembered that a company is not obliged to call a general meeting just because a shareholder has served a special notice proposing the removal of a director from the company.

Meetings and reporting

The board normally decides whether or not to call general meetings, although shareholders with 5% of the company's share capital can compel the board to call a general meeting through a members' requisition. Although the Companies Act 2006 makes clear that a company's articles cannot require anything more than an ordinary resolution to remove a director, case law (*Bushell v Faith* (1970) HL) has established that giving directors enhanced voting rights on any resolution to remove them in the company's articles is legally effective. The director concerned may therefore have enhanced voting rights which allow him or her to defeat any such resolution. If a resolution to remove a director is successful, it should be remembered that he or she may also have a contract of service with the company and substantial compensation may need to be paid.

It may be that what concerns members most is the income that their shares will produce. A declaration of a dividend is usually made in a general meeting (GM) or an annual general meeting (AGM).

Private companies are no longer required to hold an annual general meeting, although public companies must continue to do so. Any kind of business may be transacted at an AGM as long as appropriate notices have been served and appropriate resolutions are passed. If business needs to be transacted before another AGM is likely to be held, or if a private company has dispensed with holding AGMs, a GM may be convened, either by the directors or by members who together hold 5% of the company's paid-up share capital (members' requisition). Resolutions may be ordinary or special, the latter requiring 75% of the shareholders in favour. A private company can pass a written resolution, avoiding the need to hold a meeting and requiring the same percentage of the members in favour as would have been required had the resolution been put to a meeting.

The Chairman of the board will usually chair company meetings. He or she must process the business in accordance with the rules contained in the Articles, and in accordance with the general rules of the common law. Voting on resolutions can be on a show of hands, but sometimes the more formal procedure of a poll may be demanded. The Articles usually provide for situations when members can insist on a poll. The advantage of a poll is that it reflects the voting power that an individual shareholder has in the company – instead of having one vote per person as on a show of hands, voting reflects the number of shares held.

The Small Business, Enterprise and Employment Act 2015 replaced the requirement of an annual return with an annual online 'confirmation statement' where details of directors, shareholders, registered office, Persons with Significant Control (PSC) and some other matters are confirmed.

Protection of minority shareholders

The wishes of the majority should normally take effect, but not where the impact of that is to perpetrate a fraud on the minority.

The law has been alert to the problem of protecting minority shareholders and provides a number of means which can help to secure some protection. One example is shareholders' agreements, whereby the members enter into a separate contract with each other, regulating how certain aspects of the company will be run. A statutory example of protection is the right of a minority shareholder to apply to the court if the company's affairs are being carried on in a way which is unfairly prejudicial to the members or some part of them. The Companies Act 2006 has also codified the common law rules which allowed a shareholder to bring a claim on behalf of the company (a derivative action) where some wrong was done to the company and the wrongdoers controlled the company and therefore refused to take action against themselves. The 2006 Act allows a member to bring a derivative claim in respect of a cause of action arising from an actual or proposed act or omission involving negligence, default, breach of duty or breach of trust by a director of the company, although there are substantial hurdles to overcome before a court will give the member permission to continue the claim. Alternatively, a member could petition the court to wind up the company on the grounds that such a winding up would be just and equitable. These rules could be a particularly effective sanction in small private companies with, say, only half a dozen members, where the rights of one of them are being seriously prejudiced.

Company finance

In order to provide adequate resources to allow the company to develop and flourish it may be necessary for a company to borrow money or look at other methods of funding, such as government grants, as well as taking account of the capital raised by the issue of shares. The advantages of share capital are that it does not have to be repaid (unless redeemable shares have been issued) and therefore has a permanence about it, and dividends are only payable if the company makes a profit and declares a dividend.

The size of the company's share capital may well determine how much it can borrow, especially in the early stages of a newly formed and developing company. Lenders will be reluctant to advance more than the members of the business have themselves risked. Where a company needs sizeable amounts of money, it may be possible to attract funds under government-sponsored schemes.

A company can sometimes improve its financial position by siting itself where it may benefit from central and local government grants and loans. The schemes will vary from time to time, reflecting the political stance of the government of the day. Examples of schemes have included Enterprise Zones and Regional Development areas. Sometimes, the incentive to set up in a particular place will consist of very favourable rents and rate-free periods.

Borrowing

Where a company needs to borrow money, it may turn first to its own shareholders, who can provide loans called debentures. Like all forms of borrowing, the issue of debentures is only possible if the directors are empowered to borrow by the Articles. It is possible to give debenture holders greater security by creating a charge over the assets of the company.

All lenders of money worry about how they will be repaid if the company gets into difficulties. A lender may be prepared to make the loan only on condition that he or she is given some security in the event of the company being unable to pay. If that happens the lender can then realise this security, i.e. sell it, to repay what is owed. All this means that debenture holders are in a significantly stronger position than unsecured trade creditors.

One type of security is the fixed charge, which is a mortgage over a specific thing, like a building. Inevitably the company cannot dispose of, or realise, the property which is subject to the charge without the consent of the lender. This could be particularly restricting in a developing business, so the company may prefer to borrow against the security of a floating charge. This is a mortgage of assets of the company, such as the stock in trade, which changes from time to time. It is in the contemplation of both lender and borrower that the company should be free to carry on its normal business, dealing as usual with those assets. The company continues to have the flexible use of whatever is comprised in the charge. However, once the borrower defaults, the floating charge is said to crystallise. This means that it becomes a charge attached to the assets available at the moment of crystallisation and the company is henceforth prevented from dealing with those assets. The parties may have specified a number of circumstances which will give rise to crystallisation.

Floating charges are a device by which a company can raise money without affecting its ability to trade. However, they could have the effect of misleading persons who are thinking of dealing with the company, because the company might look in a more affluent and healthy financial position than is truly the case. The problem is overcome to some extent by requiring company charges to be registered with the Registrar of Companies within 21 days of being made. The company is legally obliged to effect the registration, but it would be sensible for the lender to do so too, because a charge which is not registered in time is void. Once the charge is registered, it is part of the information which is available for public inspection. Any person about to deal with a company and needing information about its financial viability could make a search at Companies House. That would reveal any charges registered, the date the charges were created, the person in whose favour the charge is made and the property over which the charge operates.

Raising money by the device of the floating charge is a procedure only available to companies and limited liability partnerships. This gives a company and a limited liability partnership an advantage over a sole trader or trading partnership, which could borrow only against the security of fixed assets.

Taxation of companies

As a separate legal entity, companies are charged Corporation Tax on profits and capital gains, assessed *before* dividends are distributed. Directors and shareholders will pay Income Tax on dividends and, for those with a day-to-day paid role, on any salaries drawn.

Personal liability – looking behind corporate structure

As noted previously, one of the key attractions of corporate structure is the limited liability of shareholders. There may be personal liabilities where personal guarantees have been given, but that is clearly a personal obligation openly entered into. There are, however, circumstances whereby the courts *will* look to the individuals rather than the company. This may be, for example, in the case of outright fraud, of wrongful trading or where a company was set up for the express purpose of avoiding a personal contractual liability (*Gilford Motor Co. Ltd v Horne* (1933) CA). This is often called, rather quaintly, lifting (or piercing) the corporate veil.

Winding up of companies

A company is an artificial body and theoretically it could continue forever with a constantly changing membership as shares are bought and sold (transfer of shares). If a company is to be dissolved, the process is referred to as 'winding up' or 'liquidation'. The directors may decide on this course, in which case the dissolution is voluntary, or it may occur by order of the court.

If the winding up is brought about because a company has financial problems, the rules of liquidation are contained in the Insolvency Act 1986 and the Insolvency Rules (England and Wales) 2016.

When a company is in financial difficulties, a full liquidation process may be unnecessary. It may be possible to save the company by using voluntary arrangements. The proposal for such a voluntary arrangement can be made by a director. The proposal will be for a composition in satisfaction of the company's debts or a scheme of arrangement of the company's affairs. An insolvency practitioner will be appointed to act as a supervisor, implementing the scheme. The value of such a voluntary arrangement is that it can be organised with the minimum of involvement of the court, as it will be approved at meetings of the creditors and the company.

Where a company is likely to be unable to pay its debts, another alternative procedure is an Administration Order. The court will make such an order if it believes that by doing so it can secure the survival of the company or that it would give time for a voluntary arrangement to come into being. There will be a court hearing. If an order is made it directs that, for the time being, the affairs, business and property of the company are to be managed by an administrator. While the order is in force there can be no order for the liquidation of the company. The administrator has three months to draw up a statement of his or her proposals, which are put to meetings of the creditors and the company.

The Enterprise Act 2002 amended the Insolvency Act 1986 to provide a procedure for the out-of-court appointment of an administrator by the holder of a floating charge, the company itself or its directors. Notice of such an appointment must be filed with the court, but no court hearing or order is required to effect the appointment. The purposes of administration are to rescue the company as a going concern or to achieve a better result for its creditors than would be likely on a winding up or to realise property to pay one or more secured or preferential creditors.

Where liquidation of the company is necessary, this may be:

- a members' voluntary liquidation, which can occur where the company is solvent;
- a creditors' voluntary liquidation, which occurs where the company is insolvent and where the wishes and interests of the creditors will prevail or
- a compulsory liquidation, which arises by order of the court, following the presentation of a petition seeking such an order.

For a compulsory liquidation, a winding-up petition may be based on a number of grounds, including the ground that the company is unable to pay its debts. This is deemed to be the case where a creditor who is owed more than £750 has made a statutory demand for this money that has remained unsatisfied for three weeks. Other evidence accepted by the court that the company is unable to pay its debts may be proof that the company's assets are less than its liabilities, or proof that a creditor entitled to execution has been unsuccessful. This means that a creditor was granted a warrant of execution to levy distress against the goods of the company, but this warrant has proved ineffective because there are insufficient goods to satisfy the demands.

Proceedings for compulsory liquidation are commenced in the county court or the High Court, depending on the amount of the company's share capital. Once a petition has been presented any disposition of company property is void unless approved by the court. This discretion is important because it means that the company can continue to trade, but it allows control over activities which may be detrimental to the interests of some of the creditors.

If the court decides to make a winding-up order, certain immediate consequences follow:

- A liquidator is appointed.
- All actions against the company must be stayed.
- A first meeting of creditors may be convened.
- The directors' duties and powers cease.
- The business of the company can carry on only to the extent that it is beneficial to the liquidation.

The liquidator must take control of all the company's property. He or she must call in all the assets, realise them and distribute the proceeds to the appropriate creditors. Some property of the company may not be available to the liquidator, e.g. where the contract by which the company was acquiring the property provides that the company's interest in the property will end in the event of liquidation or where a company has purchased goods under a contract containing a valid retention of title clause. The liquidator will also be concerned to discover whether transactions immediately prior to the liquidation were undertaken to defeat the rights of creditors or some groups of them. If so, the transaction may be void. Examples include cases where a company transfers property for no consideration, for a consideration significantly less than the real value (an undervalue transaction) or a transaction involves a preference, where a company puts one of its creditors into a more favourable position than the creditor would be under the liquidation.

Once the liquidator has realised all the assets, he or she then pays the costs and expenses of the liquidation. He or she must next pay off the creditors, who must have submitted formal proof of their debt. Secured creditors will usually realise their security to settle what is owed to them, but they may prove any outstanding balance not covered by the security. Fixed charge holders take separately, but those holding floating charges rank after preferential creditors. This tends to emphasise the superior security provided by a fixed charge. The creditors rank in the following order:

1. **Preferential creditors** – this group ranks equally among themselves after the expenses of the winding up. The categories of preferential debts have been significantly reduced by the Enterprise Act 2002. Examples of preferential debts include:

   a. Contributions to occupational pension schemes.
   b. Employees' remuneration due for the previous four months, not exceeding £800 (this figure can be adjusted from time to time by the Secretary of State).

   The Finance Act 2020, under s.98, reintroduced certain tax liabilities payable to Her Majesty's Revenue and Customs as preferential.

2 **Amounts secured by floating charges** – the Enterprise Act 2002 provides that, where there is a floating charge over the property of a company, a percentage share of the company's net assets must be made available to unsecured creditors. This provision improves the position of unsecured creditors but means that there is less money available to distribute to the holder of the floating charge. There are some circumstances in which these provisions will not apply, for example, if the company's net assets are less than a prescribed minimum threshold.
3 **Unsecured creditors** – all other creditors rank equally amongst themselves.

If there are surplus funds left over after distributing to the creditors, these are distributed to the members of the company according to their rights and interests.

Where a company has been wound up, this may reflect seriously on the competence and integrity of the directors. In that case, the courts have extensive powers to disqualify directors for specific periods. In some cases, the disqualification may be based on activities of the director which could involve criminal liability, but there is no overall requirement that the behaviour must infringe specific criminal law rules. One example where a director could be disqualified is for 'unfitness' to be concerned with the management of a company.

Once the process of winding up has finished, the company is removed from the Register and is dissolved three months later.

Limited liability partnership

Since 6 April 2001 it has been possible to incorporate a limited liability partnership. The Limited Liability Partnerships Act 2000 created this new form of organisation for businesses, which is a hybrid between the partnership and the limited liability company, although the name is potentially misleading as the form has more in common with a company.

In the same way as a company, a limited liability partnership (LLP) has a separate legal identity to that of its members. It is the LLP that enters into contracts, owns property, etc., not the individual members. This gives members of the LLP the benefit of limited liability for the debts of the partnership. There are formalities associated with setting up and running an LLP, although these are less onerous than those associated with setting up and running a company.

Creating a limited liability partnership

To form an LLP, the Limited Liability Partnerships Act 2000 requires that two or more persons carrying on business with a view to profit subscribe their names to an incorporation document. These individuals are known as designated members and have responsibility for sending any necessary information to Companies House on formation and during the life of the LLP. All the members of the LLP can be designated members, although the legal requirement is that there must be two.

The incorporation document contains the name of the LLP, the situation and address of its registered office, the name, address and date of birth of each member of the LLP, and which of the members are to be designated members. This document, the appropriate fee and a statement that the requirements of the Limited Liability Partnerships Act 2000 have been complied with must be sent to the Registrar of Companies. On registration, a certificate of incorporation is issued.

Managing a limited liability partnership

An LLP has far more in common with an unlimited liability partnership than with a limited company in terms of managing its affairs. Unlike a company, an LLP is not required to have formal constitutional documents on incorporation. It is, however, sensible for the members of an LLP to enter into a formal agreement. This agreement should deal with the sort of issues that would be covered in a partnership agreement for an unlimited liability partnership. In the absence of a partnership agreement, the Limited Liability Partnerships Regulations 2001 will apply. Under these Regulations, profits, losses and capital are shared equally and all members have an equal right to manage the LLP. Unlike a company, the members of an LLP are not obliged to make decisions in formal, minuted meetings and hold meetings of all the members for certain matters. The members' agreement should cover which members make decisions of day-to-day management and any matters requiring the agreement of the members.

Taxation of limited liability partnerships

LLPs are taxed like partnerships, with a partnership tax return and each partner being charged Income Tax on their share.

Community Interest Companies (CICs)

Although this book is concerned with business matters, note the introduction in 2005 of community interest companies (Companies (Audit, Investigations and Community Enterprise) Act 2004). This structure is for organisations which aim to utilise profits for the public good. Details of formation and registration can be found on the Companies House website.

Other business and enterprise structures

It should be noted that there is a range of other structures available. In a commercial or non-commercial context, co-operatives might be formed under the Co-operative and Community Benefit Societies Act 2014. And in a non-commercial context, there are a number of different structures available for charities and banking and savings operations, which are outside the remit of this book.

Discussion questions

1 What business structures are available to someone starting up a small property business with one employee, in addition to themselves? And what are the key features, pros and cons of the options?
2 Why was the Partnership Act 1890 introduced or, in the alternative, why are businesses not completely free to decide on their perms of operation?
3 Why would a business person who has been the managing partner of a small partnership for some years consider incorporating as a private limited company?
4 Under what circumstances might the 'corporate veil' be lifted, such that the protection of limited liability is lost?

5 The law of contract

Mark Allison

The nature of a contract

In business and commercial life, the law of contract underpins a huge range of activities as diverse as engaging staff, buying supplies, arranging insurance, leasing premises, raising loans and buying shares. These transactions all involve making a contract which must comply with the general principles governing all contracts. The modern law of contract has largely evolved from case law, although major pieces of legislation such as the Unfair Contract Terms Act 1977 and the Consumer Rights Act 2015 (in the context of consumer contracts) have radically altered some of the rules. There is no complete code of rules governing the making of a contract. The courts have generally taken a *laissez-faire* approach to the making of contracts, indicating that the parties are free to make the contract of their choice, with which the courts will rarely interfere. Although this is a guiding principle, the law will intervene in certain circumstances, as will be explained.

Where a dispute arises between the parties, litigation, arbitration or a form of alternative dispute resolution may be necessary to determine the rights of the parties. However, the role of the courts ought not to be overemphasised. Of all the millions of contracts entered into daily, only a very small proportion ever involve litigation before the courts. On the whole, the parties carry out their obligations and, by that means, discharge their contract. Even where problems do arise, some compromise or settlement may be reached between the parties without the need to take action in the courts.

Contracts are based on the idea of a bargain. Each side must put something into the bargain. A contract may be defined as 'an agreement which is binding on the parties'. Within that definition lies the first problem. The ultimate bargain arrived at between the parties may not have been 'agreed' in the commonly understood sense of that word. There are a number of reasons why this is so.

First, English law adopts an objective approach to decide whether there is an agreement between the parties. It is pointless to ask an individual who is now in dispute whether he or she intended to make a contract. Rather, the courts ask the question: 'Would a reasonable person think that there was an agreement between the parties?' Second, 'agreement' may not seem to be present where one party has little or no choice in the terms he or she is obliged to accept. There are many contracting situations where, because of the superior bargaining strength of one of the parties, the other has little or no choice but to accept the terms. Third, there are many cases where little or no negotiation takes place between the parties (e.g. sales in a self-service store), but the 'agreement' between the parties is largely created by implied terms (e.g. terms implied into contracts for the sale of goods by virtue of the Sale of Goods Act 1979 or the Consumer Rights Act 2015).

DOI: 10.1201/9781003657101-5

To return to our definition of a contract (an agreement which is binding on the parties), it is always possible for an injured party who is suffering from some breach of contract to sue to enforce it. Enforcing the contract means seeking an appropriate remedy from the court. In most cases that remedy will be an award of damages. It is only in very limited types of case that a court would order one party specifically to perform their obligations. If the court does grant such an order, it is called specific performance. This is an equitable remedy and as such is discretionary, i.e. the court may or may not award the remedy.

So, for an agreement/transaction to be a contract, there needs to be an agreement (a valid offer which has been accepted), this agreement needs to be supported by consideration and there needs to be an overarching intention to create legal relations.

Formation of the contract – agreement

It has become customary in English law to regard agreement as consisting of offer and acceptance. This is only a method of analysing the bargaining process. It is not necessary for the parties to use these words. An offer exists when one party effectively declares a readiness or intention to be bound by a set of terms without any further negotiation. In other words, the offeror (the person making the offer) has reached a stage where he or she is content for the other party to say he or she accepts, and at that point a valid, binding contract will come into existence. If the parties are not at such an advanced stage in their negotiations, no offer may yet exist. Often, what appears to be an offer may only be some kind of enticement to bargain, which the law calls an invitation to treat.

The courts have had to consider numerous instances where the argument turned on whether a definite offer was made. In the case of goods on display in a shop window (e.g. *Fisher v Bell* (1961) QBD) or goods on supermarket shelves, the court would normally find that such a display was an invitation to treat; the shopper must come forward and make an offer to buy, giving the shop the option of rejecting that offer. Similarly in auction sales, when an auctioneer calls for bids, he or she is not 'offering' to sell the goods but, rather, inviting potential buyers to put forward their offers. Equally it may be said that a website advertising goods (it may be viewed as a virtual shop) will usually constitute an invitation to treat in the absence of a clear intention to be bound by certain terms.

The problem regarding offers can also arise in relation to price lists or catalogues which are sent to customers. These are usually construed to be a mere indication of the goods available and their likely selling price and, in consequence, an invitation to treat. Any other interpretation could give rise to serious practical difficulties. Consider a case where a firm sends out a price list of second-hand earthmoving plant which it has available for sale. The circular may go to dozens of companies that could be interested in buying. If the circular was considered to be an offer and there is only one model of a particular type on the list and several purchasers come forward saying that they wish to buy it, the supplier could find itself in breach of contract over and over again, as it is unable to supply them all with the equipment. It is better to construe the price list as a mere invitation to bargain, a 'display' of what the firm has to sell and think of the approaches by potential buyers as the offers, which the seller can accept or reject as it chooses.

Where a case arises which causes doubt, the court approaches the issue of whether any offer has been made by using an objective test of the intention of the person concerned. Would a reasonable person, knowing all the circumstances of the case, think that an offer has been made? The court takes into account the words used. In one case, where a letter said: 'I may be prepared to sell', the writer could not yet be said to be ready to be bound in contract without further negotiation (*Gibson v Manchester City Council* (1979) HL). Even if the word 'offer' is used, that is not conclusive in law that an offer exists.

If it is clear that the intention is to make a firm offer, it may be made orally or in writing. It comes as something of a surprise to many people to discover that English law has very few rules about formality or written evidence in contracts. Apart from exceptional cases such as contracts for the sale of land (s.2 Law of Property (Miscellaneous Provisions) Act 1989), transactions worth hundreds of thousands of pounds can be binding even when made orally. Of course, there may be sound reasons for making a contract in a more formal way, as subsequent disputes may be capable of being resolved simply by reference to a written agreement.

An offer can be made to an individual when only that person can accept. The offer may be made to a group and, in certain cases, it may be made to the world at large (*Carlill v The Carbolic Smoke Ball Co.* (1893) CA), for example, in cases offering a reward to a finder. The offer is only effective when it has been communicated to the other party, the offeree.

Counter offers

The offeree has a number of options. He or she may wish to accept the offer, in which case he or she must exactly accept the precise offer which was made. If he or she seeks to bargain further or to introduce new terms he or she is not accepting, rather he or she is making a counter offer. The law provides that the effect of a counter offer is to terminate the original offer. So, if the counter offer is not acceptable, the person who made it cannot try to revive the original offer. He or she must make a fresh offer (*Hyde v Wrench* (1840) Ch).

This situation can be of vital importance where the person making the final offer may be fixing the terms of the deal. The offer-counter-offer situation arises frequently in business, where each side in the negotiations may be trying to make the contract on the basis of their own 'pro forma' standard documents. This gives rise to what is known as the battle of forms.

A classic example can be seen in *Butler v Ex Cello Machine Tool Co.* (1979) CA, where the seller offered to sell a machine tool for £75,000, which would be delivered in 10 months' time. The seller's offer was made on a printed form containing various conditions, one of which was a price variation clause allowing for an increase in certain circumstances. The buyers placed an order, using their own standard printed order form. On the bottom of the buyers' form was a tear-off slip, which the seller was asked to sign and return to the buyers. The seller duly signed and returned the slip. It contained the words 'we accept your order on the terms and conditions thereon'. When the tool was ultimately delivered, the seller sent a bill for over £78,000. Its claim for the increased amount was unsuccessful. Its original quotation was an offer. The buyers had made a counter offer which did not contain a price variation clause. The seller had accepted the counter offer by returning the tear-off slip. This exchange is sometimes referred to as 'the battle of the forms'.

Of course, not every communication from an offeree will necessarily amount to a counter offer. Once an offer has been made, the offeree may simply wish to seek further explanation of its terms to satisfy himself that he is getting the best deal possible. So, there may be a fine distinction between the cases. For example, X offers to sell Y his car for

£3,000. If Y replies, 'Will you take £2,700?' he is still negotiating and making a counter offer (*Hyde v Wrench*). But if Y replies, 'Is that cash or is there any possibility of HP?' he may simply be seeking to clarify the terms of the offer. A mere inquiry does not terminate an offer (*Stevenson Jacques & Co. v McLean* (1880) QBD).

Acceptance

Acceptance is defined as a final, definite and clear assent to the terms of an offer. Not only must the offeree accept the exact and precise offer made, but he or she must also communicate acceptance to the other party. Normally, the offeror must actually receive the acceptance. This tends not to create problems when the parties are face to face or using some form of instantaneous communication, such as the telephone, a telex machine or, more recently, email or messaging services. In all of these cases, the contract is made when the acceptance is received by the offeror.

The situation is somewhat different if the parties are negotiating at arm's length, i.e. not in each other's presence. This will usually be a case where they have been dealing with each other by post. The offeror can lay down rules about how the offeree has to accept, but in most cases will not have done so.

Postal rule

Then it may be appropriate to use what is called the 'postal rule'. When this applies, an acceptance made by letter is effective, and the contract comes into existence when the letter is posted (*Adams v Lindsell* (1818) KB). Inevitably this means that a legally binding contract exists for some time before the offeror actually learns that he or she is legally bound. That can sometimes be inconvenient, but an offeror could overcome the problem by stipulating as part of the offer that he or she will only be bound on *receipt* of the acceptance.

In *Holwell Securities v Hughes* (1974) CA, the offeree wanted to accept an offer under the terms of which he had to accept by a notice in writing by a fixed date. He posted a letter of acceptance, but it did not arrive until after the fixed date. The court held that the phrase 'notice in writing' indicated that the offeror must actually receive the acceptance by that date. Merely posting a letter was no use in those circumstances.

The 'postal rule' is merely a rule of convenience, and the courts would not apply it in cases where it was manifestly absurd to use the post (e.g. a case where the offeror had indicated that he or she needed a very speedy acceptance). Nor will the rule be applied if the offeror has prescribed some other method of acceptance.

Method of acceptance

The offeror can lay down quite elaborate rules for acceptance. The courts may then be faced with the problem that an offeree has purported to accept but has not followed the precise method stipulated. Generally, the courts will allow an acceptance by any method which is equally as effective as the one which was prescribed, so long as the offeror has not stipulated that acceptance *must* be made by the method prescribed. It is a question of construing whether the offeror was making a suggestion or whether that method and no other must be used. For instance, if the offeror says, 'Drop me a line to tell me if you want to buy', does that mean he or she is insisting on acceptance only by letter?

It is now generally accepted that the postal rule does not apply to email. Email is treated as a form of instantaneous communication, and acceptance therefore takes effect only when it is received by the offeror, not when it is sent. This position follows the principle established in *Entores Ltd v Miles Far East Corporation* (1955), where the Court of Appeal held that acceptance by telex was only effective upon receipt. That approach has since been extended to email by the High Court (Ch) in *Thomas v BPE Solicitors* (2010), where it was confirmed that email operates more like telex or fax than traditional post, and so the acceptance must actually arrive in the offeror's inbox to be effective. The reasoning in these cases reflects a broader consensus in academic and judicial commentary: that modern forms of electronic communication are not governed by the postal rule, but rather by principles requiring actual receipt. It is unclear whether or not the postal rule applies to e-mailed acceptances. As an email may be seen as an instantaneous form of communication (like a telex), it is suggested that there must be actual communication to complete acceptance.

Communication of acceptance

The rule that acceptance must be communicated means that silence cannot usually amount to acceptance. This is important because otherwise an offeror could impose obligations on the offeree: 'I offer to sell you my car for £3,000. If I do not hear from you before 9 pm I will assume that you have accepted'. If this were permitted, it would force the offeree into a bargain. The courts have consistently rejected such attempts to bind a party by silence, as confirmed in *Felthouse v Bindley* (1862) (CCP).

It is generally accepted that, in communications by electronic methods, receipt is deemed to take place during normal office hours unless otherwise agreed. This was affirmed in *Thomas v BPE Solicitors* (2010), where it was held that an email sent outside business hours did not amount to effective acceptance until it could reasonably be expected to have been read. As such, acceptance by electronic means must not only be communicated but also received at an appropriate time to be effective.

Termination of offers

An offer can be said to be terminated by its acceptance because, at that moment, there is a legally binding contract. There are, however, a number of ways in which an offer may be brought to an end.

The offer itself may have been made subject to a time limit. If there is no acceptance before that time, the offer will automatically lapse. Where there is no fixed time limit, the offer will remain open for acceptance for a reasonable time. What is reasonable will vary with the circumstances and the nature of the subject matter.

Once an offer has been made, the offeror may have a change of mind and wish to withdraw it. Revocation is always possible before an offer has been accepted. To be effective, the fact of revocation must be communicated to the offeree. In the case of postal communications, this means that the letter of revocation must actually be received to be effective; the postal rule does not apply to revocation. This can cause conflict where the parties are negotiating at arm's length using the post where the postal rule of acceptance is relevant. The letter revoking the offer may be posted first, but if it arrives *after* a letter of acceptance has been posted, it will be too late to withdraw the offer.

Where an offer has been made for a fixed time, the question may arise whether it can be revoked during that time. In reliance on the offer remaining open, the offeree could go to considerable trouble to put himself or herself into a position where he or she can accept, e.g. by raising finance. If the offeror revokes before the fixed time is up, the offeree may have no legal basis for complaint. Effectively this fixed offer situation is really two distinct offers: 'I offer to sell you my car for £3,000, and I offer to keep the offer open for seven days'. If the offeree wishes to enjoy the benefit of the seven-day time limit, he must accept the second offer and 'buy' the offer by giving some consideration for it. This is what happens when a person buys an option, but in most everyday situations, it would not occur to people to offer to 'buy' time in this way. The rules of consideration are considered later.

Certainty of the agreement

Even though the parties think they have reached an agreement, the courts may find that there is no legally binding contract because of a lack of certainty. Naturally, the courts do not want to disappoint people's expectations, so they approach any question of vagueness or uncertainty with the idea in mind that 'a thing is certain if it is capable of being made certain'. An agreement which appears vague at first sight may be capable of being given a sensible meaning, either because the parties have dealt together before or because there is some custom of the trade or business.

In some cases, the parties themselves may be aware that their agreement is vague or incomplete and they may provide for future resolution of these aspects by including an arbitration clause. Sometimes it is possible to fill in missing terms by reference to a statute. For example, if the parties have failed to agree on a price and goods have now been supplied under the contract, s.8 of the Sale of Goods Act 1979 provides that where no price has been fixed and no method has been agreed for determining the price, the buyer must pay a reasonable price.

The agreement

Once the processes of offer and acceptance have culminated in the formation of the agreement, the **terms** have become fixed between the parties. That has been the object of their negotiations – to determine the extent of their obligations to each other. Once agreement is reached, it cannot be altered by one of the parties without the consent of the other. If any alterations or variations to the agreement are to take place, then the relevant rules of consideration should also be borne in mind.

There can be great difficulty in establishing what has become a term of contract. A distinction has to be drawn between statements made which go to the core and essence of the transaction and those statements which merely influence the decision to enter into the contract. The former are **terms**, where a breach gives rise to a right to claim damages for breach of contract. The latter are **representations**, where rights may be available if the statement made was untrue and, in some way, induced the contract. Depending on the nature of the misrepresentation, whether fraudulent, negligent or innocent, different remedies may apply, including rescission and, in some cases, damages under the Misrepresentation Act 1967.

Consideration

English contract law is based on the idea of a bargain. There must be some exchange between the parties. One party must be doing something or giving something in consideration of what the other is doing or giving.

Consideration can be defined simply as the 'price' each party pays for the right to enforce the other party's promise. It is important to remember that consideration will not always be money, although it is quite usual for money to be the consideration provided by one of the parties. If I sell you my car for £3,000, my consideration is giving you the car and your consideration is giving me the £3,000. If we make that agreement and provide for the exchange to take place at the end of the month, then our agreement is still a binding contract *now* because our exchange *now* is an exchange of promises. In consideration of your promising to pay me £3,000 at the end of the month, I promise to hand over my car to you then. As you can see, there is no need for a physical exchange of the car for money to create a contract; the exchange of promises is sufficient. This allows parties to plan for the future and is particularly important in the commercial and business world.

This type of consideration, based on promises which are to be performed in the future, is called executory consideration. It is surprising how often people fail to realise that there is a legally binding obligation in such a case. Take the example of a shopper who orders a chair in a furniture store, delivery to be in six weeks' time, when the customer will be invoiced for the chair and will then send payment. Later the same day, having placed the order, the customer wants to back out. That would amount to a breach of contract. The customer might say that the furniture store had not even had time to send the order off, but the store has lost the profit on the sale and no doubt has been put to some trouble and expense in dealing with the customer.

Past consideration

In order to provide a valid consideration, the law insists that it must not be past. *Re. McArdle* (1951) CA illustrates the principle. Members of a family were all entitled to a share in a house on the death of their mother. While the mother was still alive, one of the sons and his wife went to live with her. The son's wife made several improvements and alterations to the property for her own comfort and convenience. When the mother died, the other members of the family promised to pay the son's wife for all the work that had been done. When they failed to pay, she could not enforce their promise as she had given no valid consideration for it. The work she had done was in the past and had not been done in consideration of their promise to pay. The promise and the act were independent of each other. To be valid, consideration must involve an interdependence between promise and act.

There are some situations which look like past consideration, but which are not true instances. Take the case of a person requesting someone to perform services, in circumstances where there is a clear expectation of payment, for example, when asking an accountant to deal with financial affairs or getting into a taxi. Once the service is performed, any subsequent promise to pay in such a case is enforceable because by asking for the service, there is an implied promise to pay a reasonable amount. The consideration is not truly past, as it was rendered at the request of the promisor and with the expectation of payment.

Adequacy of consideration

Although contracts are based on the idea of a bargain, the law does not insist that the two sides of the bargain must be equal. This is often expressed as a rule that consideration need not be adequate. The phrase 'peppercorn rent' illustrates this principle; something of value (a peppercorn) must be given, even if it is nominal. A well-known example is *Thomas v Thomas* (1842) (QB), where the payment of £1 per annum was held to be sufficient consideration to support a contract.

It is up to the parties to make the contract of their choice. One person, for good reasons of his own, may be prepared to pay far more for goods than they are objectively worth. Generally, contract law allows great freedom to the parties in making their bargain and each is expected to look after their own interests and strike the keenest bargain possible. One of the maxims of the law is *caveat emptor*, which means 'let the buyer beware'. Where the two sides of the bargain are seriously unequal, that may be evidence of fraud, duress, mistake or incapacity, in which case there may be some basis on which the court could set aside the contract. However, if the person complaining has simply made a bad bargain, the law will not intervene.

While not insisting that the two sides of the bargain must be equal, the courts will insist that each side of the bargain must have some value in the eyes of the law. This may be so even where the consideration consists of something intrinsically worthless. This point is illustrated in *Chappell & Co. v Nestlé & Co.* (1960) HL, where the chocolate company was running a promotional scheme. In return for three chocolate wrappers and a postal order for 1/6d (7½p), the chocolate company would send the customer a pop record. Chappell & Co. were – entitled to receive royalties on the selling price of all sales of the record. The question arose here whether the record was sold for 1/6d or whether the wrappers formed part of the consideration. Even though the chocolate company simply threw the wrappers away, the House of Lords found that they did form part of the consideration.

Parties to consideration

Consideration must move from the promisee, meaning that the person who receives the promise must provide consideration. However, it does not mean that the consideration must move to the promisor. In *Tweddle v Atkinson* (1861) QB, a young couple were about to be married, and their respective fathers promised *each other* that each would pay a sum of money to the son. Indeed, their agreement said that the son could sue on the agreement should either party fail to pay. One father did not pay. The son attempted to sue him but failed as he was not a party to the contract and, additionally, had provided no consideration for the promise. Note that the consideration given by the fathers did not move to the other but to the son, a third party. Consideration moved from the promisee but not to the promisor, i.e. it was for the benefit of a third person.

It should be noted that the stated difficulty of contracts not being enforceable by third parties is connected to the notion of *privity of contract*. The rule that contracts cannot be enforced by a person who is not a party to a contract has been subject to qualification by the Contracts (Rights of Third Parties) Act 1999. Section 1(1) provides: Subject to the provisions of this Act, a person who is not a party to a contract (a 'third party') may in his own right enforce a term of the contract if (a) the contract expressly provides that he may; or (b) subject to subsection (2). . ., the term purports to confer a benefit on him. If the facts of *Tweddle v Atkinson* were to be repeated today, then the son would be able to enforce the promise under s.1(1)(a) of the 1999 Act. The 1999 Act may entitle a third party to rely on an exemption clause in a contract.

Existing duties

It seems obvious that you are not providing any consideration, where you simply behave in a way in which you were legally obliged to act. This view is accepted by the courts in relation to both public and contractual duties owed.

Take the case of *Collins v Godefroy* (1831) KB, where a witness was under a public duty to attend and give evidence because he had been ordered to do so by the issue of a subpoena. The witness was promised money by the defendant in the case if he came and gave evidence. Once he had given his evidence, was it possible to say he had done so in consideration of the defendant's promise to pay? The court held that the witness gave no consideration for the promise as he was already under a duty to give evidence. He was doing nothing more than his duty.

In those cases where it can be shown that a person does more than his public duty requires, then he or she may have provided consideration for a promise to pay. This can be seen to be the basis of the decision in *Glassbrook Bros v Glamorgan County Council* (1925) HL, where a mine owner requested police to provide a 24-hour guard over his property. Although the police were under a duty to prevent crime and protect property, they were not required to mount a continuous guard. The mine owner promised to pay for the service, and it was held that the local authority could recover the money from him as the police were doing more than their public duty. They had, therefore, given consideration to his promise to pay.

The same sort of situation can arise where a duty is owed under a contract. Suppose you have asked me to supply 100 sausage rolls for your village fair and you have paid me for them. Under the terms of our agreement, I am due to deliver the sausage rolls at 12 noon on 1 June. I indicate that I am going to struggle to arrive at 12 noon and sound reluctant to make a delivery. You then promise to pay an extra £10 and, in consequence, I make the delivery. What consideration do I give for your promise to pay £10? I was already contractually bound to deliver the goods so I am doing nothing more than performing an existing duty. In an example such as this, the courts have traditionally found that I could not enforce your promise to pay £10 as I had given no further consideration to it. An old case, *Stilk v Myrick* (1809) KB, has long represented that view of the court. There, sailors were promised extra pay for sailing a ship back to port when some of the crew had deserted. Later, the ship owner refused to pay, as the sailors were already obliged by their contract to sail the ship back and had therefore simply performed the duty they already owed under the contract. Contrast, *Hartley v Ponsonby* (1857) QB, where a similar situation arose, but there were so few sailors left aboard that it was held that they had carried out *additional* work, beyond that required by their original contract, in exchange for the higher wages.

Such a decision may be entirely correct as a pure application of the law, and no doubt it prevents situations of near blackmail arising where a person could apply pressure to the other contracting party to pay more, simply by threatening not to perform. But inevitably that 'pure' approach may fail to recognise the true pressures of business life and the marketplace. In a more modern shipping case, *The Atlantic Baron* (1979) QBD, the court was able to enforce a promise of extra payment for the building of a ship by finding that the shipyard did more than they originally promised and had therefore given consideration for the extra payment.

Some judges have gone so far as to say that any promise to perform a pre-existing duty should be good consideration, because it is a benefit to the person receiving performance. When the argument is carried to these lengths, in our sausage roll example, I could recover the extra £10. The benefit to you would be that you receive performance with none of the aggravation of needing to sue for breach. The detriment to me is that I give up the possibility of being in breach of contract – and there is no doubt that this can be a very valuable right. Having agreed to sell sausage rolls to you for a fixed price, I may now find that I can sell the same goods so advantageously elsewhere that it actually pays me to be in breach with you and suffer the consequences.

Much of this thinking has been further developed by the Court of Appeal decision in *Williams v Roffey* (1990). The facts were that building contractors had subcontracted the carpentry work on a block of 27 flats to the claimant, for a price of £20,000. The claimant was entitled to interim payments under the contract. After he had completed nine of the flats and done preliminary work on the others, he had already received over £16,000. He then realised that his price for the work was far too low. The building contractors began to get anxious, as they would become liable under penalty clauses in their main contract if the work was not completed on time. They knew that the claimant was having difficulty because he had underpriced the job. They suggested a meeting with him, at which they agreed to pay him £10,000 more, on condition that the work was completed on time. In view of the extra payment, the claimant went on with the work. He completed eight further flats but then stopped work. When he tried to recover what he alleged was due to him under the contract, the building contractors refused to pay any part of the extra £10,000, on the grounds that the carpenter had given no extra consideration for it. In the Court of Appeal, the carpenter was held to be entitled to the extra money because the building contractor had obtained practical benefits in having the work completed and in avoiding the penalty under another contract.

Of course, an important feature of the case was that there was no suggestion of 'blackmail' by the carpenter. Indeed, he had been approached in the first instance by the builder who had made the suggestion to pay extra. Moreover, if the court had been determined to look for something in the carpenter's behaviour which could amount to performing more than his mere contractual duty, no doubt they could have 'found' it. For example, it could be said that the carpenter's agreement to undertake the work in a different sequence could be a detriment to him and a benefit to the builders and the consideration.

It is difficult to make a judgment yet as to the ultimate effect of the Williams case on the development of this area of law, but it is clear that the courts have found it easier to discover a valid consideration where performance of a duty owed to a third party is involved. Consider the following example. A is contractually bound to supply the engines for a ship being built by B. The engines are due to be delivered by 1st January. C has a contract with B to charter the ship from 1 June, immediately on its completion. It begins to look as if A is going to be late in making delivery of the engines. C is extremely anxious that the ship should be finished on time, as he stands to lose an enormous sum of money if his charter of the ship is delayed. C approaches A and promises him £10,000 if he will get the engines to B by 1st January. If A delivers the engines to B on time, can he then claim the £10,000 from C? The courts have been inclined to take a more relaxed view in this type of three-party situation, and in a number of cases (e.g. *New Zealand Shipping Co. v*

A M Satterthwaite & Co. (1975) PC), they have allowed A to sue for the sum promised by C. The reasoning seems to be that there is a detriment or burden to A because, by binding himself in a contract to C as well as to B, he lays himself open to two actions for breach of contract if he fails to deliver the engines. Moreover, it might have been more to A's liking to be in breach of contract with B and face whatever consequences that might have brought. Now he foregoes that possibility because of his agreement with C. C also gets what C wants, and this is a benefit to C.

In such cases it would be important to check that there was nothing in the transaction which was contrary to public policy. Latterly, the courts have shown themselves willing to control these situations by developing rules of economic duress (see what follows).

Part payment of a debt

Problems can arise in finding valid consideration where one party seeks to perform a contract by doing something less than contractually obliged. Take the example of A, who owes £100 under a contract to B, which is due for payment now. A may be in financial difficulties, so much so that B agrees to accept £75 in full settlement. What consideration has A given for B's promise to forego £25? The courts would say that the £25 was still due and B could sue to recover it despite this promise. Thus, part payment of a debt does not discharge the full debt; this rule was approved by the House of Lords in *Foakes v Beer* (1884) HL.

It would be quite different if, at B's request, A had paid £75 at a date earlier than the debt was due. That would be a detriment to A and a benefit to B, and A would then be providing consideration for B's forbearance.

In *Re. Selectmove Ltd* (1995) CA, it was considered whether the approach adopted in *Williams v Roffey* could be extended to part payment of a debt. It was argued that by paying less than the sum owed, this could constitute consideration provided there are practical benefits to the creditor. For example, the practical benefit could be that the creditor gets part of the sum owed in circumstances where they might otherwise receive no payment. However, the Court of Appeal refused to follow *Williams v Roffey* as to do so would be inconsistent with *Foakes v Beer*, a House of Lords decision. At present, therefore, the rule as to part payment of a debt, as seen in *Foakes v Beer*, remains unaffected. It is fair to say that *Williams v Roffey* has created inconsistency in the law.

The difficulty with the proposition advanced previously is that, in practice, people dealing together do frequently make concessions to each other within the framework of their legally binding agreement. These concessions or variations are usually informal and may work perfectly well so long as the parties remain on good terms. If their relationship becomes strained, one of them may wish to revert to the strict terms of their original agreement. The other will want to know if the variation can be regarded as binding. Traditionally, in order to enforce the variation, it has to be 'paid for' by some extra consideration.

Promissory estoppel

There is sometimes a way around this problem, however, by the application of the equitable principle of promissory estoppel.

Equitable principles operate to abate the rigours of the operation of the common law. Undoubtedly, there can be grave hardship where one contracting party has relied on some concession by the other, only to see it abruptly withdrawn. That was the state of affairs in *Central London Property Trust v High Trees House* (1947) KBD. The contract in that case related to a lease of a block of flats in London, where the rent was to be £2,500 per annum. Because of the war, it became extremely difficult to find tenants for the flats, so the landlord agreed to reduce the rent by half. When the war ended, the landlord wanted to revert to the original rent and was also keen to know whether he could recover the arrears of rent. Effectively, what the landlord had done was to make a promise, within the framework of an existing contract, where the tenant gave no consideration for the promise, but where obviously the tenant had relied on it. On the authority of earlier cases the High Court found a principle of equity whereby if A leads B to suppose that A will not enforce strict rights under an existing contract, then A cannot go back on a promise where it would be inequitable. Applying this principle in the High Trees case, the court would not allow the landlord to recover the arrears of rent. But the court also found that the end of the war acted as notice that the promise (made only because of the war) was no longer operative.

At first sight this case may seem to strike at the very core of the doctrine of consideration. Further examination shows that not to be the case. A number of points must be borne in mind. It was made clear in cases subsequent to the High Trees case that this equitable principle of promissory estoppel can only be pleaded by way of defence. In the colourful language of judges, it is a shield and not a sword; estoppel of this type cannot be used to create a contract. There must be circumstances which show it would be inequitable for the promisor to go back on his or her word, at least without giving notice of the intention to revert to the original terms of the contract. Of course, there may be cases where once a concession has been made, it is impossible to revert to the original terms, e.g. if the concession related to an extension of time for delivery. This raises important questions about the scope of promissory estoppel, as it seems that the effect of the promise or concession should be intended only to suspend legal rights and not to extinguish them completely. This fits well with the idea that the promisor can revert to the original terms by giving reasonable notice. Certainly, if the scope of the rule is seen to be thus limited, it gives rise to fewer legal difficulties.

It must be stressed that the development of promissory estoppel is based on equity, which is governed by a number of traditional maxims such as: 'He who seeks equity must do equity' and 'he who comes to equity must come with clean hands'. It follows from these maxims that a person seeking to take advantage of promissory estoppel must have behaved properly. This aspect was critical to the decision in *D & C Builders v Rees* (1966) CA, where two jobbing builders were owed money by Mr and Mrs Rees. Mrs Rees knew they were in a grievous financial state, and she persuaded them to accept less than they were owed in full settlement. So desperate was their plight that they accepted the smaller sum, but they were determined to see what they could do later to recover the rest. Mrs Rees claimed that they had promised to accept the smaller sum, that she had acted in reliance on their promise and that it would be inequitable to allow them to go back on their word. The court had no hesitation in finding that her own behaviour was inequitable as it verged on intimidation, so they refused to allow her to take advantage of the equitable principle.

Although promissory estoppel can be seen as an important safeguard against some of the harsher aspects of the common law rules, it must be remembered that consideration is still essential for the creation of a valid contract, although there are limited circumstances where it may not be necessary to support some modification within the contract.

Intention to create legal relations

If both parties to a contract provide valid consideration, that may be taken to indicate that they intend to make a legally binding arrangement. However, in English contract law, the courts look for a separate element of intention to create legal relations. Traditionally the courts have viewed the cases as falling into two categories: commercial arrangements where the parties are presumed to have intended to create a contract; and family, domestic and social arrangements where the parties are presumed *not* to intend to create a contract. The presumptions are sensible in most cases. Take the example of a parent who gives pocket money to a child or two friends who arrange to meet for supper. In either case, one party would never expect to be sued for breach of contract if such arrangements collapsed.

The problem facing the courts arises where one party to an arrangement is seeking to rebut the normal presumption. In family arrangements a person may want to prove that a legally binding agreement was indeed intended. He or she must then satisfy the court that this was so. Many of these cases turn on problems between husband and wife in the breakdown of marriage. In one case, *Merritt v Merritt* (1970) CA, relations between the parties were so strained that they met in the neutral territory of a car park to discuss the disposal of their matrimonial home. Once they had arrived at a decision, the wife insisted that the husband should jot down the main points on a piece of paper. The court found here that the parties did intend to create a legally binding contract, influenced no doubt by the rather formal steps that the wife had insisted on, and also by the fact that the relationship between the parties was breaking down, so that they were no longer closely bound in a family or domestic situation. Contrast *Balfour v Balfour* (1919) CA, where an agreement for payments from husband to wife for housekeeping was found not to be intended to create legal relations. In this latter case, the agreement was reached when the marriage was still intact and was thus deemed to be simply a domestic, family arrangement.

Where one party seeks to rebut the presumption, the court may also be influenced by the degree of certainty and detail with which the parties made their arrangements. In *Jones v Padavatton* (1969) CA, the details of the arrangements with regard to a house were so sketchy that it was impossible to believe that the parties could have intended to make a legally binding agreement. In other cases, it seems that as the family relationships become less close, so the strength of the presumption diminishes. There seemed to be little difficulty in establishing the appropriate intention in *Simpkins v Pays* (1955), Chester Assizes, where three people living together regularly participated in a competition in a newspaper. When one of them was ultimately the winner, she was obliged to share her winnings with the other two.

In a commercial situation, it is considerably more difficult to rebut the presumption that a contract was intended. The courts will require convincing evidence that the parties intended otherwise. In *Rose and Frank Co. v Crompton Bros* (1925) HL, the parties indicated in a written document that their arrangement was a 'gentleman's agreement, binding in honour only'. In the face of such convincing evidence, the court was obliged to find that there was no contract.

Sometimes the parties want to lift their transaction outside the framework of contract law because they do not want the trouble and expense of litigation. This can frequently be seen as a condition in competitions and is one explanation of the football pools cases, where a punter who believes he has a winning line cannot sue to enforce his win.

Provided the parties have made their intentions sufficiently clear, the court will accept their decision.

Form of the contract

As a general rule, English contract law does not require contracts to be made in any particular form. In practice many contracts are oral, which often leads people to assume that it would be difficult to sue to enforce such a contract. Certainly, there may be problems of proof, but in the end, it may be a matter of judging which party the court can more reliably believe. Where a contract is put into writing to provide the necessary proof, the writing can in turn prove to be problematical. There may be a dispute about its precise meaning, or it may fail to spell out the entire contract between the parties. A good example of written contracts is to be found in the construction industry's use of standard form contracts (see Chapter7).

Exceptionally, the law sometimes requires contracts to be made in writing. Examples include consumer credit agreements, such as hire purchase contracts (regulated under the Consumer Credit Act 1974, as amended), and contracts for the sale or disposition of an interest in land, which must comply with s.2 of the Law of Property (Miscellaneous Provisions) Act 1989. In such cases, it is usually necessary to look at the appropriate statute to see the precise form the writing must take and also to determine what will happen if the parties have not complied with the rules. Rules about the precise form of the contract can be a useful device for protecting consumers.

Consumer credit

In the consumer credit legislation, certain information contained in hire purchase agreements must be displayed in prominent information boxes, with the aim of drawing those aspects more particularly to the attention of the consumer. This requirement stems from the Consumer Credit Act 1974, as amended, and is reinforced by secondary legislation such as the Consumer Credit (Agreements) Regulations 2010. These regulations specify the form and content of credit agreements, including the requirement for clearly presented key financial terms, such as the total amount payable, interest rates and repayment details, to be displayed prominently and legibly.

Distance selling

Distance selling, e.g. contracts made in response to catalogues or press advertising with an order form or teleshopping via the medium of a television, is subject to the Consumer Contracts (Information, Cancellation and Additional Charges) Regulations 2013. By the Regulations, certain information must be given by the supplier to a consumer before a contract is formed, e.g. description of the subject matter of the contract, price, delivery costs and the existence of a right to cancel the contract.

The Regulations apply primarily to distance and off-premises contracts between traders and consumers and aim to promote informed decision-making. They also set out cancellation periods, generally giving consumers 14 days from receipt of goods (or conclusion of contract for services) in which to cancel without penalty.

The Regulations do not apply to certain types of contracts, including contracts for the sale or other disposition of an interest in land, or for the construction of a building where the contract also provides for the sale or other disposition of an interest in land on which the building is constructed.

Land

Despite the freedom about the form of the contract, there has long been special treatment for contracts for the sale of land. As early as 1677, rules were laid down in the Statute of Frauds that contracts for the sale of land must be evidenced by a note or memorandum in writing; otherwise, the contract would be unenforceable. These rules were re-enacted in more or less the same form by s.40 of the Law of Property Act 1925. Evidence in writing was meant as a safeguard against fraud, but ironically the need for such evidence has sometimes been the very means by which people could perpetrate a fraud. The rules were regarded as unsatisfactory and, when the Law Commission took a detailed look at the workings of s.40, it reported:

> As a result of judicial attempts to prevent the statute being used as an instrument of fraud, it is virtually impossible to discover with acceptable certainty, prior to proceedings, whether a contract will be found to be enforceable . . . s.40 would appear ripe for reform.

That reform took place with the repeal of s.40 by the Law of Property (Miscellaneous Provisions) Act 1989. Section 2 of the new Act provides that:

- contracts for the sale or other disposition of interests in land can only be made in writing,
- must incorporate all the terms expressly agreed between the parties and
- must be signed by both parties or their authorised agents.

Incorporation of the express terms may be by reference to another document, and the rules also make provision for the methods employed by solicitors engaged on behalf of the parties, where it is common to 'exchange contracts'. Sales of land by public auction are excluded from these rules, as are short leases (of less than three years).

Although these new rules seem designed to promote greater certainty, in that it should be easier to determine when a legally enforceable contract for the sale of land exists, one major problem remains to be tackled. It is still the case that people who are unaware of these rules may 'shake hands' on an oral agreement for the sale of land, and the purchaser may then incur expenditure on the land concerned, only to find that the vendor refuses to complete the transaction. Take the example of an oral agreement for the sale of a house, where the parties agree that the prospective purchaser can have a key for the property, and may undertake work on the property in advance of the legal transfer. If the purchaser were to install central heating and undertake extensive decorating work, it would be disastrous if the vendor then refused to transfer the house. In such a case, the vendor would seem to receive improper protection from the new rules. The aggrieved purchaser would be unable to sue for breach of contract because the s.2 rules state that a contract for the sale of land *can only be made in writing*. Lack of writing, therefore, means that there is no contract.

Problems of the sort outlined in what came before were previously solved by the use of discretionary equitable relief, in the form of the equitable doctrine of part performance. In the absence of the necessary note or memorandum in writing, this equitable principle used to permit specific performance of the contract if the behaviour of the parties was in itself sufficient to indicate the existence of a contract. Unfortunately, there is no scope now for the application of this discretion, as it depends for its efficacy on the fact that lack of writing under the old s.40 rules merely rendered a valid contract unenforceable. It did not deny the very existence of a contract, which is the result achieved under the s.2 provisions of the new Act. Clearly, the equitable doctrine of part performance has been repealed along with s.40.

The Law Commission did consider the difficulties that the new rules would create, and, in particular, they examined the plight of a would-be purchaser who had expended money and effort on the property. They considered that the present law contained a sufficient armoury of weapons by which to assist such a person, such as an action for restitution or suing on a *quantum meruit* if work had been undertaken on a property. It will take time for cases to come before the courts to show whether the Law Commission's faith in these solutions is justified.

Section 2 has generated a significant case law on its scope and meaning. In *Spiro v Glencrown Properties Ltd* (1990) CA, the vendor and prospective purchaser had both signed an agreement creating an option to purchase land, which conformed in all respects with s.2. Later, when the would-be purchaser served a signed notice to exercise the option, the vendor attempted to argue that the documents already signed merely amounted to an irrevocable offer and that it was the notice exercising the option which brought about a contract for the sale of an interest in land. As this notice was signed only by the purchaser, it did not satisfy the rules of s.2. The court dismissed that argument, preferring instead to construe the original agreement as the contract, conditional upon the exercise of the option. When the purchaser served the signed notice, he was thereby fulfilling the condition. It was the original agreement which needed to satisfy the s.2 rules, and as it clearly did, the vendor was bound.

In *Record v Bell* (1991) CA, before contracts for the sale of a property were exchanged, the purchaser asked to see entries in the Land Register to check the ownership of the property. There was a delay in the production of this information. Nonetheless, the exchange of contracts went ahead, but the continued existence of the contract was made conditional on the entries confirming what the vendor had previously stated. The condition was not contained in the contracts exchanged, but in letters between the parties. The question arose when the purchaser sought to escape from the agreement: was there compliance with s.2? It was held by the court that the condition formed the basis of a separate contract, collateral to the contract for the sale of the property. As the collateral contract was not one for the sale of land, it did not have to satisfy the requirements of s.2 and, therefore, as the condition had been fulfilled, the contract was valid and enforceable.

The courts adopted a different approach to the problem of terms missing from a written agreement for the sale of land in *Wright v Robert Leonard (Developments) Ltd* (1994) CA. Here the contract was for a flat and furnishings. The written contract made no mention of the furnishings, and before completion, these were removed by the vendor. The Court of Appeal held there was one contract which had to satisfy the requirements of s.2, but the written agreement did not reflect the agreement reached by the parties, so rectification of the contract was granted. This meant that the written agreement included a promise as to the furnishings and the purchasers were entitled to damages for the vendor's breach.

Where parties enter into a written contract for the sale of land and later seek to vary a material term of that contract, then such variation must comply with s.2. This was held in *McCausland v Duncan Lawrie Ltd* (1997) CA, where the parties altered the original completion date stated in a written contract in their subsequent correspondence. The completion date was a material term of the contract, and the variation was not in a written document or documents signed by each of the parties; the variation of the contract did not comply with s.2.

Section 2 provides that a written contract must be signed by both parties. The signature of the parties may be contained in the same document or, where documents are exchanged, each party signs a document and then sends it to the other. *Firstpost Homes Ltd v Johnson* (1995) CA explored the question of what was meant by signature. In this case the parties orally agreed on a sale of land. The purchaser prepared a typed letter for the vendor to send to the purchaser containing the names and addresses of the parties, the price and reference to an enclosed plan identifying the land. This letter was signed by the vendor, but not by the purchaser. The plan was signed by both parties. The Court of Appeal held that there were two documents, and as the letter referred to the plan, it was necessary for the letter to be signed by both parties to comply with s.2(3). It was argued that the typed name and address of the purchaser on the letter served as a signature, but this was rejected by the court.

The problem mentioned previously concerning part performance or reliance on a 'contract' for the sale of land was partly addressed by s.2(5), which states that 'nothing in this section affects the creation or operation of resulting, implied or constructive trusts'. Additionally, the doctrine of proprietary estoppel (see Chapter 11) has a role to play in rendering a contract for the sale of land, not made in writing. The cases are not clear on this latter point, but the better view seems to be that, as s.2 does not refer to proprietary estoppel and the use of such estoppel has the potential to undermine the purpose of s.2, it should not be applicable here and is only applicable, on the facts of each case, in certain narrow circumstances.

Subject to contract

It is common in dealings relating to land for the parties to be anxious to take expert advice before finally committing themselves to a binding contract. Negotiations are often carried on 'subject to contract'. Those words will usually be interpreted by the courts to mean that the parties have not yet concluded a contract. The very words deny the existence of a present contract. Of course, much depends on the precise words used, as the role of the court is to seek to construe what the parties must have intended. In one case, the parties declared their agreement to be a 'provisional agreement', but the effect of that phrase was quite different, and the court held that their agreement was binding from the start (*Branca v Cobarro* (1947) CA). It seems to follow from the cases that where the parties use the phrase 'subject to contract' in a written document, that document is *not* a contract, even if it complies in all other respects with s.2.

Agreements which are subject to contract may be contrasted with conditional contracts. For example, a contract for the sale of land may be subject to the condition that planning permission is obtained by a particular date. If the condition is not satisfied, then the parties are not bound, but there *is* a binding contract if the conditions *are* satisfied.

The contents of the contract

Once it is clear that a contract has been agreed, it may then be necessary to establish precisely the obligations which each party has undertaken. Of course, this analysis will not be necessary if everything goes well between the parties, and each performs the contract to the entire satisfaction of the other. However, the rules are important not only in problem- or dispute-solving, but also in contract planning, particularly in a business situation. The more the parties understand the rules, the better they are able to use them to ensure that subsequent problems do not occur, or where they do, that they can be solved with a minimum of expense, delay and irritation.

Where a contract consists of buying a cup of coffee from a machine or paying the fare on a bus, it may be difficult to imagine what the contents of such a contract are. But all contracts have terms. These may be express or implied. The express terms will be those specifically negotiated between the parties, either orally or in writing. Remember that the parties are largely free to fix their own terms. Contract law still adheres generally to the principle of freedom of contract. The parties must strike their own bargain. In effect, this imposes a serious restriction on the extent to which terms can be implied in a contract.

Implied terms may be incorporated from an Act of Parliament. One of the most common examples is the implied terms incorporated into sales of goods transactions by the Sale of Goods Act 1979 or the Consumer Rights Act 2015 (see Chapter 6). Without any need for specific negotiation by the parties, terms are implied relating to the fitness for purpose of the goods and the requirement that goods be of satisfactory quality. The courts have a limited role to play in judicially implying terms, but great care is needed to ensure that judges are not seen to be making the agreement for the parties. It is usually said that terms can only be implied by the judges if it is necessary to give 'business efficacy' to the arrangements made by the parties. This may occur where the parties have already undertaken a substantial part of the contract, and they come up against some problem or difficulty which they did not provide for. In such cases a judge can imply a term where it is clear that, had the parties put their minds to the problem at the appropriate time, both would have agreed to the term proposed.

Judges also have a role to play in interpreting express terms. Whilst they can utilise a measure of judgment, the Supreme Court held in *Arnold v Britton & others* (2015) that an objective approach to the meaning of a term *at the time the contract was made* must be taken; given that an objective stand must be taken, hypothetical evidence of the parties' intention is irrelevant and that commercial common-sense could not undermine the clear meaning of words used.

Status of contract terms

Terms are the obligations under a contract, but inevitably they will not be of equal importance. Take, for example, a term about the time of a delivery of goods. This may be of paramount importance where goods are perishable or where it is clear that they will be of little commercial use unless delivered on time. In other cases, the time for delivery may be far less significant.

The law has to find some way to grade or classify terms, because it must seek to offer remedies appropriate to the significance of the term which has been broken. Traditionally, the law classified terms as major terms, which it called **conditions**, and minor terms, which it called **warranties**. The terminology is unfortunate, as both of those words have sundry other meanings and uses in the law of contract.

Conditions are said to be those obligations which are of the essence, which go to the very core and heart of the contract. Breach of a condition always gives the injured party the right to sue for damages. Over and above that, the injured party may also have the right to repudiate the contract. This means that he or she is entitled to regard himself or herself as no longer bound by the contract.

A warranty is of more peripheral importance, often said to be merely collateral to the main purpose of the contract. An injured party can sue for damages for breach of warranty, but has no right to repudiate the contract.

The problem with this traditional approach to classifying terms is that it requires a decision to be made about the status of the term at the time when the contract was made. In effect this emphasises the importance of the intention of the parties, as it should be clear from what was said and written and from all the surrounding circumstances, whether they intended a particular term to be a condition or a warranty. This approach creates business certainty, as it would be possible to say at the outset what remedies would be available for particular breaches. But it takes little account of the realities of business life, where the term breached may be only a warranty, but the consequences could turn out to be very serious. This would leave the victim of the breach in a situation where he or she would be forced to go on with performance of the contract and be limited to a claim for damages.

Consider, as an example, the problem which arose in *Hong Kong Fir Shipping Co. Ltd v Kawasaki Kisen Kaisha Ltd* (1962) CA. One of the terms in a contract to charter a ship for two years provided that the ship should be in every way fitted for ordinary cargo service. When the ship was delivered at the beginning of the two-year charter period, it was unseaworthy and when it made its first voyage, repairs were needed which took four months to complete. Did the breach amount to a breach of condition? If it did, the charterer could repudiate liability under the contract and thus rid himself of the burden of this unsatisfactory vessel over the remainder of the two-year period *and* sue for damages. If it were only a breach of warranty, his claim would be limited to damages, and he would be left with the vessel for the remainder of the charter period.

In the *Hong Kong Fir Shipping* case, the Court of Appeal chose a novel solution, preferring instead to say that some terms are '**intermediate**' or '**innominate**'. Such terms defy initial classification as conditions or warranties and the parties must wait until the scale of the resulting breach is known, at which point the term can be suitably classified. Obviously, this produces a much less certain situation between the parties during the currency of the contract.

This development in judicial thinking did not, in fact, assist the charterer in the Hong Kong Fir case. The term in issue was held to be only a warranty, so the charterer had no right to repudiate the contract. As he had wrongfully repudiated, he was obliged to pay damages to the ship owner.

The same problem of classifying a term was before the Court of Appeal in another charter party case, *The Mihalis Angelos* (1971), where the term related to the date when the ship would be ready to load. In the event, it was nearly a month after that date before the ship was ready. The term was held to be a condition. The court made the point that it was influenced by earlier cases holding similar clauses to be conditions and it was vital to remember that 'one of the important elements of the law is predictability'. One important point emerging from this case is that it may already be established, either by statute or an earlier precedent, that a particular type of clause falls into the category of a condition. Then there is no scope for attempting to classify it as an innominate term. But where a term could be either a condition or a warranty, and it is not clear which the parties intended, and where the range of possible breaches and the scale of the resulting harm is very wide, then the court may classify it as innominate. This classification is theoretically made at the time when the contract is made, but the remedy available for breach of the term will only be established once the scale of that breach is known.

The appropriateness of this 'wait and see' approach is well demonstrated by the facts in the *Hong Kong Fir* case. There, the clause in question required that the ship was to be 'in every way fitted for cargo service'. These words were so wide in potential interpretation that there could be breaches ranging from failure to supply the vessel with a proper number of anchors or to put on board medical supplies to totally defective engines at the other end of the scale.

It is clear that the courts remain anxious about creating business uncertainty. The House of Lords confirmed that the traditional division into conditions and warranties is still acceptable in many cases (*Bunge Corporation v Tradex Export SA* (1981)). They have, however, left the way open to use the more flexible *Hong Kong Fir* approach in a limited number of cases where this may be justified. Clearly the problem can be largely overcome by the parties being sufficiently specific at the outset about the result of each type of breach. This is a classic example of how the rules of contract law can have a planning function.

Exclusion and limitation clauses

Although many terms in contracts will be concerned with creating obligations, it has become increasingly common for the parties to seek to limit or exclude liability which would otherwise arise under the contract. Exclusion and limitation clauses are frequently encountered in contracts involving car parks, dry cleaners, film processing, package holidays, furniture removal, insurance and the provision of professional services, for example, by a chartered surveyor. They may take forms such as 'No liability accepted for any damage caused by flooding' or 'In the event of a breach, our liability is restricted to 10 times the contract price'.

Exclusion clauses fall into two main categories: those negotiated between parties of equal bargaining strength, where such clauses can be seen to facilitate business activity and to allow the parties to allocate risks between themselves and organise which of them should insure; and those imposed on a weaker party by a stronger party who may enjoy a monopolistic position, where the weaker party has no option but to accept. Clauses falling into the latter category are, not unnaturally, unpopular with the courts. In the past, the judges strove to protect victims of harsh exclusion clauses, but some of the techniques developed by them to afford protection produced unfortunate results in business cases, sometimes upsetting the very carefully negotiated arrangements between the parties. The courts have never had a power at common law simply to strike out exclusion or limitation clauses on the grounds that they are unfair or unreasonable because contract law proceeds on the assumption that the parties are free to fix the terms of their bargain for themselves.

Judicial dislike of exclusion clauses has always meant that courts will only allow reliance on such a clause if it has effectively become part of the contract. The incorporation of the clause into the contract can occur:

1 By one party signing a contractual document containing the clause, of which he or she would then be deemed to have notice, whether it had been read or not.
2 By one party giving reasonable notice of the clause to the other. Reasonable notice may be given in a contractual document, but not every piece of paper passing between

the parties can be so classified. Generally, contractual documents are those pieces of paper on which a reasonable person would expect to find conditions printed. This would usually include documents such as airline tickets or standard company order forms. It would not include things such as mere receipts for the payment of money. In any case, such a receipt would often be handed over too late, as it is vital that reasonable notice be given before or at the time when the contract is made. Once the parties have finished negotiating and finalised their deal, it is too late for one of them to try to change it or add to it unilaterally. Where no contractual document passes between the parties, the exclusion clause can be incorporated by the reasonable display of a notice at an appropriate place, so that its contents could have been read before the contract was made.

3 By virtue of a 'course of dealings' between the parties. Where two people have dealt with each other regularly, always using the same terms, the court may be prepared to say that adequate notice has been given of any exclusion clause. This principle is only likely to be applied between business parties who have regular commercial dealings.

In cases where the court accepts that the exclusion clause is properly incorporated and forms part of the contract, the judge will next consider what the words of the clause mean and whether the wording covers the actual facts which have arisen. In the construction or interpretation of the clause, the standard policy of the courts is to construe any ambiguity strictly against the interest of the party seeking to rely on the clause. In one case involving motor insurance, a clause in the contract provided that the insurers need not pay out on a claim under the contract if the vehicle was carrying 'an excessive load'. In fact, it was a five-seater car carrying six persons, but the judge held that the word 'load' did not include people (*Houghton v Trafalgar Insurance* Co (1954) CA).

Prior to 1977, when the Unfair Contract Terms Act was passed, judges had to use these construction and interpretation techniques boldly in order to protect victims of harsh clauses. In consequence, there was a distorted result in many of the cases. Such striving on the part of judges is much less necessary now since the 1977 Act was passed, and they have been urged by the Supreme Court to give the relevant words in an exclusion clause their plain and natural meaning. Nevertheless, the two common law rules relating to incorporation and construction are still relevant, though the provisions of the 1977 Act now govern the majority of problem cases in this area with regard to contracts between businesses. There are parallel provisions for consumer contracts in the Consumer Rights Act 2015 (see Chapter 6).

Misrepresentation and mistake

The basic idea of misrepresentation is a very simple one to grasp. One party makes a false statement which has induced the other to enter the contract. This area of law is regarded as complex, largely due to the number of overlaps with other areas of law, so that it is seldom possible to view a misrepresentation in isolation. The overlaps occur either when misrepresentation induces a mistake or when the statement made, depending on its relative importance, may be classed as a representation or a term or because the law of contract and the law of tort have separately developed ideas about negligent statements.

In complex contractual negotiations, both parties will make many statements to each other about the subject matter. We have already considered the problem of classifying terms of a contract. Terms may result from statements made which now form part of the contractual obligations. Alternatively, the statement may be of less importance and may not become part of the contract. Nevertheless, that statement may weigh with one party in influencing him or her to make the contract. He or she is induced by the representation. Some statements may simply have been designed to create a good bargaining atmosphere, and these may give rise to no liability at all. These examples are sometimes referred to as 'sales puffs'.

The court must determine into which category a statement falls. When deciding whether statements made prior to contracting are terms or representations, they are influenced by factors such as:

- How early in negotiations was the statement made?
- Was an oral statement subsequently included in a later written contract?
- Did the person making the statement have special knowledge on the subject matter of the contract?

The answers to these questions will not necessarily produce a definite solution. Each case will turn on its own facts.

Where a false statement is classed as a representation, it is possible for it to be made innocently, negligently or fraudulently, depending on the maker's state of mind. The classification is important because each type of misrepresentation gives rise to different remedies. It is also important to distinguish terms from representations because of the different remedies available. Non-compliance with a term results in breach of the contract. Misrepresentation is a false statement of fact standing outside the contract and not forming part of the obligations and does not, therefore, involve a breach of contract. The difference in the remedies available used to be so great that the distinction between a term and a representation assumed immense importance. Since the changes in the law introduced by the Misrepresentation Act 1967, it is now far less significant.

Figure 5.1 shows how any particular statement can be graded. Take an example like 'The car has only done 20,000 miles' or 'This lorry can carry 30 tons'.

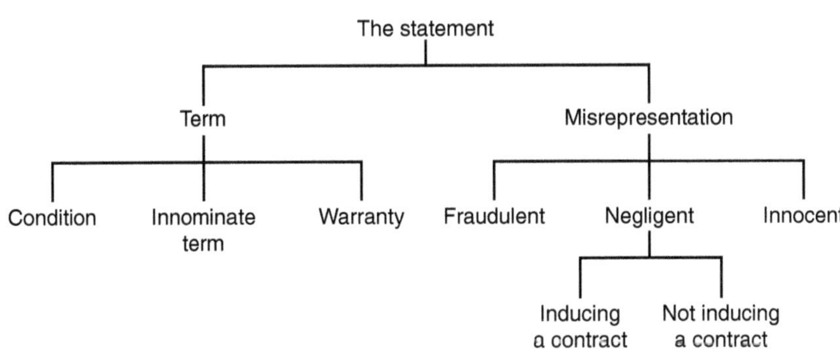

Figure 5.1 The grading of terms and misrepresentations.

Misrepresentation is defined as a false statement of fact, made by one party to the other, before or at the time of contracting, which is one of the causes that induces the contract. Within the definition, a number of points of difficulty occur:

1 There must be a statement, but the courts have been prepared to imply a statement from behaviour. 'A nod or a wink, or a shake of the head or a smile' from the seller, intended to induce the buyer to believe the existence of a non-existing fact, would be sufficient.
2 The statement must be factual. This excludes statements of intention and statements of opinion. Statements of opinion give rise to particular problems, as they can be made by someone who has more knowledge and information in the particular bargaining situation. This makes the statement seem especially authoritative. It seems as if the maker of the statement knows facts which justify him or her in holding that opinion. In one case a vendor, trying to sell property which was let, described the tenant as 'a most desirable tenant'. The vendor knew that the tenant was in arrears with his rent and earlier rent had only been extracted with considerable difficulty. In these circumstances, knowing the facts as he did, the court held that the vendor could not reasonably hold such an opinion. They regarded his statement as actionable misrepresentation (*Smith v Land and House Property Corporation* (1884) (CA)). In that case, the vendor had superior knowledge. He alone knew the problems about the tenant. Compare the situation in *Bissett v Wilkinson* (1927) (PC), where a vendor was selling land in New Zealand. In response to an enquiry from the purchaser, the vendor said that the land could carry 2,000 sheep. The vendor had never kept sheep nor had the land ever been used for that purpose, so he had no facts on which to base this opinion. It was only an opinion and not actionable as misrepresentation.
3 Since there must usually be a false statement, it follows that keeping silent cannot usually constitute misrepresentation. If I want to sell my second-hand car, and I say nothing about its condition or performance, it will be up to a prospective purchaser to enquire specifically about anything that concerns or worries him or her. This is part of a general principle in contract of 'let the buyer beware' (*caveat emptor*). I would be under no duty to point out faults and defects. Of course, if asked specific questions, then I should answer honestly; otherwise, my statements could amount to misrepresentation. There are some situations, however, where the law imposes a positive duty to speak. In these cases, silence can constitute misrepresentation. These situations are as follows:

    a Where one party tells a half-truth but remains silent about some fact which distorts the positive statement made. An example is where a prospective purchaser asked about the farms on an estate, enquiring whether all the farms were let. The vendor replied that they were (a positive statement), but failed to go on to say that all the tenant farmers had given notice to quit (silence distorting the positive representation) (*Dimmock v Hallett* (1866) CACh).
    b Where the relationship between the parties is fiduciary (a relationship of trust), e.g. solicitor and client or banker and customer. If one then makes a contract with the other, there is a positive duty to disclose known material facts. This rule is significant when considering undue influence.
    c Where the contract is one involving utmost good faith. The commonest example of such a contract is insurance. Here the facts are all within the knowledge of one

party. On the strength of what he tells the other, that other must decide whether or not to accept the risk and how much to charge as a premium. The law imposes a positive duty in these cases to reveal all material facts. Insurance contracts are discussed in greater detail in what follows.

4 The false statement must actually help to induce the contract. There will be many occasions when false statements are made, but they do not act as an inducement. This may be because the other party does not hear the statement, does not believe it or is not in any way influenced by it, but makes his decision to enter the contract for quite separate reasons. In *Vahey v Kenyon* (2013) CA, a house vendor did not fully disclose the extent of flooding which had taken place during his occupancy. The contract was held to be void for fraudulent misrepresentation. The statements about flooding were made with full knowledge of them being false. The statements were material in Mr Vahey's purchasing the property and/or not seeking a significant discount.

Where all the relevant features of misrepresentation can be proved, the injured party will want to know if he or she can escape from the contract, and also whether he or she has any claim for damages. The answer depends largely on the type of misrepresentation involved. Was it a statement made fraudulently, negligently or innocently? Once the maker's frame of mind is established, the remedies can be considered.

Fraudulent misrepresentation

This is defined as a false statement made knowingly without belief in its truth or recklessly, careless whether it is true or false. In the very limited number of cases where this can be proved, the behaviour of the person making the false statement would amount to the tort of deceit. Unfortunately, it is very difficult to convince a court that a statement falls into this category. As Lord Esher commented in *Le Lievre v Gould* (1893) CA: 'A charge of fraud is such a terrible thing to bring against a man that it cannot be maintained unless it is shown that he has a wicked mind'.

Negligent misrepresentation

This category of misrepresentation was introduced by s.2(1) of the Misrepresentation Act 1967 and is one of the factors which have effected a significant improvement in this area of the law. A statement is negligent if it was made in circumstances where the maker had no reasonable grounds for believing it to be true. He or she may have thought that they were making a true statement, but the court will evaluate whether the basis for this belief was reasonable. This may be a difficult question to resolve, and much may be at stake. An important commercial case, *Howard Marine and Dredging Co. Ltd v Ogden & Sons Ltd* (1978) CA, illustrates the problem.

In order to tender for a large excavation contract with a water authority, Ogdens needed to hire or buy barges to move spoil out to sea. They approached Howard Marine and had extensive discussions and negotiations with Howard's manager. He thought that two particular barges would be suitable for Ogdens' needs, and he told Ogdens that each barge could carry 1,600 tonnes. He was basing his statement on his recollection of information contained in *Lloyd's Register*. In the company's files about the barges, the manager could have discovered that the *Lloyd's Register* figure was wrong. The actual carrying capacity of each barge was nearer 1,000 tonnes. Ogdens were successful in getting the excavation contract, and they immediately hired the two barges from Howard. As soon as they began using them, they realised the vast discrepancy between the manager's statement and the true carrying capacity. Ogdens refused to pay any further hire charges for the barges and wanted to claim damages from Howard. Obviously, the court had to classify the statement made. First, they had to consider whether it was of such importance that it could be regarded as a term of the contract. Given the fact that the statement was made quite early in the negotiations, and not subsequently included in the charter party (the written hire contract), the court found that it was not a term. What kind of misrepresentation was it? By a majority verdict only, the Court of Appeal found that this was a negligent misrepresentation and awarded damages under s.2(1) of the Misrepresentation Act 1967. One of the judges thought that it was reasonable for the manager to rely on *Lloyd's Register*, which was normally famous for its detail and accuracy, but the other two judges thought that the manager should have referred to, and checked in, the authoritative file on each barge.

Liability for negligent misrepresentation under s.2(1) of the 1967 Act can only arise if the statement causes the other party to enter into a contract. There is considerable overlap in this area with the law of tort, and negligent statements are a problem area. Just before the 1967 Act was passed, the courts had an opportunity to consider what liability, if any, should arise from negligent statements. In the case of *Hedley Byrne and Co. Ltd v Heller and Partners Ltd* (1964) HL, the facts involved a company that wanted assurances about the financial stability of a prospective customer. They asked the customer if it was in order to approach the customer's bank for a credit reference. The bank negligently informed the company that the customer was financially sound. Subsequently, the customer went into liquidation, and the company suffered losses in excess of £15,000. As they were unlikely to get their money from the customer, the company was anxious to sue the bank. On these facts, it is clear that the bank's statement had not induced any contract between the bank and the company, so s.2 of the 1967 Act could not cover such a case. That section requires that a person must have entered into a contract. Could the company recover the damages against the bank in the tort of negligence? The law on negligence had not developed at that stage to cover such misstatements causing pure financial loss, but the House of Lords in the *Hedley Byrne* case showed themselves ready, in appropriate circumstances, to develop the law and award damages.

In the mid-1960s, this created a sudden new wealth of possibilities where a person was a victim of a negligent statement. It has become common to refer to negligent statements which result in a contract as negligent misrepresentations and those which do not result in a contract as negligent misstatements. While it may be to a claimant's advantage to have more possible causes of action, these twin developments in respect of negligent statements make the law more complex than it need be. See further treatment of negligent misstatements in the law of tort in Chapter 9.

Innocent misrepresentation

Since the 1967 Act, this is limited to those cases where the maker of the statement honestly believed that his or her statement was true, *and* he or she had reasonable grounds for believing it to be true.

Where the misrepresentation is innocent, the court has discretion under s.2(2) of the Misrepresentation Act 1967 to award damages in lieu of rescission. See what follows.

Remedies for misrepresentation

Once it is clear which type of misrepresentation is involved, the next problem is to establish the appropriate remedy. All types of misrepresentation have the same effect on the contract, which is to make the contract voidable at the option of the party misled. This means that the injured party can seek to rescind the contract, to have it set aside as if it had never existed. The parties would then be restored to their original positions as far as the contract is concerned. This is known as the remedy of rescission, and in some circumstances, it can be achieved on a self-help basis without any court assistance or intervention. Inevitably such help will often be required, especially where the party at fault will not cooperate.

There will be times when a court is unable to order rescission. In principle, the remedy is possible for all forms of actionable misrepresentation, whether fraudulent, negligent or innocent, but it will not be available in the following circumstances:

1 Where the court is unable to restore the parties to their original position. This could be the case where goods sold have been consumed or inextricably mixed with other goods.
2 Where the injured party has affirmed the contract. This means that the injured party is fully aware of the false statement but is prepared to go on with the contract.
3 Where the injured party has delayed too long in seeking rescission. There is no precisely defined time limit, but a long delay once all the facts are known may in itself be an affirmation of the contract.
4 Where third-party rights have intervened. This would typically involve a situation where goods have passed to an innocent third party who has acquired them for value and in good faith.
5 Where the court decides to exercise its discretion under s.2(2) of the 1967 Act to award damages in lieu of rescission. This applies primarily in cases of non-fraudulent misrepresentation, and particularly where the misrepresentation is innocent. The court may consider that rescission is not an appropriate remedy and instead award a monetary sum.

Damages

When considering the availability of damages for misrepresentation, it must be remembered that the false statement in question may have induced the contract, but it does not form part of the contract obligations. If the statement was regarded as sufficiently important by the parties to form part of the contractual obligations, it will be classified as a term. If it then turns out to be false, the appropriate remedies for breach of contract will be available.

As misrepresentations stand outside the contract, it follows that there can be no damages available in the law of contract. Where the statement is fraudulent, damages have always been available in the tort of deceit.

Prior to the recognition of negligent misrepresentation, a misrepresentation made without fraud was innocent, for which no damages were available. Now, if a claimant can prove negligent misrepresentation under s.2(1), the court has the power to award damages. It seems clear that these damages will be assessed on a tort basis rather than a contract basis.

Royscot Trust Ltd v Rogerson (1991) CA shows a most favourable application of s.2(1) in the assessment of damages. The court held that damages under s.2(1) are to be assessed on the same basis as for fraudulent misrepresentation, even where the misrepresentation was merely negligent. This allows recovery for all losses directly flowing from the misrepresentation, regardless of foreseeability.

Where the claim is for negligent misstatement within the *Hedley Byrne* principle, damages for the tort of negligence are available. The difference between tort damages and contract damages is significant, as it could result in quite different calculations. Tort damages are designed to put the claimant into the position he or she would have been in but for the tort; contract damages are designed to put the claimant into the position he or she would have been in if the contract had been properly performed.

It should be noted that in a claim under s.2(1) while a representee must show that a misrepresentation has occurred and damage has been suffered, it is then for the person making the representation, the representor, to prove that they had reasonable grounds for believing the statement to be true and did so believe up to the time that the contract was formed. This makes it easier for the representee as the burden of proof is placed on the representor to establish reasonable grounds (see previously mentioned *Howard Marine and Dredging Co. Ltd v Ogden & Sons Ltd* CA, where Howard Marine was unable to show reasonable grounds for the statement as to the capacity of the barges).

Damages in lieu of rescission

Even if a false statement has been made, the remedy of rescission is sometimes far too extreme to produce a fair result. If I sell you my car and tell you that it has done 30,000 miles, and you subsequently discover it has done 31,000 miles, it can often be quite satisfactory to award damages in recognition of the false statement. It may seem to give undue protection to the purchaser to allow him or her to back out altogether, although this will often be a question of degree.

Under s.2(2) of the 1967 Act, the court has a discretion to award damages in lieu of rescission for non-fraudulent misrepresentation, i.e. for negligent or innocent misrepresentation. This discretion only arises if the right to rescind still exists at the time the matter is before the court. If the right to rescind has already been lost (e.g. by affirmation or lapse of time), then s.2(2) cannot apply, and no damages under this section will be available.

Limiting or excluding liability for misrepresentation

In the same way as it is possible to limit or exclude liability for obligations which would otherwise arise under a contract, it is also possible to limit or exclude liability for misrepresentations which induce a contract. The relevant rules are contained in s.3 of the Misrepresentation Act 1967.

Section 3 permits the exclusion or limitation of liability for negligent or innocent misrepresentation (but not for fraudulent misrepresentation), provided that the clause satisfies the requirement of reasonableness. This test of reasonableness is interpreted in accordance with s.11(1) of the Unfair Contract Terms Act 1977. The party seeking to rely on the clause bears the burden of proving that the exclusion or limitation is fair and reasonable in all the circumstances, taking into account the bargaining positions of the parties and whether the clause was properly brought to the attention of the other party.

Mistake

As a broad rule, English law admits few mistakes as affecting the validity of a contract. Often, when one party says, 'I made a mistake', what this really means is, 'I failed to watch out to make sure I was making a good bargain'. In such cases, the maxim *caveat emptor* will apply (let the buyer beware).

In some instances where the parties are mistaken, no contract ever comes into existence between them. The extent of their misunderstanding is so great that it prevents them from reaching true agreement. In one case the parties believed they had an agreement relating to a cargo of cotton coming to this country on a ship called *The Peerless*, which was sailing from Bombay. By coincidence there were two ships, both called *The Peerless*, both carrying cotton and sailing from Bombay. As it was impossible to say which ship the parties had in mind, it was held that they had never reached a true agreement (*Raffles v Wichelhaus* (1864) CE).

Sometimes, the state of confusion is caused by one person deliberately seeking to lead the other into making a mistake. This can occur where one person misrepresents his or her identity. Such fraudulent behaviour would be likely to give rise to remedies for misrepresentation. But in cases where it can be shown that the identity of the person was crucial, the mistake about identity may be so important that there is no real agreement. The contract would be void, meaning that no contract has been formed.

Other factors affecting validity and enforceability

A party seeking to enforce a contract may be unable to do so in cases where it was induced by duress or undue influence or in cases where there is some illegality. These issues affect the validity or enforceability of the agreement and will be considered separately in later sections.

Duress

It is hard to imagine that any contracts are made nowadays as a result of threats of physical violence. The old rules about duress would only permit relief if the contract came about because of threatened or actual violence. Originally the result of such behaviour was that the contract was void, i.e. there was no contract at all. Subsequently, the rule was modified. Now, if such facts occur, the injured party could seek to have the contract set aside. The modern effect of duress is thus to render the contract voidable; the contract is valid unless and until it is avoided.

These rules are of such limited significance that they would not be worth mentioning if it were not for the development by the courts in the late twentieth century of principles relating to economic duress. The cases which have developed this theme tend to be complex, but the usual scenario is that one party threatens the other with breach of contract in order to gain some new concessions from the other. For example, one party sees that he has not negotiated a particularly good price for building a ship, so he threatens to abandon the contract unless the price is renegotiated. This will put the other party into an impossible position if he has already contracted to charter the new ship to another party. Unless he agrees to the higher price, he will find himself in turn in breach of his charterer. He could resort to litigation against the shipbuilder, but that is costly, uncertain and very time-consuming. The potential loss of business and goodwill may be so great that he reluctantly agrees to pay the higher price. At a later date, would a court allow him to recover the extra sum paid?

The courts have not yet developed a particularly coherent set of principles to solve such a problem. There must be more than mere commercial pressure. To succeed, it is necessary to show:

- That there was a threat which was illegitimate, such as a threat to break a contract;
- That the innocent party's will was overborne, so that the pressure was a significant cause of the decision to agree and
- That there was no reasonable, practical alternative available.

The leading guidance remains the case of *Pao On v Lau Yiu Long* (1980) (PC), which set out a number of factors to assist the court:

1 Did the victim protest?
2 Did the victim have any other course of action open to him, e.g. could he have sued?
3 Did the victim receive independent advice?
4 Did the victim take immediate steps to try to avoid the new deal?

This developing area is one which is likely to cause considerable difficulty for the courts, as these situations may arise between business people who are of apparently equal bargaining strength, where the courts have a traditional reluctance to interfere with the contract as agreed between the parties.

Consider *Carillion Construction Ltd v Felix (UK) Ltd* (2001) TCC, where economic duress was held to have been applied by Felix, the defendant cladding sub-contractor, under the threat of discontinuing supplies which were critical to the completion of the project by the claimant main contractor.

Undue influence

Rules of undue influence were developed in equity because of the narrowness and harshness of the old common law rules about duress. Where there is behaviour falling short of such open violence or threats of harm, equity may grant relief if a contract has been brought about by undue influence. There is no precise definition of this phrase, but it usually occurs where there is a dominant/subservient relationship between two people and the dominant party seeks to take advantage of or exploit the other.

Undue influence was traditionally classified as actual or presumed. A relationship of influence is automatically presumed to exist in some relationships, e.g. solicitor and client or parent and child. In other instances, it is not presumed but may be proved to exist.

In *Royal Bank of Scotland v Etridge (No. 2)* (2002) HL, presumed undue influence was explained by Lord Nicholls. A claimant must prove undue influence:

> proof that the claimant placed trust and confidence in the other party in relation to the management of the claimant's financial affairs, coupled with a transaction which calls for explanation, will normally be sufficient, failing satisfactory evidence to the contrary, to discharge the burden of proof . . . [T]he court may infer that, in the absence of satisfactory explanation, the transaction could only have been procured by undue influence.

This inference of abuse of influence must then be countered by a defendant.

Where a contract is highly favourable to the dominant party, it may be avoidable on grounds of undue influence. If so, the contract may be set aside where the transaction calls for explanation and the dominant party is unable to rebut the allegation of undue influence, usually by showing that the innocent party was independently advised.

Some of the cases in this area involve the big banks, where they have obtained signatures on guarantees and mortgages from relatives or debtors, often visiting them in their homes and giving them no chance to receive independent advice. The court must weigh very carefully whether such a transaction was only agreed to because of the bank bringing undue influence to bear.

National Westminster Bank Ltd v Morgan (1983) HL is a useful illustration of these rules in practice. Banks must beware of transactions where, for example, a husband seeks a loan for which a bank requires security involving property of the wife. This security may take the form of a charge on the matrimonial home of the husband and wife. In obtaining the signature of the wife to give the bank the necessary security, this may be due to undue influence exerted over the wife by the husband. Such undue influence may affect the validity of the transaction between the wife and the bank if the wife gains no financial advantage from the transaction. In this situation the bank is said to have constructive notice of the undue influence. To avoid this outcome, banks ought to advise that independent legal advice should be sought before entering into the transaction.

Illegal contracts

The phrase 'illegal contracts' is used here to cover those situations where contracts are either void or illegal at common law or by statute. A wide range of situations is thus encompassed, for which the law offers numerous solutions. Sometimes the illegality involves an immoral contract, such as a contract to commit a crime or to pervert the course of justice. On some occasions Acts of Parliament designate particular types of contracts as unlawful. In these cases, the consequences of making such a contract will frequently be governed by the relevant statute, which may expressly provide whether the agreement is unenforceable or void.

There is a general power available to the courts to interfere with contracts where it can be seen that such contracts are contrary to public policy. Such a potentially wide-ranging doctrine must be applied with caution, because the judges must not forget that the parties are largely free to fix their own bargain. Inevitably the definition of 'public policy' must remain somewhat flexible as it will need to shift and change to reflect the changing needs of society. As Burrough J remarked in *Richardson v Mellish* (1824) CCP, public policy is 'a very unruly horse and once you get astride it, you can never know where it will carry you'.

Where contracts involve some considerable element of moral wrongdoing, it has been usual to describe them as illegal at common law. Examples would be contracts promoting sexual immorality or contracts made to facilitate illegal activity. In such cases, neither party will be able to enforce the contract. However, if one party is less guilty of the two, he or she may be able to recover money paid or property transferred, particularly where they are able to show repentance; for example, by withdrawing from the agreement before its illegal purpose is carried out. In such cases, the courts may allow restitution to prevent unjust enrichment.

Contracts in restraint of trade

Where less moral wrongdoing is involved, for example, in contracts in restraint of trade, the approach of the courts is to declare void either the contract as a whole or its offending parts. A contract in restraint of trade is one whereby one party promises or agrees to restrict his or her individual freedom to contract. These clauses are frequently found in employer-employee contracts or in contracts for the sale of a business. They are regarded as important because so many disputes in this area are litigated in the courts.

It is useful first of all to establish why such clauses are regarded as contrary to public policy. Surely, one might argue, if the two parties agree, that should be the end of the matter. That would be to take a pure freedom of contract approach. However, viewed from the point of view of the public interest, it is seen as wrong and contrary to public policy that a person's skill and experience should be rendered effectively useless and sterile by a contract. The courts have, therefore, been active in controlling those contracts in restraint of trade which they regard as being unreasonable. The modern rule is that such a clause will only be enforced if it is reasonable both in the interests of the parties and in the interests of the public.

When judging what is reasonable, each case will tend to turn on its own individual facts, so precedents are of little value other than as examples. The courts have always made clear that it is not appropriate to incorporate a restraint of trade in every contract of employment. An employer cannot include such a clause merely to prevent competition. There must be a legitimate business interest to protect, typically involving confidential information, trade secrets or customer connections.

Take the case of an employee who is engaged as a travelling salesman within a 25-mile radius of London. It may be permissible to include a clause in his contract that when he leaves his present employment, he will not solicit any of the employer's customers within that 25-mile radius for a period of one year. It would clearly not be reasonable to restrict the salesman throughout the UK, as he only poses a threat to the employer within a very small area.

When evaluating the reasonableness of a restraint clause, the courts will pay close attention to factors such as the area of the restraint and the time limit involved. The restraint should be no wider and no longer than is strictly necessary to protect the employer's business interests.

It will be clear that there is no precise science about clauses in restraint of trade. An employer who wishes to put such a clause into an employee's contract will have to tread carefully in the drafting to make sure that it goes no further in the restrictions it imposes than will be found acceptable by the courts. If the employer 'gets it wrong' in the drafting, he or she will pay a high price as the clause will be held to be void and will be struck from the contract. The only help the courts will give is to construe a restraint of trade clause in the light of the circumstances existing when the parties made their contract, in order to determine what they must have intended. By interpretation of the clause, it may be possible to limit in scope what otherwise appears to be too far-reaching. An example of this approach can be seen in *Home Counties Dairies Ltd v Skilton* (1977) CA, where a milkman, once he left his present employers, was to be restrained from selling milk or dairy produce for a period of six months. The phrase 'dairy produce' was so wide that, on a literal interpretation, the man could not even go to work in a supermarket selling yoghurt. The court held that the phrase had to be construed in the light of what the parties must have intended and should therefore be more confined in its meaning.

The courts tend to concentrate on the issue of reasonableness as between the parties, but it should be borne in mind that rules of public policy are in issue, and public interest is always a factor which could sway a court's decision. This can clearly be seen in the case *Pharmaceutical Society of Great Britain v Dickson* (1970) HL. The Society intended to change the rules for its members, many of whom were owners of small chemists' shops. As a result, they would be severely restricted by their professional rules from dealing in non-traditional lines. For example, their shops would have to stop selling goods like paperback books or music cassettes. For many small chemists these were high-profit lines which effectively kept the business going. One chemist sought a declaration that such a change in the rules would be an unreasonable restraint of trade. The House of Lords agreed and considered public interest to be a paramount consideration. Such a change in the rules could well have caused many small chemists' shops to close and thus deprived large sections of the public of an important amenity and public service.

Over the years the courts have been vigilant in finding restraint of trade clauses hidden in some unusual situations. In one case, the rules of a pension fund were held to operate as an unreasonable restraint (*Bull v Pitney Bowes* (1966) QBD). The principles have also been applied in cases involving licences for racehorse trainers (*Nagle v Fielden* (1966) CA) and the transfer system for footballers (*Eastham v Newcastle United* (1964) Ch). These cases clearly demonstrate that the courts will not allow the parties to achieve through a back door what they could not do by a direct method.

Where the courts are prepared to interfere and find a restraint unreasonable, it is void and struck from the contract. Where it is only one clause of a much larger contract, the remainder of the contract may be perfectly valid and enforceable. In some cases, especially borderline situations, the inevitable effect of declaring a restraint to be void is that the employer is left with no protection for legitimate business interests. An example of this can be seen in *Commercial Plastics Ltd v Vincent* (1965) CA, where a research chemist had been working on PVC but only on a limited aspect of its application. The restraint in his contract was drafted to cover too general a field of work on PVC and so the whole restraint was held to be void. The judges felt sure that the employer had genuine interests to protect and, with a properly drafted clause, would have secured such protection.

This rigorous approach of the courts stems from the principle that the parties must make their own terms and cannot expect the judges to do this for them. In one limited circumstance the judges may be able to give some assistance. Where the restraint is couched as a series of alternatives, some of which are valid but some of which are too wide and thus unreasonable, the court may be able to sever parts which are excessive. Consider a restraint for '12 months within a 10-mile radius of our shops at Newcastle or York'. If the employee has never worked in the York shop, he or she probably poses no threat to the employer there and that aspect of the restraint could therefore be too wide and unreasonable. Using an approach known as 'the blue pencil test', the courts will simply cross out the words 'or York' and leave a valid restraint behind. This approach was recently confirmed and refined by the Supreme Court in *Tillman v Egon Zehnder Ltd* (2019), where the Court emphasised that the words to be severed must be capable of being removed without modifying the remaining contract. If the offending words can be cleanly severed and the restraint remains sensible and enforceable, the court may uphold the remainder of the clause. Of course, this will be no help in a great many cases where the restraint consists of one over-wide restriction, e.g. a restraint for '12 months within 100 miles of our shop in York'. The judges have no power here to cut this down in scope if it should turn out to be too wide to be reasonable.

It is obvious that the rules about restraint are not clear or precise. The main value of putting a restraint clause into a contract is its deterrent effect. It acts as an early warning to an employee that he will not get away with interfering with his employer's interests. But even if no restraint is imposed or a restraint has been struck out as unreasonable, it should be remembered that employees owe duties to their employers at common law under their contracts of service.

Discharge of the contract

Discharge by performance

In the normal course of events, both parties to a contract expect that the contract will be properly carried out. Where both parties perform their obligations under the contract exactly and precisely as promised, then the contract is discharged by its complete performance. It follows that if either party does not exactly and precisely perform, there is a breach of contract. Contractual obligations arise by agreement, and it is right that both parties should carry out their obligations in full. A necessary result of this rule is that a person who only partly performs his or her obligations and is to be paid upon completion of these obligations cannot be paid. This is clearly demonstrated in a case involving the installation of central heating. The job was done, and the customer had not paid. But the heating failed to give the promised temperatures in the house and the boiler emitted vile fumes. As the installer had not exactly and precisely performed, it was held that he was not entitled to be paid (*Bolton v Mahadeva* (1975) CA).

That case does demonstrate how harshly the rule of exact and precise performance can operate. Not surprisingly it is subject to a number of exceptions:

1 Substantial performance – A party may be paid for partial performance where he or she has substantially performed the contract. So, if one party has performed except for some trivial details, he or she will be entitled to be paid, less the value of the other party's counter-claim in respect of those trivial breaches. This was affirmed in *Hoenig v Isaacs* (1952) CA.
2 Acceptance of part performance – A party may be paid where the other party is prepared to accept part performance. In these cases, however, the other party must have a genuine choice whether to accept part performance or not. So, where I am due to deliver 10 tons of coal and I arrive with only eight tons, you may agree to accept eight tons. But in the central heating case, the claimant had no such choice. The work was only partly done, but it could not be undone. Of course, in those cases where the other party does agree to accept part performance, the obligation to pay will be correspondingly adjusted.
3 Prevention of performance – A party who has partially performed may be paid as much as the work or performance is worth in cases where the other party will not permit him or her to complete this performance. In this case, of course, the person who will not permit performance is in breach, and the party who has partly performed may either sue for damages in respect of that breach or claim a reasonable sum for the value of the work already done. A quantum meruit claim may be possible in such circumstances.
4 Divisible contracts – Where a contract is divisible, a party who has performed some parts may be paid for them. So, in a contract of employment with a salary paid monthly, each complete month could be regarded as a divisible contract. In building contracts,

any agreement about progress payments will make it a divisible contract. See Chapter 7 with regard to construction contracts.

Although the rules about precise performance are quite strict, there are, nevertheless, some defences which are regarded as acceptable excuses for non-performance or partial performance. For instance, the parties may have agreed to some lesser performance. The validity of their agreement will be the subject of rules about discharge by agreement.

Or again, the parties may be unable to perform their contract as originally contemplated because of some external factor or event outside their control. The validity of such an excuse is considered under discharge by frustration.

The parties themselves may have taken account of the need to change the rules about performance, e.g. in a building contract, more time may be needed if a contractor unearths antiquities, and their contract may make provision for altered performance.

The time for performance of the contract

It is often said that 'time is of the essence' in a contract, meaning that the time stipulated for performance is a major term, and any breach of that term would entitle the other to repudiate the contract. This will often be true, but each case will turn on its own facts. Have the parties expressly provided for a time for performance? If not, then the contract must be performed in a reasonable time. If a precise time has been fixed, the surrounding circumstances may make it clear that it must be adhered to. If a wedding dress is ordered to be completed in time for a specified wedding date, time is clearly of the essence. But if a new armchair is ordered with an undertaking to deliver it next week, it may not be so disastrous if it is not delivered on time. Inevitably, in that situation, the time may come when the patience of the customer starts to wear thin. Then, although time was not originally of the essence in the contract, the customer can make it so by serving notice on the seller. The customer may indicate that, unless he has the chair by the end of the week, he regards the contract as at an end. This reflects the general principle confirmed in *United Scientific Holdings Ltd v Burnley Borough Council* (1978), where the House of Lords held that time may become of the essence where reasonable notice is given following a delay.

Tender of performance

Sometimes it is only possible to perform the obligations under a contract with the cooperation of the other party. Take the example of a sale of goods where the seller agrees to deliver. He needs the other party to be there to receive the goods. If he offers to perform, i.e. tenders performance, by taking the right amount of goods of the correct description and quality, and the other party will not accept them, then the seller can argue that he has done all he can, and he can then consider himself discharged from his obligations. In *Startup v McDonald* (1843) CCP, a delivery of oil was due to be made 'within the last 14 days of March'. It was delivered at 8.30pm on 31st March, a Saturday. Being late in the evening, the buyer refused to accept delivery. It was held to be a valid tender of performance within the terms of the contract so the seller could recover damages for the non-acceptance of the goods.

Where the obligation under the contract is to pay money, there is only a valid tender if money of the exact amount is offered, made up in such a way as to constitute 'legal tender'. Of course, it is quite common for the parties to have agreed expressly to payment in some other way, e.g. by cheque or bank transfer. If no such agreement exists, the other party is entitled to expect money.

Discharge by agreement

What is created by agreement can be discharged by agreement. Although this statement is largely true, discharging a contract by agreement depends on the operation of the rules of consideration. Effectively, if the parties agree to discharge a contract, they are making a new contract. That new contract will be supported by valid consideration if each party is giving up something of value by discharging the first contract. If I agree to sell you goods for £10 and I have not yet delivered and you have not yet paid, then we can discharge that contract by a new agreement. My consideration for the new agreement will be giving up the right to receive £10 and your consideration will be giving up the right to receive the goods.

Obviously, it is not so straightforward to achieve a discharge by agreement if one party has already performed. If I have already delivered the goods, my part of the bargain is complete. It would be possible to discharge our agreement by drawing up a deed (a contract under seal), but this would be absurd to contemplate except in commercial cases involving very large sums of money. In these cases, the rules about waiver play an important part. In the example already given, I may agree to forego payment of the £10 due. What would prevent me from changing my mind at a later stage and suing for the money? It could be argued that I am prevented from going back on my earlier promise because of the defence of promissory estoppel. As this is an entirely discretionary principle, the other party may feel less than secure in relying on it. I would also be prevented from suing for my £10 if we had made a valid variation of our original contract, supported by new consideration.

Discharge by frustration

When two parties make a contract, their natural expectation is that both of them will perform it. However, in some cases, one party may find it impossible to perform their part of the bargain due to events quite beyond their control. The law of contract basically adopts a harsh principle that if contractual obligations are undertaken, then any non-performance for whatever reason amounts to breach.

However, since the nineteenth century, the courts have developed rules relating to 'frustration' or subsequent impossibility, in an attempt to achieve justice between the parties when things go wrong in an unforeseen way. If one party chooses not to perform because he or she now regards the contract as a bad deal, he or she will be in breach. If the parties themselves have provided an express term to cover some particular event, then the terms of their agreement will take effect. The common law rules developed by the courts provide that if there has been a frustrating event, it will have the effect of discharging the future obligations of the parties under the contract. The resulting financial arrangements between the parties will then be dealt with, when appropriate, by the Law Reform (Frustrated Contracts) Act 1943, which remains in force and unchanged in material respects.

The first problem is the difficulty of establishing which events should be regarded as frustrating events. Take, as an example, the closure of the Suez Canal. If my ship is locked in the canal as a result of the closure, then any contract I have made to hire the ship to you would surely be frustrated. But if I had a contract to transport your goods, and the closure of the canal simply means that I must use a longer and more costly route, then this is by no means certain to be found to frustrate the contract. In such a case my major complaint would be that I am not going to make a profit on our contract. This factor alone will not influence the court. Contracts inevitably involve risk, and the parties must look out for their own interests.

Davis Contractors Ltd v Fareham Urban District Council (1956) HL is a good example of this point. In July 1946, the contractor had agreed to build 78 houses for £92,000, the work to be completed within eight months. Owing to quite unexpected circumstances, and through no fault of either the contractor or the council, there was a serious labour shortage and necessary building materials were extremely difficult to obtain. As a result, the job took almost two years to complete, and the contractor incurred an extra expense of £17,500. He wanted to argue that the very long delay had frustrated the original contract, and as that would avoid the original contract, he could then claim payment for a reasonable sum. The House of Lords rejected the contractor's argument. Lord Radcliffe observed: 'The possibility of enough labour and materials not being available was before their eyes and could have been made the subject of special contractual stipulation. Frustration is not to be lightly invoked as the dissolvent of a contract'.

Examples of situations in which the court is prepared to find frustration include:

1 *A forecast event fails to take place.*

   A classic example was the cancellation of the coronation of Edward VII. Many people had made contracts to hire rooms to view the coronation procession. In one such case, *Krell v Henry* (1903) CA, the contract for the hire of a room overlooking Pall Mall was held to be frustrated by the cancellation of the coronation. It is evident in a case like this that although the hirer could still use the room on the day in question, it is impossible for the parties to achieve the substantial object of their contract. There is an interesting contrast to be made with the case of *Herne Bay Steamship Co. v Hutton* (1903) CA, where the contract indirectly arose out of the coronation activities. The Royal Naval Review was to take place at Spithead, and the claimant chartered a boat to cruise around the assembled foreign fleets in the harbour and to see the naval review. Because of the King's illness, the review had to be cancelled. But the contract for the charter of the boat was held not to be frustrated, because it was still possible to enjoy a day's sailing and see all the foreign fleets. In other words, the substantial object of the contract had not been destroyed.

2 *The subject matter of the contract is physically destroyed.*

   In *Taylor v Caldwell* (1863) QB, the contract provided for the hire of a music hall which burned down before the date for performance of the contract. The fire was accidental, and the contract was held to be frustrated. A similar rule for goods which perish after the contract is made but before the date for delivery is contained in s.7 of the Sale of Goods Act 1979. The contract would be frustrated provided neither party was at fault. The same principle also applies where the person who was to perform the contract dies or becomes too ill to perform. This rule is only appropriate where the contract calls for the services of one particular person. So, if I hire a famous concert pianist to give a recital, then the death or illness of the pianist could frustrate the contract. Where there is an ordinary ongoing contract of employment, short-term illnesses would not usually frustrate the contract. There will inevitably be borderline cases, where a long-term illness rendering the employee unfit to perform his or her former duties may amount to frustration. This sort of example should not give rise to too many difficulties when it is remembered that such an employee's contract could be terminated by giving appropriate notice.

3 *Subsequent intervention makes performance of the contract illegal.*

This situation can occur when parties have made a perfectly valid contract, but before they perform it, some change in the law makes that type of contract unlawful. So, in *Metropolitan Water Board v Dick Kerr & Co.* (1918) HL, the defendants had undertaken to build a reservoir for the Water Board. They were already at work when they were ordered to stop by the government, which directed them to build a factory instead. The court held that the contract with the Water Board was frustrated. It is true that the defendants could have resumed work on the reservoir very much later, but because of changing costs of materials and labour this would have then been a substantially different task from the one originally undertaken. Obviously, the reservoir example can be compared with the council house case (Davis) discussed earlier. In the one case the rules of frustration operated, and the defendant was released from what had become a bad bargain. In the other case, Davis Contractors were held to their original contract. How can such a distinction be justified? It seems that in most cases the courts will abide by the traditional view that you must perform as you agreed, but there will be exceptional situations where the increased cost of performance is so great that it can be said to make the contract into something radically different from what was contemplated by the parties. No doubt this allows the courts to find frustration in extreme cases, but it does demonstrate how difficult it is to pin down the scope of the doctrine with any precision.

4 *Frustration affecting contracts concerning land.*

It used to be thought that the rules of frustration would not operate in the case of leases or contracts for the sale of land. Two important cases emphasise how significant the rules could be in such instances. In *National Carriers v Panalpina (Northern) Ltd* (1981) HL a warehouse was let on a 10-year lease. With five years of the lease still to run, the street giving access to the warehouse was lawfully closed off by the local authority because there was a listed building in a dangerous condition. It could only be demolished with the consent of the appropriate minister, which took almost two years to obtain. During all that time, the tenant was unable to use the premises for warehousing as no lorries could gain access. He stopped paying the rent and claimed frustration of the lease. The House of Lords rejected his claim. Lord Russell said:

> Under the bargain . . . the land has passed from the lessor to the lessee with all its advantages and disadvantages . . . I ask myself why should justice require that a useless site be returned to the lessor rather than remain the property of the lessee?

In *Amalgamated Investment Co. Ltd v John Walker & Sons Ltd* (1976) CA, the facts involved the purchase of a piece of land which the purchaser intended to redevelop. He contracted to pay £1.7 million for the land. On the day after he signed the contract, officials at the Department of the Environment listed a building which was presently on the site as a building of special interest. This made it extremely difficult to go ahead with development plans, and the site without development potential was worth only £250,000. Despite the scale of the financial disaster, the purchaser was held to be bound by the contract. Although it was now apparent that he had made a bad bargain, the contract was not frustrated.

5 *Government-imposed restrictions rendering performance impossible or radically different.*

The COVID-19 pandemic prompted unprecedented government interventions, including mandatory lockdowns and restrictions on business operations. These actions have led parties to invoke the doctrine of frustration, arguing that such supervening events have rendered contractual performance impossible or fundamentally different from what was originally contemplated.

In *Bank of New York Mellon (International) Ltd v Cine-UK Ltd* (2021) QBD, commercial tenants sought to withhold rent payments, claiming that the enforced closures due to COVID-19 restrictions frustrated their leases. The High Court rejected this argument, emphasising that there is no concept of 'temporary frustration' in law; frustration must result in a contract being brought to an end permanently. The tenant's appeal was dismissed in the Court of Appeal.

Similarly, in *Financial Conduct Authority v Arch Insurance (UK) Ltd & others* (2021), the Supreme Court examined whether business interruption insurance policies covered losses stemming from the pandemic and associated lockdowns. While this case primarily addressed insurance contract interpretation, it underscored the complexities of applying traditional contractual doctrines to the unprecedented circumstances of a global pandemic.

From the cases and examples we have looked at, it is possible to state only a very broad principle: frustration occurs when performance of the contract is impossible or when the parties can no longer achieve the substantial object they had in mind. This principle has been reaffirmed in recent case law arising from the COVID-19 pandemic. Although government-imposed restrictions due to unforeseen events like lockdowns have severely disrupted contractual performance, the courts have remained reluctant to discharge contracts on these grounds alone. As illustrated in *Bank of New York Mellon (International) Ltd v Cine-UK Ltd* (2021), frustration will not be found where the impediment is temporary or performance merely becomes more onerous. The doctrine continues to apply only where the change is so fundamental that it renders performance radically different from what was originally agreed.

It is essential that the supervening event should be outside the control of either of the parties. In practice, this is extremely rare. If it comes about through some act or election of one party or as a result of negligence, then it cannot be frustration. Both of these points are well illustrated in *Lauritzen AS v Wijsmuller BV* (1990) (known as *The Super Servant Two*) CA, where the defendants had contracted to transport a drilling rig from the Japanese shipyard where it was being built to Rotterdam. The contract provided that the carriage would be undertaken between 20th June and 20th August and would be effected using either *Super Servant One* or *Super Servant Two*, both of which were large transportation units. In fact, it was always the intention of the defendants to use *Super Servant Two* to fulfil the contract, but, unfortunately, it sank in January. In the meantime, the defendants had made other contracts to use *Super Servant One*, so that vessel was not available to be substituted. The defendants argued that the sinking of the *Super Servant Two* had frustrated the contract. The Court of Appeal held that this was not a supervening event such as to frustrate the contract, because the defendants could have performed without breach by using *Super Servant One*, and it was by their own election that they chose not to do so.

Inevitably, frustration often occurs after the parties have embarked on performance. Their valid contract automatically ends at the moment of the frustrating event, and the parties are therefore discharged from further obligations under it. This can produce a very untidy and unfair situation, as it means that obligations which fell due before the frustrating event are still owed. Take the example of a contract to import and sell £10,000 worth of a particular type of toy, delivery to take place in November, with £2,000 payable on the making of the contract, and the remaining £8,000 to be paid within one month of delivery. The contract is made in June and £2,000 is duly paid. In October the government bans the import of the toy in question. At that point the contract is frustrated. But the buyer has paid £2,000 for which he has received nothing. The early approach of the courts to sorting out these problems was to decide that the loss lay where it fell. In an attempt to solve the financial plight of the two parties to a frustrated contract, the Law Reform (Frustrated Contracts) Act 1943 was passed.

The Act makes no attempt to define what frustration is. It simply deals with the consequences. Its two main objectives are to allow for the recovery of money paid and for compensation for partial performance. These provisions remain in force and were notably relevant in contractual disputes arising during the COVID-19 pandemic, particularly where government-imposed restrictions rendered performance impossible or radically different from what was agreed. These will be considered separately.

1 *Recovery of money paid* – The Act provides that all sums paid before the frustrating event shall be recoverable and all sums due before the frustrating event cease to be due. Applying this to the sale of toys example, the £2,000 deposit could be recovered. Of course, this may not be particularly fair to the seller, who may have incurred considerable expense in going about the performance of the contract. The Act recognises this by providing that where expenses have been incurred before the frustrating event, and those expenses were 'in, or for, the purpose of performing the contract', the court may allow a party to recover in respect of those expenses, to the extent that the court thinks is just and equitable. In the toy example, the seller may have made phone calls and sent telexes and may have had transport expenses and insurance costs. The judge may say that he can keep £500 of the £2,000 deposit. Now the buyer is £500 out of pocket. That is an inevitable price which has to be paid when a transaction has gone wrong. The rules of frustration only operate where neither party is at fault, so inevitably the courts are trying to make the best of a bad situation and no solution will ever be perfect. They can only aim to achieve as fair a compromise as possible. An exercise of the courts' discretion in relation to expenses was seen in *Gamerco SA v ICM Fair Warning (Agency) Ltd* (1995) QBD.
2 *Recovery of compensation for partial performance* – We have already seen that a frustrated contract will have commenced life as a valid contract, under which the parties may have started to perform their obligations. Suppose I have contracted to build you a house for £60,000 payable on completion. If I have completed three-quarters of the work when the government interferes and puts a ban on private building work, can I recover the value of the work I have done? The Act provides that if one party obtains a valuable benefit by reason of work done in the performance of the contract by the other, the court may allow the party who has performed to recover a sum which does not exceed the value of the benefit conferred, and which is just in all the circumstances of the case.

The scale of disaster when such frustrating events occur can be seen in *BP Exploration Co. Ltd v Hunt* (1982) HL. Hunt owned an oil concession in Libya. He did not know if oil would be found there or, if so, how much. He wanted, somehow, to finance the exploration of the oil field but to hold on to a share of any profits if large quantities of oil were found. He contracted with BP to share the field. BP would bear all the exploration costs and, if oil was found, Hunt would repay them from his share of the oil revenues. The exploration work cost a staggering $87 million. A large amount of oil was found, and extraction had only just begun when the Libyan government confiscated the land and nationalised the concession. BP had done work in performance. Had that work conferred a valuable benefit on Hunt? Merely exploring for oil gave Hunt no valuable benefit, but its discovery greatly increased the value of his concession before it was nationalised. The judge calculated Hunt's valuable benefit to be the value of the oil already removed and the compensation he would receive from the Libyan government. This formed the upper limit of what he could have awarded to BP. He then had to consider what figure, up to that upper limit, he thought was just in all the circumstances. Although the judge thought that the 'valuable benefit' to Hunt was over $80 million, he awarded $35 million to BP, as it had been an agreed basis of the contract that it would bear the exploration costs. The exploration might have found no oil at all, and in that case, it would have had to bear those costs in total with no claim at all against Hunt.

The 1943 Act may not apply if the parties themselves have included a term to cover the results of a frustrating event. It will be a question of construing or interpreting their term to discover if they intended it to operate in such extreme cases. This question arose in several commercial contracts during the COVID-19 pandemic, where parties had included force majeure clauses or 'material adverse change' provisions. In those instances, the courts generally upheld such clauses were clearly drafted and applicable, rather than applying the doctrine of frustration.

The rules of the Act have no application to contracts of insurance, so if I insure against sickness, pay a year's premium and then die after the first month, there can be no recovery of part of the premium. In some sale of goods situations, the appropriate rules are found in s.7 of the Sale of Goods Act 1979, rather than the 1943 Act.

Discharge by breach

A breach of contract may occur because one party does not perform his or her obligations at all, performs them late or performs them in an unsatisfactory way. One of the major difficulties when looking at discharge by breach of contract is trying to establish exactly which breaches have the effect of discharging obligations under the contract. Where such a major breach can be identified, the injured party may nonetheless want to go ahead with the contract and simply ignore the breach or, more likely, seek compensation for it.

All breaches of contract, however large or small, give the injured party the right to sue for damages. In some cases, it may be possible to exercise that right through set-off. Where a painter contracts to decorate the outside of your house for £1,000, to be paid on completion of the work, he may commit breaches by not burning off where specified, or by applying only one undercoat instead of two. In such a case, the injured party could solve the problem by deducting something from the £1,000 due. In some cases, a contract may even provide a schedule of damages for specified breaches, thereby anticipating those which are likely to occur and providing a solution. Such an example is the case where a builder undertakes to complete a job by 30th June, and if he fails to do so, to pay £100 per day (agreed or liquidated damages) in respect of each day's delay thereafter.

A much more difficult question still remains – when does a breach put an end to the contract itself or permit the innocent party to choose to treat the contract as at an end? And what exactly is meant when we say the contract is 'at an end'?

As a general rule, where the party at fault is in breach of a major term (a condition) or in breach of an innominate term where the effects are serious, the innocent party can treat that behaviour as repudiatory, and regard himself or herself as discharged from further performance. Where the breach is of a minor term (a warranty), the injured party is restricted to an action for damages and must continue with performance of the contract.

The various categories of terms have already been examined. Let us assume that when the breach occurs, it is clear which type of term has been broken. Take the case where you want to ship goods under refrigeration. You enquire whether the ship has a refrigeration plant and are assured that it has. On that basis you contract to ship 10 tons of soft fruit under refrigerated conditions. When you arrive to load the fruit, you discover that the ship has no refrigeration equipment. This is likely to be a breach of a major term, a condition. You now have a choice. You can treat it as a repudiatory breach and regard yourself as discharged from further performance, i.e. the obligation to pay the shipping charges. You can also sue for damages in such a case. No doubt you would take this course of action if you were able to find alternative shipping facilities immediately or if you could sell the fruit in a local market. Where that is not possible, you may decide that you have no option but to ship your fruit in the unrefrigerated ship, thereby choosing to affirm the contract, but, of course, you can still sue for damages for the breach. As you have affirmed, you have to pay the shipping charges, but if your fruit deteriorates as a result of the breach, you can be compensated for the damages.

This example assumes that, even though the breach was major, you were left with a choice of courses of action, either to affirm and go on or consider the contract at an end. Sometimes that will not be the case, as the very nature of the breach itself may put an end to the contract. Take the case where you engage a central heating contractor to install a system for you for £2,000, the work to commence on 1st June. If the contractor informs you in May that he has found more lucrative contracts and does not intend to come to do your job, he is effectively repudiating the contract in advance. You could now wait until June to see if he comes. If he fails to turn up, then this is a case of breach by non-performance. Or you could sue immediately, making clear that you are choosing to treat the contract as at an end. Where these situations occur in practice, the difficulty may be that the injured party needs to respond quickly to minimise the harm and inconvenience, and he may therefore act precipitately without the benefit of proper advice. Yet, once it is clear that the other party does not intend to perform, it might be dangerous to wait and see if he will change his mind, because in the meantime, the injured party could lose his rights through some supervening frustration.

Enforcing the contract and remedies for breach

The law provides remedies for breach of contract, but the extent to which purely legal remedies can be truly effective is limited. A remedy is necessary because things have gone wrong. An ideal solution would be for the parties themselves to plan in advance for situations which may arise. They can then impose solutions which are satisfactory to both of them.

This contract planning role is very significant in the business and commercial world. The standard form building contract is an example where many contingencies are provided for, without the need to go to court. One advantage of such a system is that it creates a climate within which the parties can stay on reasonably good terms, which is vital if they need to do business with each other frequently. Sometimes the contract planning role can also avoid those situations where court action would probably be useless. There are business situations where the customer is asked to pay a deposit. Not only does this create a situation where the customer is more likely to perform, but it also provides a ready small sum of compensation against non-performance. An example would be 'same-day' photographic processing, where deposits are common.

One problem of seeking to enforce a contract through the courts is that the primary and usual remedy for breach is damages. The courts are prepared to order the other party to carry out their obligations, but only in a very limited number of cases. If they grant a remedy, it is called specific performance. Such an order is not readily available because the court may then need to keep the defendant under constant supervision, and it has neither the desire nor the means to do so. Moreover, in many cases, money will be adequate to compensate the claimant. The types of contract where an order of specific performance may be granted usually relate to sales of land and leases. No two pieces of land are identical, and it is thought that money cannot adequately compensate. That is quite a difficult idea to accept when the 'land' in question may be a semi-detached house on an estate surrounded by dozens of properties which are seemingly identical! The argument about the unique quality of the property can also be used, although less frequently, in relation to the sale of goods, for example, a unique painting. If the goods are in some way unique or not readily obtainable, a court may be prepared to grant an order for specific performance.

As the power to award specific performance is equitable and discretionary, a number of points will be borne in mind by the courts. They will not grant an order if it would operate very harshly on the defendant. Nor will they grant it where a claimant has delayed too long in seeking relief – delay defeats equity. This rule is not precise, and what counts as excessive delay will depend on the facts of each case. The order will not be granted if constant supervision would be required, nor in cases involving personal service or employment (*Warner Bros v Nelson* (1937) KBD), as it is thought that this would be tantamount to slavery, as well as being impractical to enforce. It used to be suggested that an order of specific performance would not be granted in cases involving breach of contract to pay money, as damages would surely be an adequate remedy. This may not be so, and the problem was highlighted in the case of *Beswick v Beswick* (1968) HL, where a coalman sold his business to his nephew, who was to pay for it by paying the uncle £5 per week for his life and thereafter the money was to be paid to the coalman's widow. After the death of the coalman, the payments to the widow were not made. She sued as the personal representative of her late husband and was awarded specific performance of the contract. This was the only way to achieve justice in this case.

There is another form of equitable relief, the injunction, which can be useful in a limited number of contract situations. This is particularly the case where a party to a contract has promised *not* to do a particular thing. A common example is a clause in restraint of trade, where an employee promises not to solicit his or her former employer's customers or a seller of a business promises not to compete. Where a breach of such a clause is proved, the court may grant an injunction, but it should be remembered that the relief is entirely discretionary. In recent years, courts have shown a greater willingness to grant injunctive or other equitable relief in urgent or highly specific commercial contexts, provided the remedy sought is proportionate and just. For instance, in *Classic Maritime*

Inc v Limbungan Makmur Sdn Bhd (2019), the Court of Appeal reaffirmed that specific performance will not be ordered where damages are adequate but acknowledged the importance of preserving contractual expectations in exceptional cases.

Awarding damages

In theory, an award of damages can be made in respect of every breach of contract, however large or small, but the amount of damages is linked to the loss suffered. The aim of the court when awarding damages is to put the injured party into the position they would have been in if the contract had been properly performed. This compensates him or her for the loss of expectation. Somehow, the court has to arrive at an appropriate figure for the damages. This is described as the measure of damages. The figure can only be worked out once it is established for which harm the claimant can recover. He or she can only recover in respect of losses which 'arise naturally', according to the usual course of things, and losses which must have been in the contemplation of the parties when they made the contract. This is the rule of remoteness of damage, which was laid down in the case of *Hadley v Baxendale* (1854) CE. The claimant's losses must also have been caused by the breach of contract.

The two aspects of the remoteness rule can be seen at work in the case of *H Parsons Ltd v Uttley Ingham & Co. Ltd* (1978) CA, where the defendants manufactured and installed an animal feed dispensing hopper for Parsons, who was a pig breeder with a class 1 pedigree herd. The installation was carried out after the hopper had been transported by road. During the journey the driver had jammed the hopper ventilator with a rag to stop it rattling and he did not remember to remove it before the hopper was erected. Pig food was poured into the hopper and at first all seemed well. Inevitably, because the ventilator was jammed shut, the pig food eventually turned mouldy. Of course, there was a breach of contract here, and Parsons could undoubtedly claim damages in respect of the mouldy pig food. Unfortunately, the pigs had eaten the food and developed a disease which killed off a large proportion of the herd. In consequence, not only had Parsons lost more than 250 pigs, but he had also sustained the loss of a breeding season. Should he be allowed to recover in respect of the loss of the pigs? The scale of the problem can be clearly seen in this case, where the defendants accepted liability with regard to the pig food and were prepared to pay damages of £1,802, but Parsons was trying to claim about £30,000. Parsons' claim was successful. The defendants must have contemplated that if they installed a defective hopper which made the food mouldy, the pigs could become ill. It was not relevant that the degree of harm (the death of the pigs) was greater, as long as the type of harm (illness) could have been contemplated.

It is clear that the remoteness rules operate very flexibly, but the more the defendant knows about the claimant and his or her requirements, the more he or she is likely to be able to contemplate by way of potential harm. The remoteness rule was considered in *Transfield Shipping Inc. v Mercator Shipping Inc.* (2008) by the House of Lords. It appears that if a loss is wholly out of the ordinary, being disproportionate or unpredictable, and there is a general market understanding as to what can be claimed, then the loss is too remote unless there is an assumption of responsibility for the loss on the part of the defendant. So, the rule in *Hadley v Baxendale* will usually apply, but in unusual cases, as described in the preceding sentence, there will need to be an assumption of responsibility for loss to be recoverable.

Once the courts are clear about the harm or loss which falls within the remoteness rule, they must then assess an appropriate measure of damages. Sometimes this will be quite straightforward, as in cases for the sale of goods where the seller has failed to supply. The buyer must now look elsewhere for the same goods and may only be able to buy at a higher price. The measure of damages is the difference between the two prices. Or a buyer may refuse to accept delivery of goods properly tendered under a contract. There, the measure may well be the loss of profit on the deal. Or a holidaymaker may cancel a booking for a country cottage late in the day, and the damages will be the loss of profit on the booking.

Sometimes the courts will give damages on the basis of how much it costs to cure a defect in work done instead of the difference in value. This is the usual measure in relation to defective or incomplete building work. But the cost of cure will not be given if it is excessive when viewed against the benefit accruing to the innocent party. In *Ruxley Electronics & Construction Ltd v Forsyth* (1995) HL, a contractor had agreed to construct a swimming pool to a depth of 7 feet 6 inches for £17,797, but the pool constructed was only 6 feet 9 inches in depth. The pool was usable and evidence showed that the lack of depth had not decreased the value of the pool. However, to increase the depth of the pool to 7 feet 6 inches would require the demolition of the existing pool and cost £21,500. The House of Lords held that the award of cost of cure damages of £21,500 was out of proportion to the loss suffered by the pool owner and would not be given. Difference in value damages would be appropriate (which in this case would have been nominal as there was no difference in value between the two pools), but the House of Lords restored the original trial judge's award of £2,500 for loss of amenity.

Where there has been a breach, the injured party cannot just sit back and wallow in their misery. The law imposes a duty to take reasonable steps to mitigate these losses. A claimant cannot recover for any harm which could have been prevented by taking reasonable steps in mitigation. If a buyer refuses to take delivery of goods under a contract, the seller must take reasonable steps to find another buyer. In the previously mentioned holiday cottage example, he must take reasonable steps to re-let the property. An ironic result of taking reasonable steps to mitigate may be that the claimant actually increases his losses. In the holiday cottage example, he may place advertisements in newspapers and still not manage to re-let the property. He can then include the cost of the advertisements in his claim for damages.

Of course, the claimant is the person who has been wronged, so there is a limit to what he or she can be expected to do under the duty to mitigate. As the defendant has committed the breach, he or she is hardly in a strong position to 'call the tune'. This is well illustrated in *Pilkington v Wood* (1953) CA, where a solicitor gave incorrect advice to a client, in consequence of which the client found that his house had a defective legal title and was virtually unsaleable. When the solicitor was sued for damages, he argued that the claimant had failed in his duty to mitigate, as he could have solved the problem of the defective title by suing the person who had originally sold him the house. This was hardly asking the claimant to take a reasonable step, as such litigation would have been expensive and the outcome by no means predictable. The claimant was right not to have embarked on such a costly and uncertain procedure.

It used to be a feature of damages for breach of contract that nothing would be awarded for distress, injury to feelings or disappointment. One justification for this rule may have been that it is very difficult to quantify such losses. The rule has been subject to an exception over recent years in cases where the claimant was contractually entitled to expect pleasure and enjoyment, not pain and distress! However, the sums awarded for distress and disappointment have not tended to be large. *Jarvis v Swans Tours Ltd* (1973) CA shows the sort of case where an award for disappointment may be appropriate. The claimant, a solicitor, had booked a holiday in Switzerland on the strength of a brochure which claimed that he would be joining a house party and a special resident host. He would enjoy a welcome party and special yodelling evening, plenty of *gemütlichkeit* (good atmosphere) and, generally, he would be 'in for a great time'. In fact, by the second week of his holiday, there were no other guests at all, and in every other respect the holiday failed to live up to the brochure claims. Even the yodeller was a local man singing in his working clothes. When awarding him damages for disappointment the judge said: 'Mr Jarvis has only a fortnight's holiday each year. He books it far ahead and looks forward to it all that time. He ought to be compensated for the loss of it'. The court awarded him £125.

It was thought that a contract must have as its *main object* the provision of pleasure or the avoidance of distress for damages for disappointment or distress to be recoverable. In *Farley v Skinner* (2001), the House of Lords decided that for the award of such damages, the contract must have *as an important or major object* of the contract the provision of pleasure or peace of mind. In that case the claimant employed a surveyor to inspect and report on a private dwelling house that the claimant wished to purchase. The defendant surveyor was expressly asked to report on aircraft noise. In the defendant surveyor's report, it was negligently stated that the property was unlikely to suffer greatly from such noise. In reliance on the report the property was purchased. On occupying the property, the claimant discovered that the property was severely affected by aircraft noise. The House of Lords held that the claimant was entitled to damages for disappointment. It would appear that damages of this type will not usually be available for a negligent survey, as under an ordinary surveying contract, there is no promise to provide peace of mind. The important feature of *Farley v Skinner* was that the surveyor was expressly instructed to investigate aircraft noise, which became an important or major object of the contract. This approach has remained the position in subsequent cases, with the courts maintaining a cautious stance and continuing to confine such awards to contracts where emotional satisfaction or peace of mind is clearly a significant part of the bargain.

Liquidated damages

It has already been emphasised that by effective contract planning, the parties can provide in their contract for the sums to be paid in the event of specified breaches, a common example being late performance. Such contractually determined damages are called liquidated damages. The advantages of such a clause are clear. Problems can be resolved quickly and privately, without resort to the courts and, hopefully, in a manner which will preserve good relations between the parties. The parties also know in advance the risk and cost involved in particular breaches. On the other hand, such clauses can cause problems, particularly where the party agreeing to the clause lacks bargaining strength. In that case, the result of the clause might be unfavourable to him or her. Either party may miscalculate the real effects of a specified breach and thus be unhappy to be regulated by what turns out to be an unfair clause. Against that, of course, must be set the fact that such clauses are freely negotiated, at least in theory!

The courts themselves are alert to the possibility that such a liquidated damages clause could operate harshly. If there is a dispute about the clause, it is open to the court to consider whether it represents a genuine attempt by the parties to estimate their losses, in which case the clause will be enforced even if the estimate is wrong, *or* if a clause was inserted by one party to frighten or terrorise the other into performance. In the latter case, the clause is classed as a penalty clause and is unenforceable. It is irrelevant what name the parties give to their clause. The court will determine which category it falls into, using a number of fairly flexible tests:

1. Does the clause provide for the same amount to be paid in respect of different breaches, some of which may be serious while others are trivial? If so, it will be hard to show that the clause was inserted as a genuine pre-estimate of loss.
2. Does the clause provide for damage for breaches involving non-payment of money at a level far higher than the sum unpaid? If you are due to pay £25 per week and a clause is inserted saying that for every late payment, a £100 fine will be due, that looks like a penalty.
3. Does the sum provided for by the clause appear to be harsh and unconscionable, given the breaches which are likely to occur?

Although the generally accepted principle of civil damages is simply to compensate for loss, recent cases have seen development beyond this. In *Cavendish Square Holding BV v Talal El Makdessi* (2015), the Supreme Court re-examined *Dunlop Pneumatic Tyre Co. Ltd v New Garage & Motor Co. Ltd* (1915) HL and held that the concept of a 'genuine pre-estimate of loss' was unhelpful and the true test for unenforceability was whether the clause in dispute was a secondary obligation which imposed a detriment on the contract-breaker out of all proportion to any legitimate interest of the innocent party in the enforcement of the primary obligation. This case may qualify the attitude of the courts in future cases concerning liquidated and specified damages in construction contracts. Heard at the same time was *ParkingEye Ltd v Beavis* (2015) SC, where the concept of 'commercial justification' for apparently disproportionate damages for breach of contract was held to be a valid basis. This case involved parking charges of up to £100, i.e. the amount payable if no parking ticket had been purchased or it had expired. This clearly did not relate to the actual loss caused by the breach, but was upheld in the Supreme Court.

The revised test in *Cavendish* has since been confirmed and applied in later decisions. In *Triple Point Technology Inc v PTT Public Company Ltd* (2021), the Supreme Court considered the operation of a liquidated damages clause in the context of delayed software delivery. The clause continued to apply up to the point of termination, clarifying an earlier ambiguity from Court of Appeal reasoning and confirming the enforceability of such clauses even where part-performance had occurred. The case underscores that where the clause protects a legitimate business interest, such as predictable financial exposure to delay, it will be upheld, even if it does not directly reflect actual loss.

These developments reinforce the idea that the courts are concerned not only with compensation but also with commercial fairness and the proper allocation of risk, especially in complex contracts such as those involving construction, software or other specialist services.

Suing on a quantum meruit

Quantum meruit means 'as much as it is worth'. Generally, the rules about *quantum meruit* have no particular significance in the law of contract, because the contract itself should fix the amount of any payment due or the method of calculating it. In some cases, however, even the law of contract provides for the payment of a reasonable price. This could happen where minors have been supplied with necessaries. Another example is where goods have been sold, delivered and accepted, but no price has been fixed (s.8, Sale of Goods Act 1979).

There are times when the parties may think that they have a valid contract. They may have acted in reliance on that belief in performing obligations, only to find that the contract has not taken effect. In such cases, if one party confers a benefit on the other, in circumstances where both believe that the benefit will be paid for, then the courts will allow the performer to recover a reasonable sum under the *quantum meruit* rules. This turned out to be a useful solution in *British Steel Corporation v Cleveland Bridge and Engineering Co. Ltd* (1984) QBD, where the parties planned to enter into a contract, but it had not been finalised between them. In anticipation of that contract, one party began to perform. When a contract failed to materialise, the performer was able to be compensated for the work done on a *quantum meruit* basis.

The principles in *British Steel* remain good law and were reaffirmed more recently in *Benedetti v Sawiris* (2013), where the Supreme Court emphasised that the aim of *quantum meruit* is to prevent unjust enrichment, not to reward the claimant's expectations. The court confirmed that a reasonable sum is assessed by reference to the market value of the benefit conferred, rather than what the claimant believed they were entitled to receive. This means that even if a party performed expecting a higher sum, they will only recover what the benefit was objectively worth.

Limitation periods

The remedy of damages, as the principal relief for breach of contract, must be sought in time; otherwise, any claim is statute-barred by the Limitation Act 1980. The imposition of time limits is sensible because evidence eventually becomes stale and unreliable and any party at fault must ultimately have some certainty about whether they will be sued or not.

In claims for breach of contract, the time limit is basically fixed at six years from the date when the cause of action accrued. That usually means six years from the date of the breach of contract. The 6-year rule is extended to 12 years where the contract was made by deed.

These rules are fairly rigid except where a claimant fails to discover that he or she has a cause of action because of the defendant's fraud. In these cases, time only begins to run when the claimant discovers the fraud or when he or she could with reasonable diligence have discovered it. 'Fraud' for these purposes could include cases where a defendant has deliberately concealed relevant facts from the claimant. This is what happened in *Applegate v Moss* (1971) CA, where a builder constructed a house on foundations which did not comply with the contract terms. The house was finished in 1957, but it was 1965 before the defects began to manifest themselves. The house was declared unsafe for habitation. Although eight years had elapsed, and normally, a claim would have been statute-barred, the time was extended to run from the date when they learnt that the foundations were defective because of the builder's concealment of facts relevant to the claimant's claim – he had covered the foundations up before they could discover how inadequate they were.

These strict time limits do not apply to equitable remedies such as specific performance. But in exercising any discretionary equitable relief, the court works on the maxim that delay defeats equity. It will be a question of fact in each case whether a claimant has unduly delayed.

In construction and professional negligence disputes, the Limitation Act 1980 continues to operate alongside the Latent Damage Act 1986, which may extend time for claims involving latent defects, especially in tort, to three years from the date of knowledge of the material facts (subject to a long-stop of 15 years from the act or omission complained of). However, this does not generally apply to contract claims unless a concurrent duty in tort is established. This distinction was upheld in *Platform Funding Ltd v Anderson & Assoc.* (2019) QBD, reaffirming that limitation periods must be carefully considered where claims could be brought in both contract and tort.

Limitation periods in certain situations have been extended to 30 years by the operation of the Building Safety Act 2022. See chapter 7.

Interim orders

Pending the final outcome of litigation, there are provisions which can be taken to prevent the defendant from destroying records or funds.

A **freezing order** (formerly known as a Mareva injunction, after the case *Mareva Compania Naviera SA v International Bulkcarriers SA* (1975) CA) freezes assets to prevent them from being disposed of, ensuring their availability in the event of a court judgment. If an injunction is granted, the defendant will be ordered to make a statement disclosing his or her assets. There will be court hearing with the opportunity for the defendant to argue that the order ought not to have been made or is too wide and should be limited in amount. Freezing orders are now governed by CPR Part 25 and Practice Direction 25A and are available in both domestic and worldwide form, depending on the location of the assets.

A **search and seize order** (formerly known as an Anton Piller order, after the case *Anton Piller KG v Manufacturing Processes Limited* (1975) CA) is a serious and even more aggressive injunction which can only be made in the High Court. These are often carried out in the early morning, known as a 'dawn raid'. The aim is to preserve evidence or property relevant to court proceedings. The defendant is ordered to let the applicant's solicitors enter the premises, search them for anything named in the order and to take photographs and samples. An expert will search computers. Often the defendant will also be ordered to disclose where things are. To obtain a search and seize order, you need an extremely strong *prima facie* case, very serious damages, clear evidence of possession of something incriminating, a real possibility of destruction of that thing and a likelihood that the harm caused by the order will be proportionate to the object of the order. Courts also now require that such applications comply strictly with the safeguards set out in the Practice Direction 25A and relevant case law, particularly the need for full and frank disclosure, minimal invasion of privacy and the supervision of the search by an independent solicitor acting as a supervising solicitor.

Interest on debts

Note that there is a statutory right to interest on overdue commercial debts (i.e. *not* when the debtor is a consumer), under the Late Payment of Commercial Debts (Interest) Act 1998 (currently 8% plus Bank of England base rate). Note that if there is a lesser rate of interest agreed by contract, then the express terms will be applied. The Act has since been supplemented by the Late Payment of Commercial Debts Regulations 2013, which also provide an automatic fixed sum for recovery costs (e.g. £40–£100 depending on the size of the debt) and allow reasonable further recovery costs to be claimed if the fixed sum does not cover the creditor's expenses.

Discussion questions

1. How do the courts distinguish between frustration and a mere increase in difficulty or cost of performance?
2. What determines whether a breach of contract is sufficiently serious to justify termination by the innocent party?
3. Should the remedy of specific performance be more widely available in modern contract disputes?
4. Has the Supreme Court's decision in *Cavendish v Makdessi* changed the law significantly on penalty clauses?
5. In what circumstances can a claim be brought on a quantum meruit basis, and what limitations are placed on the recovery?
6. Should damages for mental distress be more widely available in contract law?

6 Special contracts

Mark Allison

Introduction

This chapter covers the following special contracts:

- Agency
- Legislation relevant to estate agents
- Statutorily implied terms regarding goods and services
- Insurance

Agency

The agency relationship lies at the heart of commercial and business life. Agents are used to negotiate the sale of goods or services or to negotiate and conclude the sale of goods or services on behalf of a principal. The relationship of principal and agent arises whenever the principal authorises an agent to make contracts on his or her behalf. Generally, there will be a contract between the principal and the agent, with the latter being paid for his or her services.

Agency is usually created by express or implied agreement, but agreement may be lacking where the agent binds the principal under his or her apparent authority. The usual rule is that agents may only bind a principal in contracts which are within the scope of their authority.

Where the agency has been expressly created, the scope of the authority may have been clearly established. More often, however, it is a question of implying whatever authority is customary in the circumstances. So, for example, a board of directors that passes a resolution appointing a managing director of the company impliedly authorise that director to do all the things that fall within the usual scope of the office of managing director.

Where an agent is appointed for a particular purpose, he or she will also have authority to carry out any necessary incidental acts. Where the agency arises as part of a trade, profession or business, the agent has implied authority to perform such acts as are usual in that trade, profession or business. As a general rule, an estate agent has no implied authority to take a deposit in respect of the property to be sold, nor has he or she the right to sign a contract of sale.

Apparent authority

It is clear that agents can always bind principals if they make a contract of a type for which they have actual authority. Additionally, the acts of an agent will bind the principal if they

DOI: 10.1201/9781003657101-6

are within the scope of the agent's apparent authority. Simply, this means that an agent may have exceeded his or her real authority, but, in the eyes of the third party with whom he or she is negotiating, there was nothing to raise any doubts about the agent's authority to act and to bind the principal. In other words, there was the appearance of authority. The law is inclined to hold the principal liable for the acts of the 'agent' in these cases of apparent authority for reasons of equity and business confidence.

Although architects and engineers are frequently appointed under standard conditions of engagement which lay down in detail the duties to be carried out, the conditions rarely set out the extent of the agent's authority. The scope of their agency may thus depend upon implied or apparent authority.

In some cases where an agent has no authority to act but has done so, it is possible that the principal will subsequently ratify the transaction and thus become liable to the third party.

Powers of attorney

Wherever an agent needs to be given authority to execute deeds, e.g. where he or she is given power to sign a transfer of land, then the agent's authority must be conferred by deed. The deed conferring such authority is called a power of attorney.

Under the Powers of Attorney Act 1971, a general power of attorney must be made by deed and must be executed in accordance with that Act.

Duties of the parties

Where the agency relationship is created by a contract, the duties of the parties will be rather similar to the common law duties between an employer and an employee. The chief duties of the principal are to pay the agent the agreed commission or remuneration and to indemnify him or her against liabilities reasonably incurred when carrying out his or her duties.

Paying agreed commission or remuneration

Problems about when payment is due have arisen mainly in relation to estate agents' fees. This is not surprising when the various phrases used in their contracts are examined. Some contracts provide for the fee to be paid if the agent introduces a purchaser. Or the contract might provide for payment if the agent secures an offer. Sometimes the fee is only due on completion of the sale.

The courts had always been inclined to construe these commission clauses very strictly (*contra proferentem*), because it was thought that estate agents enjoyed some advantage over their clients, by being able to impose standard terms on them. Despite the passing of the Estate Agents Act 1979, this aspect of the law, about when commission is payable, was not clarified. Section 18 merely requires that the agent should give the client particulars of the circumstances in which the client will become liable to pay remuneration to the agent.

This area remains governed largely by the terms of the individual agreement and common law interpretation. Courts have continued to affirm that the entitlement to commission must arise clearly from the agreed terms. In *Devani v Wells* (2019), the Supreme Court held that even where an agreement to pay commission lacked some formal terms (such as the trigger for payment), it could still be enforceable if the essential terms could be implied and there was sufficient certainty.

Indemnity and lien

The principal must indemnify the agent against all liabilities and expenses reasonably incurred in the execution of his or her authority. An agent would not, however, be entitled to an indemnity in respect of illegal transactions, or in any situation where liability is caused by his or her own breach of duty.

An agent has a lien (a right to hold on to goods belonging to another) over property belonging to the principal until the principal settles all outstanding dues, such as remuneration and expenses, with the agent. Often, this will be a particular lien, where it will relate only to the property in respect of which the dues have been incurred.

Duties of the agent

The agent is bound to carry out duties with due care and skill, and the standard of care required will vary. If agents hold themselves out as being skilled in some particular field, then they must carry out the agency duties with the skill of a reasonable practitioner in that field.

The agent cannot normally delegate performance of the duties, unless the contract with the principal expressly or impliedly provides for delegation.

Whilst undertaking these duties, agents must not put themselves into a position where their own interests conflict with the duty they owe to a principal. Strict rules apply if agents have compromised themselves by making a secret profit from the transaction. Not only must the agent hand over the secret profit to the principal, but also, he or she is liable to be dismissed for such behaviour and would also lose the right to any commission earned on that transaction, and in some cases could be prosecuted under the criminal law.

Relationship between principal and third party

Once an agent has created a contract between the principal and the third party, his or her role is effectively over. The task has been to act as an intermediary, but in the normal course of events, the agent is not a party to the ultimate contract. That contract is made between the principal and the third party.

The rights of the principal to enforce the contract against the third party now depend on whether the existence of the principal was known to the third party when he or she was making the contract. If the third party was unaware that the agent was acting on behalf of a principal (i.e. the principal was undisclosed), then the agent may be personally liable or able to enforce the contract, subject to the usual exceptions.

As for the liability of the principal to the third party, he or she is fully liable under the contract. If the principal gives the agent money to settle with the third party, but the agent fails to do so, the principal is still liable to the third party.

Relationship between the agent and third party

As a general rule, once an agent has brought about a contract between the principal and the third party, their role is complete and they should not be liable on the concluded contract, nor should they be able to sue to enforce it. The agent's role is one of communication between the parties.

However, where the agent has contracted in his or her own name, or where the identity of the principal is not disclosed, the agent may become personally liable or acquire rights under the contract, depending on the circumstances and the intention of the parties.

Termination of the agency relationship

In many cases, the agency relationship exists within the framework of a contract, and where this is so, any method of terminating a contract will terminate the agency relationship. So, for example, it may be possible to discharge the relationship by either party giving notice.

A Lasting Power of Attorney (LPA) lets the donor appoint one or more people as attorneys to help make decisions on the donor's behalf. There are two types of LPA: health and welfare, and property and financial affairs. If the donor subsequently becomes incapable, for example, due to dementia, the authority of the attorney under the LPA is not terminated.

Under the Mental Capacity Act 2005, which governs LPAs, an LPA must be registered with the Office of the Public Guardian before it can be used. The attorney's powers continue even after the donor loses capacity, but are subject to legal duties, including acting in the donor's best interests and following the principles of the Act.

The Commercial Agents (Council Directive) Regulations 1993

These Regulations extend the common law rights of commercial agents against their principals. The Regulations only apply to a commercial agent, defined as a self-employed intermediary having continuing authority to sell, or negotiate the sale, of goods on behalf of a principal. The activities of an agent marketing only services, or engaged for a one-off transaction, are therefore not covered by the Regulations.

Although originally derived from EU law, the Regulations remain in force in the UK following Brexit, now forming part of retained UK law under the European Union (Withdrawal) Act 2018. UK courts have confirmed that pre-Brexit Court of Justice of the European Union (CJEU) case law interpreting the Directive remains binding unless and until overturned by a higher UK court or legislation.

Where they do apply, the Regulations include the following provisions:

1 The duties owed by agent to principal, and vice versa. These largely reflect the position at common law.
2 The remuneration payable to the agent in the absence of provisions in the agency agreement.
3 The agent's entitlement to commission, both during and after the agency agreement has been terminated, including due dates for payment.
4 Notice periods for terminating the agreement.
5 The agent's right to an indemnity or compensation payment following termination of the agreement.

The right to compensation or indemnity under Regulation 17 continues to be interpreted with reference to key case law such as *Lonsdale v Howard & Hallam Ltd* (2007) HL, which confirms that compensation is the default remedy in the UK unless an indemnity has been expressly agreed.

The Estate Agents Act 1979

The Act empowers the National Trading Standards Estate Agency Team (NTSEAT, run from Powys County Council) to prohibit an unfit person from undertaking certain or all estate agency work, as defined by the Act, or to warn this person that future breaches of obligations imposed by the Act will result in prohibition. Before exercising its powers under the Act, NTSEAT must instigate an inquiry into the person's fitness. It may instigate such an inquiry if the person has:

- been convicted of an offence involving fraud, dishonesty or violence, an offence under the Estate Agents Act or other specified offences;
- discriminated in the course of estate agency work;
- failed to comply with obligations imposed by the Act or
- carried on a specified undesirable practice.

The Act provides that a person who is bankrupt shall not engage in estate agency work except as the employee of another.

The Act requires estate agents to inform clients of the circumstances in which they will become liable to pay remuneration or make other payments (e.g. advertising costs and expenses). Clients must also be informed what the amount of remuneration or other payments will be or what the method of calculation will be.

The Act also requires estate agents to disclose personal interests that they have (or will have following the proposed transaction) in land, before entering into negotiation for its acquisition or disposal.

The Act lays down requirements in relation to clients' money held by an estate agent. Such money is held on trust for the person entitled to it. It cannot be used to pay off the estate agent's creditors if he or she becomes bankrupt. The Act requires an estate agent to keep properly audited accounts.

Clients' money must be kept in a special client account. Payments out of the account can only be made to persons and for purposes specified by the Act. For example, where an estate agent holds money belonging to the purchaser, he or she is entitled to make authorised payments to the vendor.

Unfair Trading and Misleading Marketing Regulations

The Consumer Protection from Unfair Trading Regulations 2008 prohibit all traders from using unfair commercial practices in their dealings with individual consumers and specifically, estate agents are prohibited from engaging in commercial practices that are unfair to sellers, buyers, potential sellers or potential buyers of residential property.

The Business Protection from Misleading Marketing Regulations 2008 prohibit traders in all sectors, including estate agents, from using misleading practices in their business-to-business advertisements, which catches misleading marketing used to advertise property for sale.

These provisions give rise to criminal liability and those agents in breach of the regulations could be at risk of prosecution resulting in a fine or imprisonment on conviction.

The Consumer Protection Regulations have been amended by the Consumer Protection from Unfair Trading (Amendment) Regulations 2014, which clarified the scope of misleading omissions and aggressive practices and introduced a limited right of redress for consumers where a misleading or aggressive practice has led to a transactional decision. Enforcement remains the responsibility of local authority Trading Standards and, in some cases, the Competition and Markets Authority.

Regulation of Property Agents Working Group

The government proposed that property agents be regulated by an independent regulator, with mandatory qualifications and a code of practice. The Regulation of Property Agents Working Group, chaired by Lord Best, was tasked with advising on this proposition. Their Final Report, published in July 2019, included proposals for the following:

1 The scope of a new system of regulation
2 A new licensing regime
3 A framework for codes of practice
4 Mandatory qualifications
5 Transparency and use of leasehold and freehold charges
6 The set-up, functions and relationships of a new regulator
7 Assurance and enforcement under the new system

Following the publication of the report, progress on implementing these recommendations was limited. In March 2024, the House of Lords Industry and Regulators Committee urged the government to publish a full response to the Working Group's report and to introduce legislation establishing a new regulator to improve standards in the sector.

In May 2024, the government acknowledged the need to address the regulation of property agents but indicated that other legislative priorities had delayed action. The subsequent General Election in July 2024 further postponed developments in this area.

In October 2024, the newly appointed Housing Minister, Matthew Pennycook, reaffirmed the government's commitment to implementing the Working Group's proposals, stating that reforming the sector included raising professional standards among agents to protect consumers and defend the reputation of good agents from the actions of rogue operatives.

As of April 2025, while the government has reiterated its intention to regulate property agents, no specific legislation has been introduced. The industry continues to anticipate further developments in this area.

Sale of goods

A contract for the sale of goods is defined by the Sale of Goods Act 1979 as one where: 'The seller transfers or agrees to transfer the property in goods to the buyer for a money consideration called the price'. The Act covers a wide range of transactions from selling loaves of bread to buying cars. It must be read in conjunction with all the basic rules of contract, as the Act is not a complete code in itself. Its significance lies in the fact that it sets out:

- the terms to be implied into a sale of goods transaction,
- the extent to which those implied terms can be excluded, subject to the Unfair Contract Terms Act 1977,

- the rules for determining when ownership and risk pass from the seller to the buyer in the absence of express or implied agreement and
- the remedies for the buyer and seller where breaches of contract have occurred.

The Act does not apply to hire purchase transactions, contracts for work and materials (e.g. installing central heating), or contracts for the supply of services, where separate statutory provisions apply.

For consumer contracts, the Sale of Goods Act 1979 has been largely superseded by the Consumer Rights Act 2015, which now provides the main statutory framework for contracts involving the sale of goods, digital content and services between traders and consumers. However, the 1979 Act continues to apply to business-to-business contracts and to consumer sales made before 1 October 2015.

Implied terms

It is quite common for people to enter into contracts to buy goods without ever speaking to the seller, e.g. when buying from a vending machine, or in a self-service store. If asked what the terms of their contract were, they might reply 'none'. However, such transactions are covered by the Act, and the following terms are automatically included for the buyer's protection.

In consumer transactions, these protections are now found in the Consumer Rights Act 2015. In business-to-business contracts, the relevant provisions remain those set out in the Sale of Goods Act 1979.

Quality of the goods

Section 14 provides that goods sold in the course of a business must be of satisfactory quality (i.e. fit to be sold). To be of satisfactory quality, the goods must meet the standard that a reasonable person would regard as satisfactory, bearing in mind any description applied to them, the price paid and other relevant circumstances. The Act goes on to say that quality includes the state and condition of the goods, whether the goods are fit for all purposes for which goods of the kind in question are commonly supplied, their appearance and finish, freedom from minor defects, safety and durability. This implied term is a condition of the contract, but does not apply if:

a Defects were specifically drawn to the buyer's attention before the contract was made. If a buyer buys a washing machine at a reduced price, having been told that the reduction is because the cabinet is dented, he cannot subsequently try to reject the washing machine because of this damage.
b The buyer examined the goods before the contract was made and the defects ought to have been revealed by that examination.

One of the practical problems of applying the rule about satisfactory quality is the question of how long the goods should remain of satisfactory quality. How long should a pair of shoes last? If they leak at the first wearing, clearly they would not seem to be of satisfactory quality. But what if they leak after six weeks? How often did the buyer wear them during that time? In what weather conditions? Whether goods are satisfactory is a question of fact and, in disputes, it would need to be answered on the individual circumstances. Expert evidence may be sought.

In consumer contracts, the equivalent protection is now found in s.9 of the Consumer Rights Act 2015, which requires goods to be of satisfactory quality, assessed by the same criteria. The provision continues to be interpreted with reference to case law under the Sale of Goods Act 1979.

Fitness for purpose

Section 14 further provides that where goods are sold in the course of business and the buyer makes known the purpose for which he or she requires the goods, then the goods must be reasonably fit for their purpose. What must a buyer do to 'make known the purpose' for which he or she wants the goods? In most cases, where goods have one common purpose, and that is what the buyer wants to use them for, then the answer is nothing in particular, as he or she is regarded as impliedly making known the purpose. It is only where the buyer wants to use goods for some unusual purpose that he or she must make that purpose known in order to benefit from this condition. The protection of this rule about fitness for purpose will not apply if the buyer did not rely on the seller's skill and judgment. This might be shown by the buyer specifically asking for goods by their brand name. In such a case, the inference is that the buyer knows that such goods will satisfy this purpose, i.e. by relying on his or her own skill and judgment.

One obvious advantage of these clear duties imposed on the seller of goods is that the buyer has rights against someone who can be identified and located. In practice, the seller may try to indicate that he or she is not liable and that the buyer will have to take the matter up with the manufacturer. The seller is under a strict liability to sell goods which are of satisfactory quality and reasonably fit for their purpose.

In cases covered by s.14 these conditions are important consumer protections, and they limit the operation of the usual principle of *caveat emptor* (let the buyer beware). That principle is still significant in cases not covered by s.14, e.g. private sales of second-hand goods, because s.14 specifically states that in circumstances outside the rules set out previously, there is *no* implied condition or warranty about the quality or fitness for any particular purpose of goods supplied under a contract of sale.

In contracts made with consumers, the equivalent duty is now found in s.10 of the Consumer Rights Act 2015, which provides that goods must be fit for the particular purpose for which the consumer requires them, provided that purpose is made known to the trader. This protection is subject to the same limitations, including whether the consumer relied on the trader's skill and judgment.

Sales by description

Section 13 provides that if goods are sold by description, there is an implied condition that they must correspond with that description. There is obviously a sale by description in any case in which the buyer has not seen the goods, but they have been described to him or her, e.g. goods selected from a catalogue or by reference to a price list. The Act specifically states that there can still be a sale by description even where the goods are displayed and selected by the buyer. However, although some of the older case law suggested that even a minor discrepancy between the contractual description and the goods (for example, in *Moore & Co v Landauer & Co* (1921) CA regarding how tins of peaches were packed), gave rise to a right to reject for breach of s.13, more modern case law suggests that what must be determined is whether a particular item in a description constitutes a substantial ingredient of the identity of the goods, and only if it does to treat it as a condition. Note also the effect of the provisions in s.15A of the Sale of Goods Act 1979 (see later on).

As regards consumer contracts, specifically, the equivalent provision is now found in s.11 of the Consumer Rights Act 2015, which similarly requires that goods match their description. The section continues to be interpreted in line with principles developed under s.13 of the Sale of Goods Act 1979.

Implied terms about title and right to sell

A person buying goods is naturally anxious to become the proper legal owner of them. Section 12 provides that there is an implied condition that the seller has the right to sell the goods, or will have when the time comes to pass ownership. The term normally not only relates to the title but also covers other reasons where there might be no right to sell, such as a prohibition on sale due to trademark issues (per *Niblett v Confectioners' Materials Co. Ltd* (1921) CA). This condition can never be excluded.

When goods are supplied to a consumer, this protection is now provided by s.17 of the Consumer Rights Act 2015, which states that the trader must have (or will have) the right to supply the goods at the time ownership is to pass. Like s.12 of the Sale of Goods Act 1979, this term is treated as a fundamental condition of the contract.

Sale by sample

The mere fact that a buyer sees a sample during the course of negotiations does not necessarily make the transaction a sale by sample. What is necessary is that the sale should be made by reference to that sample. The parties may well indicate this, e.g. when the buyer orders goods 'equal to sample'. In such a case, s.15 then implies conditions:

a that the goods supplied will correspond with the sample;
b that the goods will be free from any defect making their quality unsatisfactory, which would not be apparent on a reasonable examination of the sample.

Whilst ss.13, 14 and 15 imply conditions into contracts for the sale of goods, the remedy of rejection of goods for breach of such a term is curtailed in non-consumer transactions. Where a buyer does *not* deal as a consumer, s.15A states that if the breach is so slight that it would be unreasonable for a buyer of goods to reject them, then the breach is not to be treated as a breach of condition but as a breach of warranty.

Under a consumer-facing contract, the equivalent provision is found in s.13 of the Consumer Rights Act 2015, which implies similar conditions into contracts where goods are supplied by reference to a sample. The Act preserves the buyer's right to reject goods in the event of breach, without the limitation found in s.15A of the Sale of Goods Act 1979.

Implied terms about payment and delivery

Although the parties may not have negotiated specifically about any of the previous terms, it is likely that they will have thought about payment and delivery. This is reflected by s.10, which applies 'unless a different intention appears'.

Section 10 states that stipulations about the time of payment are not 'of the essence', which means that any agreement as to when money is due does not go to the root of the contract. In practice, a buyer who fails to make payment on time can be sued for damages for non-payment, but this behaviour will not allow the whole contract to be repudiated.

As to the time of delivery, s.29 makes it clear that if no time is fixed, delivery must be within a reasonable time if the seller is obliged to send the goods. It is important to note that there is no general rule that a seller must send the goods. Whether it is for the seller to send or the buyer to collect is a question to be determined by the contract. In other words, what was expressly or impliedly agreed between the parties? In many cases, the operation of these rules will be straightforward. Suppose that a buyer purchases a deep-freeze cabinet from a shop which promises, as part of the bargain, 'free delivery within 20 miles'. The buyer lives within the 20-mile radius and agrees on a day when the goods will be sent by the seller. In this case, by the terms of the contract, the seller is bound to send the goods. If he or she fails to send them on the appointed day, the provision about the delivery date may be a major term, the breach of which would entitle the buyer to repudiate the contract. Alternatively, the failure may be a breach of a minor term, merely entitling the buyer to sue for damages. The solution will depend on the weight attached to the term by the parties themselves. Their contract may make it clear that delivery dates are vital and must be complied with.

Unless otherwise agreed between the parties, delivery of the goods and payment of the price are concurrent conditions (s.28). It is very common practice to exclude the operation of this rule, e.g. in cases where goods are bought on credit and a bill is rendered at the end of the month.

In consumer contracts, similar rules apply under the Consumer Rights Act 2015. Section 28 provides that goods must be delivered without undue delay and, unless otherwise agreed, within 30 days. A failure to deliver within this period may entitle the consumer to reject the goods and terminate the contract. Provisions on payment timing are also governed by the general principles of contract law unless expressly modified by agreement.

Exclusion of implied terms

Notwithstanding the extensive 'writing-in' of terms into contracts for the sale of goods, the Sale of Goods Act (SGA) provides by s.55 that any or all of these implied terms may be excluded, subject to the Unfair Contract Terms Act 1977 (UCTA). When considering the relevance of the UCTA to the sale of goods transactions, a distinction must be drawn between those buyers who 'deal as consumer' and those who do not. The UCTA identifies dealing as a consumer occurring when:

1 A person dealing as a consumer *does not* make the contract in the course of business.
2 The seller *does* make the contract in the course of a business.
3 The sale is of goods of a type ordinarily supplied for private use or consumption (UCTA 1977, s.12).

The Unfair Contract Terms Act 1977 has been largely replaced by the Consumer Rights Act 2015 in contracts between consumers and suppliers. The relevant provisions govern the fairness of contract terms and consumer notices. Under the 2015 Act, terms that seek to exclude or restrict liability for the implied rights, such as those relating to quality, fitness for purpose, and conformity with description, are unenforceable. For non-consumer (i.e. business-to-business) transactions, exclusions and limitations continue to be governed by UCTA, with enforceability depending on the reasonableness test under s.11 of that Act.

One party dealing as consumer (UCTA 1977, s.6)

The rule in such cases is that there can be no exclusion of any of the terms implied by SGA 1979, ss.12–15 (i.e. no exclusion of the terms relating to title, sales by description, satisfactory quality, reasonable fitness for purpose or sales by sample). Any attempt by a seller to exclude or limit this liability in these areas will be ineffective.

Where the transaction involves a consumer, the position is now governed by the Consumer Rights Act 2015, which provides equivalent protections. Under that legislation, any attempt to limit or exclude the statutory rights relating to the conformity of goods with the contract, including quality, description and fitness for purpose, is automatically unenforceable (see ss.31 and 57 CRA 2015).

Neither party dealing as consumer (UCTA 1977, s.6)

Although no exclusion or limitation of SGA 1979, s.12 (title to the goods) is allowed, the seller may limit or exclude this liability under SGA 1979, ss.13–15, if the contract term by which the limitation is imposed satisfies the requirement of reasonableness. That is, the clause must be fair and reasonable when judged in light of the circumstances known, or reasonably foreseeable, to the parties at the time the contract was made.

UCTA 1977 contains guidelines for the application of the reasonableness test, some of which are set out in the following, which suggest that a court should consider:

- the relative strength of the bargaining position of each party;
- whether the buyer received an inducement to agree to the term;
- whether the buyer knew, or ought reasonably to have known, of the existence of the term and
- whether the goods were manufactured, processed or adapted to the special order of the buyer.

Passing ownership and risk

A contract for the sale of goods is defined by the Sale of Goods Act 1979 as one where the seller transfers or agrees to transfer the property in goods for a money consideration called the price (s.2). In this definition the word 'property' is used to mean ownership. It is important to realise that ownership can pass at a different time from possession. When making the contract the parties could agree to ownership passing immediately, but no delivery of the goods (i.e. transfer of possession) may take place until a later date. It is important to establish the exact time at which the buyer becomes the owner because:

1. The buyer can only effectively transfer ownership to a third party after becoming the owner.
2. As a general rule, risk passes with the ownership, irrespective of whether the goods have been delivered. This means that if goods ought to be insured, the buyer should do so, despite the fact that he has not yet received them. If the goods are lost or stolen

at this stage, the loss will fall on the buyer. But the seller who remains in possession is under a duty to take reasonable care of the goods.

The parties are free to fix for themselves when ownership will pass, but if their contract does not deal with this point, SGA 1979 provides rules to cover the various situations that can arise. The following points should be noted:

1 The time when ownership can pass depends on how the law categorises the goods in question. If they are unascertained (goods mixed with others and not yet separated or identified as fulfilling the buyer's contract) no property can pass until they become ascertained (goods identified in accordance with the contract after the contract is made). For example, if a builder orders a quantity of timber from a timber merchant who has huge stocks, the goods are unascertained until the merchant selects timber of the correct quantity and specification. Once he makes the selection, the goods become ascertained. Goods may also be specific. These are goods identified and agreed upon by the parties at the time when the contract is made.
2 Ownership of specific or ascertained goods passes at the time agreed by the parties. If they do not make their intention clear, the rules of SGA 1979 s.18 will be applied. These can be summarised as follows:

   a In the case of a sale of specific goods which are ready to be delivered, ownership passes when the contract is made.
   b When specific goods have to be put into a deliverable state (e.g. they must be repaired, cleaned, tested or weighed), ownership passes when this has been done and the buyer has notice to that effect.
   c In a sale of unascertained goods, ownership passes once appropriate goods are set aside in fulfilment of the particular contract, provided that at this stage the goods have become ascertained.

It will be seen that it is better for the parties themselves to fix when ownership will pass, as the operation of these rules leaves a buyer with little control over the timing.

The SGA 1979 s.19 allows sellers to retain ownership of goods until the buyer has paid for them, even though the seller has parted with possession of the goods. This can be particularly useful when a buyer becomes insolvent, as the seller can insist that he or she still owns the goods in question and can take them back. Otherwise, the seller would simply rank as an ordinary creditor in a bankruptcy or liquidation, with little hope of being paid in full. Reserving ownership of goods which are now in the buyer's possession is not without practical problems. One of the first significant cases on this point was *Aluminium Industrie Vaassen BV v Romalpa Aluminium* (1976) CA. Indeed, it was from this case that 'retention' clauses began to be known as Romalpa clauses. AIV had sold aluminium to Romalpa, on terms that ownership would only vest in Romalpa once the goods had been paid for; that if the foil was used in manufacture, the end product would belong to AIV; and that if this end product was then sold, the proceeds would be due to AIV. This clause was held to be valid and was enforced by the court. Their reasoning was based on the fact that Romalpa were not owners but merely bailees under a bailment of the foil. A bailment is a fiduciary relationship, a relationship of trust which would allow the bailor to recover goods and also the right to trace the proceeds of sale.

This case appeared to hold out important advantages for retention of ownership clauses, but it must be viewed in the context of later decisions, where the court reached a different conclusion. In *Borden UK Ltd v Scottish Timber Products Ltd* (1981) CA, the material supplied under the contract was resin to be used in the manufacture of chipboard. An 'ownership retention clause' was inserted. When Scottish Timber went into receivership, Borden wanted to trace their resin (which was now in manufactured chipboard) and trace the proceeds of the sale of chipboard which contained their product. They failed on both counts. There could be no bailment here as a bailment implies a right of redelivery. Borden was supplying small quantities at short, regular intervals and must have realised that their product was being incorporated immediately into the manufacturing process. If the clause was intended to create some kind of charge over the chipboard, it needed to be registered, and it was not.

The difference between the cases seems to rest on the fact that no mixture took place in Romalpa, whereas that was inevitable in Borden. Romalpa clauses can still be effective, as long as the seller makes clear that the buyer is a bailee and no mixing occurs; however, it should be noted that claims to trace sale proceeds in subsequent cases have not been successful. Simpler retention of title clauses, where a seller does not attempt to claim products manufactured from their materials or their proceeds of sale, tend to be more effective.

Where the transaction is between a trader and a consumer, the position is governed by the Consumer Rights Act 2015. Under ss.28 and 29, risk remains with the trader until the goods are delivered into the physical possession of the consumer or a person authorised to receive them.

Remedies

The remedies available for breach depend on which party is in breach and what type of breach has been committed:

1 *Seller in breach* – where the seller commits a major breach, e.g. by supplying goods which are defective, and not therefore of satisfactory quality, the buyer ordinarily has the right to repudiate the contract (put an end to it), usually by rejecting goods. The buyer also has the right to sue for damages. In practice, the buyer may choose not to reject the goods, and indeed in some cases the Act specifies that they cannot be rejected (see previously mentioned s.15A). The loss of the right to reject the goods by virtue of having accepted them (s.35 Sale of Goods Act 1979) is likely to happen when the buyer has purchased specific goods (goods which were identified by both parties at the time when the contract was made), where there was a reasonable chance to inspect the goods and where they have now been accepted. Acceptance can occur through express intimation or through keeping the goods beyond what is regarded as a reasonable time, without indicating rejection to the seller. In these cases, the right to reject is lost, and the buyer must be content with an action for damages.

   For consumer sales, the Consumer Rights Act 2015 offers a tiered system of remedies. Where goods fail to conform to the contract, the consumer is entitled to a short-term right to reject within 30 days (s.22), followed by a right to repair or replacement. If that fails, or is not provided within a reasonable time and without significant inconvenience, the consumer may claim a price reduction or final right to reject the goods (ss.23–24).

2 *Buyer in breach* – the most likely situation is that the buyer has failed to pay for the goods. The seller's remedy will then depend on whether he or she is still in possession

of the goods. If so, the right of lien can be exercised over the goods (i.e. holding on to them) with a view to ultimately exercising the right to resell them. If delivery has already been made to the buyer and ownership of the goods has passed to the buyer, then the seller may sue for the price of the goods. If ownership has not passed to the buyer, for example, because there is a reservation of ownership clause in the contract between seller and buyer, the requirements of which have not been fulfilled, then generally the seller may not sue for the price and will instead claim damages for non-acceptance.

Supply of goods and services

The Supply of Goods and Services Act 1982 (SGSA) relates to contracts under which ownership in goods will pass, where the contract also provides a service element. The Act also regulates contracts for services only. Obvious examples of contracts falling within the Act are builders who supply materials and do the work, and garages who supply parts and undertake the repair.

The Act deals separately with the supply of goods and the supply of services. Where the goods are supplied under a work and materials contract, a number of terms are implied:

1 An implied condition that the seller has a right to transfer the ownership of the goods (s.2).
2 Where goods are transferred by description, there is an implied condition that the goods must correspond with their description (s.3).
3 The goods transferred must be of satisfactory quality and reasonably fit for their purpose (s.4). These conditions apply only where the goods transferred are supplied in the course of business.

These conditions are couched in terms very similar to those used in the Sale of Goods Act 1979. Very similar terms are incorporated by ss.7–9 in contracts for the hire of goods. These are significant nowadays, as hiring cars, office machinery and heavy plant is very common in business.

As usual, the possibility of excluding these implied terms is limited by the operation of the Unfair Contract Terms Act 1977. Generally, in consumer transactions, these protective conditions cannot be excluded, but in business transactions, an exclusion clause may be acceptable if it satisfies the requirement of reasonableness under the 1977 Act.

Where the contract is between a trader and a consumer, the relevant protections are now provided by the Consumer Rights Act 2015. In such cases, Part 1 of the Act implies similar terms into contracts but uses updated language and introduces specific consumer remedies where these terms are breached.

Supply of services

The 1982 SGSA does not define what is meant by a service, except to make clear that the services provided by professions such as architects and surveyors are included. Obvious examples are activities like car servicing, boiler maintenance, dry cleaning and watch repairing. In any of these cases the customer will feel aggrieved if the job is not done effectively, if it costs too much or if it takes too long. In essence, those are the three areas where the Act now prescribes implied terms.

It is possible to exclude the terms implied by these sections, but again, only within the limits provided by the Unfair Contract Terms Act 1977. It should particularly be remembered that where, by negligence, a person causes death or personal injury, liability cannot be excluded or limited by a contract term (s.2).

For consumer contracts, the Consumer Rights Act 2015 now applies. Section 49 of the Act provides that services must be performed with reasonable care and skill. Section 52 adds that, where no time has been agreed, services must be performed within a reasonable time. Price must also be reasonable where not agreed in advance (s.51). These mirror the protections previously found in the 1982 Act and remain subject to similar rules on exclusion, fairness and reasonableness under Part 2 of the 2015 Act.

Exercising reasonable care and skill

Section 13 states that if a supplier of a service is acting in the course of a business, there is an implied term that the service must be carried out with reasonable care and skill. One interesting feature of this section is that it does not prescribe whether the term is a condition (major term) or a warranty (minor term). The standard laid down is apparently imprecise, but it means that a builder must exercise the care and skill of a reasonably competent builder, a surveyor must exercise the care and skill of a reasonably competent surveyor and so on. The standard is thus set objectively. It is no use asking whether the builder did his best, or even whether he tried reasonably hard, given his own limitations! He must be measured against a mythical, reasonably competent builder.

Section 13 merely codifies the existing common law rules, i.e. similar duties would previously have been implied into such contracts by the courts. As soon as the idea of 'reasonable care and skill' is introduced, it is clear that there is the possibility of a claim in the tort of negligence, too. Usually, it is more straightforward to sue in contract, as the relevant duty is owed by virtue of the contract and need not be specifically proved.

For agreements involving consumers, the relevant rule is now found in s.49 of the Consumer Rights Act 2015. This section restates the same obligation that services must be performed with reasonable care and skill and applies it specifically within the consumer context. While the wording is similar to the 1982 Act, it forms part of a broader statutory scheme that includes additional remedies and protections tailored to consumer transactions.

The time for performance

Section 14 provides that there is an implied term that a supplier acting in the course of business, where no time is fixed by the contract, must perform the service within a reasonable time. Establishing what is a reasonable time is a question of fact, which can sometimes be implied from the nature of the contract. For example, if a firm is engaged to come and spray fruit trees when they are in blossom, there is only a short span of time within which the contract can be performed.

In contracts with consumers, the same expectation is reflected in s.52 of the Consumer Rights Act 2015. Where no specific timeframe has been agreed, the service must be completed within a reasonable period. The standard is still fact-dependent, but the Act provides a clearer basis for redress where delays occur.

The cost of the service

Section 15 provides that where the contract does not fix the charge or provide a method of fixing it, the customer must pay a reasonable charge. What is reasonable will be a question of fact.

This principle is echoed in s.51 of the Consumer Rights Act 2015, which states that if the price is not agreed in advance, the consumer is required to pay only what is reasonable. As with time for performance, the assessment depends on the nature of the service and the expectations of a typical consumer in that context.

Hire purchase, credit sales and conditional sales

It is often the case that a buyer can only afford to purchase goods if payment can be spread over a period. Historically, hire purchase developed at the end of the nineteenth century as a means of allowing people to buy more expensive durable goods like pianos and sewing machines by spreading payments out over a period, and yet achieving a measure of protection for the seller who remained the owner of the goods until the last payment was made. If a buyer fell into arrears with payments, the seller could 'snatch back' the goods. Where a buyer was found to be over-committed financially, he or she could usually terminate the agreement, but often on terms which were disadvantageous. Recognising these evils and the potential inequality of bargaining power between the buyer and the seller in such transactions, Parliament passed a series of Hire Purchase Acts, the rules from which are now consolidated in the Consumer Credit Act 1974.

A number of different ways exist for a customer to buy on credit. Traditionally the law provided several types of transactions, which give flexibility about the terms incorporated and which vary in the protections afforded to the customer. The supplier could choose to use the following:

1 A hire purchase agreement, where the customer hires goods for a period, with the opportunity to exercise an option to purchase at the end of the period. In this type of agreement, the supplier remains the owner of the goods until the option is exercised. The hirer may terminate the hiring agreement without exercising the option, although this would not usually be wise, as it would prove to be a costly way to hire goods. As the goods remain in the ownership of the supplier, he or she has certain rights to seize back the goods if the hirer defaults on payment. In this type of transaction, terms about satisfactory quality and reasonable fitness for purpose are implied in a form similar to the implied terms of the Sale of Goods Act 1979. Hire purchase agreements are controlled either by the Consumer Credit Act 1974 or by the rules of common law.
2 A credit sale agreement. These extend credit to the buyer of goods, e.g. where it has been arranged to pay by instalments, but the ownership in the goods passes to the buyer immediately. They are governed mainly by the Sale of Goods Act 1979.
3 A conditional sale agreement, where ownership of the goods remains vested in the seller, but the buyer has no right to terminate the agreement. Because credit is extended to the customer, the Consumer Credit Act 1974 may apply.

Of course, these types of contract are not the only ways in which credit may be given. There has been a dramatic rise in the use of credit cards – either those issued by particular stores for use in their own shops, or those issued by credit card companies for use in any accredited outlet of the company. In either of these cases the sum owed can represent many different contracts made by the debtor. Sometimes, by the terms of his or her agreement with the credit card company, the debtor must repay at regular monthly intervals and may not exceed a fixed credit limit. In other cases, the debtor may owe up to a fixed credit limit and may choose whether to settle in full at the end of the month, in which case no interest is paid, or part of what is owed may be paid and interest run up on the balance.

Inevitably, most organisations which are prepared to extend credit do so only for a return. That return is the interest they charge. The law has long believed that rules are needed to protect people from extortionate bargains, and such rules are all the more vital when seen as part of 'consumer protection'. The rules were dramatically altered by the introduction of the Consumer Credit Act 1974. As its name implies, the Act is of special significance to 'consumers', and the protections of the Act range from licensing people and organisations which offer credit to insisting on sufficiently full information being available, so that a customer can establish the cost of the credit. To give the Act 'real teeth', there are criminal as well as civil sanctions.

For the business community, making contracts to purchase supplies or equipment, the Act may be of limited significance. Its main provisions only operate where the credit is extended to an individual or other unincorporated body. Formerly, only transactions below a limit of £25,000 were covered, but this limit was removed with the Consumer Credit Act 2006. Where the Act applies, it can offer vital protections, including a key provision in s.75, which makes the credit provider jointly liable with the supplier where goods or services fail to meet implied terms such as satisfactory quality or fitness for purpose. This can be especially important when the original supplier is insolvent, unreachable, or otherwise not worth pursuing.

Consumer Rights

Much of the chapter has already referred to the Consumer Rights Act 2015, which now provides the central legal framework for contracts between a trader and a consumer. This legislation brought together and replaced several earlier statutes, including parts of the Sale of Goods Act 1979, the Supply of Goods and Services Act 1982 and the Unfair Terms in Consumer Contracts Regulations 1999, with the aim of simplifying and strengthening consumer protection law.

The Act introduced a number of significant changes and clarifications, including:

- clearer rules on what happens when goods are faulty, including tiered remedies such as the short-term right to reject, repair or replacement and final rejection,
- legal standards for the supply of services – including that services must be provided with reasonable care and skill, within a reasonable time and for a reasonable price (if not agreed),
- new statutory rights and remedies for digital content, such as downloads, streamed media and e-books,
- a single framework for assessing and challenging unfair terms in consumer contracts and notices,
- rules ensuring that letting agents in England display a full tariff of their fees,

- greater powers for public enforcers such as Trading Standards, including the ability to seek redress for affected consumers and the requirement to give written notice for routine inspections, and
- provisions addressing anti-competitive practices by businesses, alongside consumer-specific enforcement mechanisms.

The Act applies only where a trader is supplying to a consumer, and its terms are non-excludable, meaning the statutory rights it provides cannot be overridden by contract. It also introduced a clearer system of remedies, helping consumers understand not only what their rights are, but what happens when those rights are breached.

Other methods of customer protection

Many of the protections discussed in this chapter can only be fully effective if a person is prepared to bring a court action to enforce his or her rights. But there is often fear of the cost of litigation and, especially with smaller cases, recognition that the time, trouble and effort involved is disproportionate to the size of the claim. In recognition of these barriers, an important trend in recent years has been the growth of alternative dispute resolution (ADR) mechanisms (see Chapter 2).

An interesting trend in customer protection is the establishment of ombudsmen, including the Property Ombudsman and the Financial Ombudsman Service. An ombudsman is an independent official who has been appointed to investigate complaints that people make against organisations and, if justified, make recommendations for redress. In many schemes, these recommendations are binding on the business concerned, but not on the customer, who retains the right to pursue the matter in court if dissatisfied.

However, the best customer protection comes from dealing with trustworthy organisations with high standards who wish to maintain their reputation.

Insurance contracts

When a businessperson considers the risks inherent in running a business (e.g. that workers may injure third parties, the workers themselves may be injured, business premises may burn, a customer may be injured by a defective product), there are clearly potential claims against the business. It is true that some organisations are sufficiently large to bear their own risks. This tends to be the case only where the insurance premiums charged are so high that these large organisations find it more economical to meet claims, if and when they arise, out of their own funds.

In all normal cases a firm would be wise to secure insurance cover, and in some circumstances the law makes insurance compulsory, e.g. employers' liability and motor insurance.

Two vital principles are fundamental to all insurance contracts.

Utmost good faith

A person who seeks insurance usually knows all the relevant facts. On the basis of what is revealed the insurers can decide whether to accept the risk (i.e. give insurance cover) and how much to charge by way of a premium. Because of this inequality in the bargaining position of the parties, the law imposes a duty on a person seeking insurance to disclose all the material facts to the insurer. The duty of utmost good faith (*uberrimae fidei*) only requires disclosure of facts which are material and which are known, or ought to have been known, by the insured.

a *When are facts material?* The answer seems to be when they would influence a prudent insurer as to whether or not he will accept the risk, or as to the amount of the premium. With motor insurance, previous motoring convictions are obviously material. In the case of most types of insurance the fact that another insurer has refused the risk is material. In life insurance, previous illnesses may be material.
b *Knowledge of the facts.* The duty on the insured is quite strict. He or she must reveal facts that are known or presumed to be known. This duty cannot be avoided by evidence of ignorance. However, if the facts within the knowledge of the insured actually reduced the risk, or were matters of common knowledge (such that the insurer would be presumed to know them), no disclosure was necessary.

These rules about disclosure recognise the possibility that an insured party may be silent despite being under a duty to speak. They are an exception to the general contract rule that silence does not usually amount to misrepresentation. This is normally the case because of the principle of *caveat emptor* (let the buyer beware). Usually, it is up to the parties to negotiate their own bargain and if one of them requires information, he or she must ask the other for it. The law does not generally require the other party to volunteer it. Of course, a person seeking insurance may have made positive statements which turn out to be false. In such a case the normal rules of misrepresentation would operate and the insurance company's remedy would depend on the type of misrepresentation, which can be innocent, negligent or fraudulent.

However, the duty of utmost good faith has now been modified by the Insurance Act 2015, which applies to most contracts of insurance entered into after 12 August 2016. Under this Act, the insured must make a fair presentation of the risk, disclosing material facts in a manner that is reasonably clear and accessible. The Act softens the strictness of the former rule by introducing proportionate remedies, depending on whether the non-disclosure or misrepresentation was deliberate, reckless or simply careless.

Where there has been a serious failure to disclose material information, the insurer may still avoid the policy, but only in more limited circumstances. It is no longer the case that the insurer can avoid the policy merely because a fact was material and undisclosed, regardless of fault. The requirement that the insured must have known, or ought to have known, the fact remains, but it is now judged in a fairer, more structured way.

The old law could produce harsh results. In *Woolcott v Sun Alliance of London Insurance* (1978) CA, for example, the plaintiff did not complete the proposal form himself. He was buying a house with the aid of a mortgage from a building society, whose mortgage application form indicated that the building society would insure the property. The house was subsequently damaged by fire. When the plaintiff sought to claim on the block policy held by the building society, the insurers avoided the policy and refused to pay. The reason was that the plaintiff had been convicted of robbery in 1960 and sentenced to 12 years in prison, a fact the insurers deemed material, but which had not been disclosed. Under the modern rules, the outcome in such a case might differ, depending on the insured's actual knowledge and the fairness of the presentation made.

Insurable interest

If people could insure goods or lives in which they had no legitimate interest, the result would be a wager or gambling agreement. It is a necessary ingredient of an insurance contract that the insured must have an insurable interest, i.e. some foreseeable financial loss or liability.

a *Life insurance* – it is accepted that everyone has an insurable interest in their own life and a person may be insured against death or personal accident to an unlimited extent, so long as the premiums can be paid. In certain circumstances, a person may have an insurable interest in the life of someone else, e.g. a wife has an unlimited insurable interest in her husband and vice versa; a creditor has an insurable interest in the life of a debtor, up to the amount of the debt.
b *Property insurance* – the owner of property has an unlimited insurable interest up to its full value. In other cases, where a person is not the owner, the law allows insurance to the extent of that person's interest or liability. For example, a building society could insure a building for the amount loaned on a mortgage. Where goods are concerned a buyer can insure before delivery is taken or warehouse workers can insure other people's goods entrusted to them for storage.
c *Liability insurance* – where a person can foresee liability (e.g. arising out of their occupation of premises or use of a vehicle), he or she has an insurable interest. The law does not allow insurance against criminal liability, as that would be contrary to public policy. So, a motorist cannot insure against being fined for careless driving.

The insurance contract

Subject to the two principles discussed previously, the usual rules about the formation of a contract apply. The following points should be noted:

1 *Agreement* – this usually results from the person seeking insurance completing a proposal form (the offer), which is then accepted by the insurers.
2 *Consideration* – insurance contracts are not usually made by deed, so consideration is necessary. This usually consists of paying a premium in return for the promise of a payment out whenever the risk occurs.
3 *Form of the contract* – as a general rule, contracts of insurance do not need to be in writing. An oral contract could be valid and enforceable. In practice it is usual to record the terms of the agreement in a policy. Should any dispute arise over the risks which are covered, the words of the policy must be construed (i.e. interpreted) to see whether they cover the facts of the case. It is usually said that the words of a policy should be read so as to give effect to the intention of the parties. This suggests that both play a part in drawing up the policy, whereas the truth is that it will usually be a standard form contract drawn up by the insurers. Mindful of this, the courts will construe any ambiguity against the insurers. The Consumer Rights Act 2015 applies to insurance (whereas the Unfair Contract Terms Act 1977 did not).
4 *Terms of the contract* – one feature of insurance contracts is that all terms, major or minor, tend to be called warranties. This is particularly confusing because of other meanings which the word can carry, e.g. in the sense of a guarantee. In any contract situation it is always a difficult question to determine whether a statement which has

been made is so important that it becomes part of the contract (i.e. becomes a term), or is of lesser importance so that it is merely a representation. Persons seeking insurance may make a number of statements in answer to questions on the proposal form, e.g. stating that they have no motoring convictions or that they have never had an operation. Insurers have been in the habit of making such statements form the basis of the contract, by including a note on the proposal form (which is then signed by the insured) to the effect that all the answers will be regarded as warranties. This practice puts the insurers into a most powerful position in relation to the insured. Should there be anything wrongly answered on the form, then, even if the insured has been acting honestly, and even if it is immaterial, an insurer can avoid liability on the whole contract by treating the error as a breach of warranty.

This strict position has now been modified by the Insurance Act 2015, which provides that a breach of warranty no longer automatically discharges the insurer from liability. Instead, the insurer's remedy will depend on the nature and effect of the breach and, in particular, whether it was connected to the loss. This more proportionate approach limits the harsh consequences that once followed from technical or innocent errors.

5 *Duration of the policy* – often an insurance policy runs for a fixed period, usually a year. There are cases where insurance relates to a specific event (e.g. bad weather insurance for a country show) and the protection lapses once the date of the event has passed. Where the policy is of a continuing type, e.g. life assurance, the insurer must go on accepting the premiums even though the risk inevitably becomes greater year by year. With other types of policy, e.g. motor insurance, the expectation is that it will be renewed, but neither party is obliged to make a fresh contract. If renewal does occur, then the duty to disclose material facts becomes relevant again, because each renewal creates a fresh contract.

Types of insurance

Employers' liability insurance

Although sensible employers have always tended to have insurance cover for protection against claims by injured workers, problems have arisen in the past when some employers had no such insurance and were unable to meet claims for damages from their employees because they had become insolvent. Although an employee injured in an accident at work may have claims under the industrial injuries scheme, the law also provides that damages can be claimed from a negligent employer. That claim is now protected by the requirement of compulsory insurance imposed by the Employers' Liability (Compulsory Insurance) Act 1969. The following main points about the Act should be noted:

a All employers (except local authorities, statutory corporations and nationalised industries) are covered by the rules.
b Employers must maintain insurance against liability for injury or disease sustained by employees in the course of their employment.
c The level of cover required is £5 million in respect of claims from one or more employees arising from any one occurrence.
d The insurance must be made with an authorised insurer. This term is defined by regulations.

e The employer's duty is to maintain an 'approved policy'. The effect of this rule is to make it illegal for the insurer to limit its liability in certain ways.
f The rules are enforced by criminal sanctions. The Act provides for fines of up to £2,500 for failure to comply. In practice, updated penalties under delegated legislation may now exceed this amount.
g The insurance certificate must be displayed and, if required, produced to an inspector of the Health and Safety Executive.

Limitations on the cover provided under the Act are as follows:

a It only protects employees who work under a contract of service or apprenticeship.
b Only those incidents arising 'out of and in the course of the employment' are covered. The same phrase is used in the industrial injuries scheme (for Industrial Injuries Disablement Benefit), where its interpretation and application have not been without difficulties.

Public liability insurance

The employers' liability insurance rules only oblige an employer to arrange insurance cover against injury to its employees. Liability may be owed to a third party who is injured by some activity of one of the employees, e.g. if one of the labourers negligently hits a pedestrian with a plank of wood. In such a case, the principle of vicarious liability means that the pedestrian is likely to bring an action for negligence against the employer.

It is not compulsory for an employer to insure against this type of liability (usually called public liability insurance), although it is very common in practice. Despite not being required by law, such insurance is often contractually required, particularly in the construction industry and other sectors involving work in public spaces.

It is not only employers who need public liability cover. Anyone may cause death, injury or damage while going about his or her daily life. For example, a pedestrian may cause a motorist to swerve and crash, while a property-owner could be liable if a chimney were to fall on a passerby. However, public liability cover is often included within broader insurance policies, such as under a householder's or occupier's policy.

Insurance against damage to property

This type of insurance usually covers fire damage but will often extend to damage caused by explosion or lightning. The policy may then be further extended to cover other risks to property caused by, for example, storm, flood or impact. This type of insurance is not compulsory by law, and it is essential that a policyholder ensure that all the desired protection is included in the policy. For example, storm and flood coverage has become a contentious issue, particularly in areas prone to climate-related events.

It is now quite common to extend fire insurance protection to cover consequential loss. Should a fire render a building unusable, a firm's business activities could be severely disrupted. Insurance cover can be provided to cover loss of profits and additional expenses necessarily incurred, e.g. renting alternative property.

Contractors' all risks insurance covers physical damage to the materials used for construction. The policy should be in the joint names of the employer and contractor, as is now commonly required under standard building contract conditions.

Professional indemnity insurance

Professional indemnity insurance provides cover for awards of damages or settlements for which a professional practice, for example, an architect or surveyor, is liable, resulting from a claim made against them for any act, error or omission arising out of the conduct of their business. The extent of the business activities should be detailed in the policy, as cover will usually only apply to work carried out within the declared scope of professional services.

Motor insurance

In the case of motor insurance, certain aspects of insurance cover are compulsory by law. As a minimum a motorist must have a policy which covers the following:

- Liability for death or personal injury to third parties, including passengers or damage to property caused by or arising out of the use of the vehicle.
- Hospital charges (up to a fixed maximum) in respect of the treatment of an injured party and emergency treatment fees.

It is unusual for such limited cover to be taken out by a motorist. He or she would normally also want cover against the following:

- Fire and theft in relation to his or her own vehicle
- Damage to the vehicle and to the person

The insurance companies may not find motor insurance to be profitable. Premiums charged naturally vary with the level of assessed risk, and most insurers will take into account the type of vehicle, its value, the age and experience of the driver, the locality in which he or she lives, the driver's claims record and any driving convictions.

When a third party is injured by an uninsured motorist, or a 'hit and run' driver, the victim may claim from the Motor Insurers' Bureau (MIB). Every insurer that provides motor cover in the UK must be a member of the MIB and contribute to the fund used to meet such claims. This scheme ensures that innocent victims are not left without redress where no valid policy is in place.

Claims under an insurance policy

Once the need to make a claim under a policy arises, the question of the construction of the contract becomes important. Is the event which has occurred actually covered by the policy? Naturally the insurers will have based their premium on a fixed amount of cover to be provided, and they will not wish to pay out more than is necessary.

If the loss which has occurred is covered by the policy, then, in the case of most types of insurance (except life and personal accident), the principle of indemnity operates. This means that the amount paid out by an insurer should be sufficient to cover the loss sustained, but should not allow the insured person to make a profit. However, even in those cases where the indemnity principle is applicable, the following restrictions on it should be noted:

1 The policy itself may provide for a maximum sum insured. This can provide a complete indemnity if the sum insured is adequate, but a property may have been under-insured,

and the policy may contain an average clause which will then operate. The object of such a clause is to 'penalise' the person who has under-insured, since this practice is unfair from the insurer's point of view, preventing a person from paying a proper premium in return for the risk that is borne. Of course, under-insurance is not necessarily caused by deceitful motives. A person may genuinely fail to realise that the rebuilding of a property will cost much more now because of inflation than when it was purchased some years earlier. Generally, average clauses work on a pro rata basis. If you under-insure by 50%, your claim will be settled on the basis of 50% of the loss sustained. For example, a house which could cost £100,000 to rebuild at present-day prices is insured for £50,000. If it is totally destroyed by fire, £50,000 is the maximum payable. If damage amounting to £25,000 is caused (one-quarter of the true insurance value) the insured will be paid one-quarter of the sum insured, i.e. £12,500.

2 The policy itself may be subject to an excess. It is quite usual for the contract to specify that the insured will bear an agreed proportion of each claim, for example, the first £250. In such cases, the insurance does not provide a full indemnity.

3 An insurer may not be wholly liable if the property has been insured with more than one company. In such a case the contribution principle will operate, whereby other insurers who are covering the same risk share in the cost of paying out. The indemnity principle is thus preserved, as the total settlement figure will only cover the insured's loss and not allow this person to make a profit. As with indemnity, the contribution principle is not appropriate in the case of life insurance. Double insurance consists of covering the same risk by more than one policy. This is not the same as reinsurance. As soon as an insurer has taken on a risk, there is always the likelihood that it may have to pay out. In cases where a single accident may be very costly, e.g. the loss of a telecommunications satellite, the insurer may wish to spread the risk. In such a case, it can reinsure part or all of the risk with another insurer. If the risk then occurs, the insured claims against the original insurer in full. The original insurer then claims under its own reinsurance policy.

4 When the risk occurs, an insured person may have a right of action for damages against the person responsible for the injury or for damaging property, as well as the possibility of claiming under the offending person's insurance policy. If a motorist has a vehicle negligently damaged by another motorist, the subrogation principle will apply. If the victim decides to claim from his or her insurers, they may, in turn, claim against the negligent motorist. The practical effect of this would be that one insurance company would be claiming from another. Given the thousands of claims which they deal with, this process would be expensive and time-consuming. It is, therefore, common for insurers to make agreements amongst themselves not to claim from one another under the subrogation rules. In the long run, given the amount of business involved and the administrative savings made, this will usually be a perfectly equitable arrangement. This type of agreement between insurance companies results in the 'knock for knock' arrangements found in motor insurance. The practice does not affect the rights of the insured, who still receives compensation as per the terms of the policy.

Discussion questions

1 When can a principal be bound by the actions of an agent who has acted without actual authority, and why might the law choose to uphold such arrangements?

2 In what ways do the Consumer Rights Act 2015 and the Sale of Goods Act 1979 protect buyers, and how do these protections differ depending on whether the buyer is a consumer, is in business or when both parties are dealing in their private capacity?
3 Why might a term in a contract for the sale of goods be treated as a condition in one case but not in another? What practical consequences follow from this distinction?
4 How has the Insurance Act 2015 changed the way courts and insurers approach non-disclosure and breaches of warranty?
5 Why is it important for contractors and professionals working in the built environment to maintain adequate insurance cover, even when not required by law?
6 In what ways does the law attempt to balance fairness between insurers and the insured when claims arise, particularly where indemnity, excesses or under-insurance are involved?

7 Construction contracts

Jennifer Charlson

Standard forms of contract

It would be uneconomic to draft complex new terms for every construction contract, so standard forms of contract have developed in the construction industry. These have been subject to interpretation by the courts, and updated versions, responding to changes in practice and the law, are issued regularly. A further advantage of standard forms of contracts is that regular users are familiar with their provisions.

The primary issues that a construction contract usually incorporates are:

- the scope of work undertaken by the contractor and the procedure for variations;
- the time for completion and interim control of the progress of the works;
- the mechanism for payment of the contractor;
- supervision of the works on behalf of the employer;
- insurance against a range of risks and
- dispute resolution provisions.

The construction industry relies on standard forms of contract, which should achieve a fair allocation of risk between contractors and clients. Nevertheless, a standard form of contract needs to be customised to a specific project, and often the contract is amended to change the balance of risk.

The following sections detail current Joint Contracts Tribunal (JCT), NEC (formerly New Engineering Contract) and Royal Institute of British Architects (RIBA) forms of contract and introduce Association of Consultant Architects (ACA), Infrastructure Conditions of Contract (ICC) and FIDIC (International Federation of Consulting Engineers) contracts.

Joint Contracts Tribunal

Since 1931, the JCT has produced standard forms of contract and guidance notes for use by the construction industry. The JCT is made up of eight members who represent a wide range of interests from the construction industry: the British Property Federation, Build UK Group, the Contractors Legal Group, Local Government Association, the Royal Institute of British Architects, the Royal Institution of Chartered Surveyors and the Scottish Building Committee. JCT develops, publishes, revises and disseminates suites of standard forms of contract, including main and subcontracts and designer agreements. The current suite is the JCT2024 Edition.

DOI: 10.1201/9781003657101-7

The JCT Design and Build Contract is often used for large and complex construction projects, including sports stadia, shopping centres and office blocks. It is used by both the public and private sectors. The current edition is explained as an example JCT main contract.

The JCT Design and Build Contract 2024 (DB 2024) can be used when the Contractor completes the design and then carries out and completes the Works based on the Employer's Requirements. The Employer employs an agent to administer the conditions. The DB 2024 comprises the agreement, recitals, articles, contract particulars and attestation, which customise the contract for the project and conditions which can be used for every project.

The project-specific details include the Employer's Requirements, the Works, the Contractor's Proposals, the Contract Sum, Date for Completion of the Works, Liquidated Damages and Retention Percentage. Each section of the conditions is identified here:

1 *Definitions and Interpretation*.
2 *Carrying out the Works*. The Contractor's obligations are detailed, including completing the design for the Works and the required standards for materials, goods and workmanship. The Contractor must give notice to the Employer of any Relevant Events and the Employer must then give a fair and reasonable extension of time by fixing a later Completion Date. The Contractor will be liable to the Employer for damages for delay, often liquidated damages, if practical completion is achieved later than the adjusted Completion Date. The Contractor is responsible for remedying defects during the Rectification Period.
3 *Control of the Works*. The Contractor shall not subcontract the design or the Works without the Employer's consent. The Contractor must immediately comply with the Employer's instructions. Each party agrees to comply with the applicable CDM Regulations.
4 *Payment*. The procedure for interim payments and notices is detailed. The Employer is entitled to deduct and retain a specified percentage of the money due to the Contractor. The Contractor is entitled to reimbursement of loss and expense incurred as a consequence of the Employer's actions. However, the Contractor should notify the Employer of Relevant Matters. Following practical completion, the Contractor submits a Final Statement and the Employer makes the final payment. The Contractor is entitled to suspend performance for non-payment by the Employer.
5 *Changes*. The Valuation Rules for measurable work and daywork are detailed.
6 *Injury, Damage and Insurance*. The Property, Personal Injury and Professional Indemnity Insurance and Joint Fire Code provisions are set out.
7 *Assignment, Performance Bonds and Guarantees, Third-Party Rights and Collateral Warranties*. Performance Bonds and Guarantees provide financial security to the Employer. Third-Party Rights and Collateral Warranties provide contractual rights to Purchasers, Tenants and Funders against the Contractor and subcontractors.
8 *Termination*. The grounds and procedure for termination by the Employer or the Contractor are described.
9 *Settlement of Disputes*. The dispute resolution mechanisms recognised include mediation, adjudication and arbitration. Litigation may be selected as an alternative to arbitration.

JCT contracts have stood the test of time and are well-established and understood.

New Engineering Contract

The first edition of the NEC was published in 1993. NEC is a wholly owned subsidiary of the Institution of Civil Engineers. This suite of standard forms of contract adopted a new and simple drafting style to promote a high standard of project management across the construction industry. NEC argues that each of its suite of standard contracts has the following characteristics:

- Its use stimulates good management of the relationship between the two parties to the contract and, hence, of the work included in the contract.
- It can be used in a wide variety of commercial situations, for a wide variety of types of work and in any location.
- It is a clear and simple document, using language and a structure that are straightforward and easily understood.

Sir Michael Latham (1994) recommended that the NEC suite should become the national standard for both private and public sectors. The Engineering and Construction Contract has been used on high-profile projects including the Channel Tunnel Rail Link, London Olympics 2012 and Crossrail.

The NEC4, the current edition of The Engineering and Construction Contract (EEC), is outlined as an example. This contract comprises Core, Main, Secondary and Dispute Option Clauses and Contract Data to be provided by the Client and Contractor. The Core Clauses are:

1 *General.* The obligation to act in 'a spirit of mutual trust and co-operation' is the first clause. Terms are then identified and defined. Early warning is to be notified, registered and meetings held.
2 *The Contractor's main responsibilities.* The Contractor is to provide the Works and design parts of the Works in accordance with the Scope. The Contractor submits the name of each proposed Subcontractor to the Project Manager for acceptance. The Contractor obeys instructions given by the Project Manager or Supervisor.
3 *Time.* The Contractor is to do the work so that the Completion Date is achieved. The Contractor is to submit detailed programmes to the Project Manager for acceptance. Acceleration can be proposed.
4 *Quality Management.* The Contractor is to operate a quality management system which complies with the requirements stated in the Scope. The Contractor is to correct Defects.
5 *Payment.* The Contractor submits an application for payment and the Project Manager then assesses the amount due at each assessment date. The Project Manager certifies a payment within one week, and payment is made within three weeks of each assessment date. The Project Manager makes an assessment of the final amount due and certifies a final payment.
6 *Compensation events.* Compensation events are listed and the mechanism for their notification is set out. The Project Manager may instruct the Contractor to submit quotations for compensation events or proposed instructions. The Project Manager assesses compensation events.
7 *Title.* The provisions for the passing of title of Plant and Materials to the Client are described.

8 *Liabilities and insurance.* The Client's and Contractor's liabilities and insurance requirements are identified.
9 *Termination.* The termination reasons, procedures and amounts due are detailed.

One pricing mechanism needs to be selected from the following main options:

- Option A: priced contract with activity schedule
- Option B: priced contract with bill of quantities
- Option C: target contract with activity schedule
- Option D: target contract with bill of quantities
- Option E: cost reimbursable contract
- Option F: management contract

Twenty-three Secondary Options are offered for consideration, but it is not necessary to use any of them. Two of the more significant options are:

- X7 *Delay damages.* A daily liquidated damages rate for delay can be agreed.
- X16 *Retention.* A retention percentage can be agreed.

There are three resolving and avoiding dispute procedures. The choice depends on whether the Housing Grants, Construction and Regeneration Act 1996 (HGCRA) applies and whether a Dispute Avoidance Board is the method of dispute resolution.

In addition, the Contract Data, including the Works, Scope, completion date and total of the Prices customises the contract to the project.

NEC contracts have a modern approach and their use has increased in recent years.

RIBA Building Contracts

RIBA publishes Building Contracts which are suitable for either domestic or commercial projects:

- Domestic Building Contract 2018
- Concise Building Contract 2018

Endorsed and supported by the HomeOwners Alliance, the RIBA domestic building contract is an agreement between the client and the contractor. It can be used on all domestic projects on the client's home, including renovations extensions, maintenance and new buildings.

The RIBA concise building contract is an agreement between the client and the contractor. It can be used on all types of commercial building works in both the private and public sectors. Its features include:

- Collaboration provisions: advance warnings, joint resolution of delay, proposals for improvements and cost savings
- Includes optional provisions dealing with official secrets, transparency, discrimination and bribery as normally required by public sector clients
- Flexible payment options
- Provision for contractor design, with 'fit for purpose' liability option
- Optional provisions for a contractor programme

- Optional provisions for client-selected suppliers and subcontractors
- Mechanisms for dealing with changes to the project, which allow for agreement and include specified timescales
- Option for commencement and completion in stages
- Terms compliant with the Unfair Terms in Consumer Contracts Regulations 1999
- Guidance notes on use and completion are included

Other standard forms of contract

The ACA publishes suites of partnering (PPC2000, TPC2005 and SPC2000) and alliance (FAC-1 and TAC-1) contracts. The ICC is a standard suite of infrastructure contracts, including 2018 versions of ICC Design and Construct and ICC Target Cost, which is jointly sponsored by the Association of Consultancy and Engineering (ACE) and the Civil Engineering Contractors Association (CECA).

Founded in 1913, the International Federation of Consulting Engineers (FIDIC)'s aim is to promote and implement the consulting engineering industry's strategic goals on behalf of its international Member Associations. FIDIC has published widely used (including the World Bank) international standard forms of contract for over 50 years. The current 2017 second editions include Construction Contract (Red Book), Plant and Design-Build Contract (Yellow Book) and EPC/Turnkey Contract (Silver Book).

Choice of form of contract

The publishers of standard forms of contract are in competition with each other for business. JCT, RIBA and NEC contracts are the most used contracts. The *RIBA National Construction Contracts and Law Report 2022*'s findings included the following:

- Contracts organisations have used most often: JCT –59%, RIBA – 15% and NEC – 13%
- Contracts used during the last 12 months: JCT – 71%, RIBA – 28% and NEC – 31%

The JCT DB 2016 and NEC4 EEC contracts exhibit the following significant differences:

- The programme is not a contract document for JCT, whereas the programme is fundamental to the NEC ethos.
- In relation to extensions of time and loss and expense, JCT uses relevant events and relevant matters with time and money identified as separate concepts. NEC's compensation events address both time and money simultaneously.

Each suite of contracts is eager to promote its use. The choice of a form of contract depends on many factors, including the type of work to be performed and the designer's familiarity and preference.

Payment provisions

Part II of the Housing Grants, Construction and Regeneration Act 1996 (HGCRA) applies to construction contracts. Please refer to Chapter 2: Dispute Resolution for further details about the HGCRA. Sections 109–113 detail the Payment requirements:

- 109 *Entitlement to stage payment.* A contractor is entitled to stage payments if the duration of the work is to be longer than 45 days.

- 110 *Dates for payment.* There must be an adequate mechanism for determining what and when payments become due and provision of a final date for payment.
- 110A *Payment notices: contractual requirements.* Payment notices must be given by the specified deadlines.
- 110B *Payment notices: payee's notice in default of payer's notice.* If a payment notice from the payer, as required by s.110A is not given, the payee may give the payer a notice.
- 111 *Requirement to pay notified sum.* Generally, the payer must pay the notified sum on or before the final date for payment.
- 112 *Right to suspend performance for non-payment.* Subject to giving seven days' notice stating the grounds, there is a right to suspend performance for non-payment.
- 113 *Prohibition of conditional payment provisions.* A provision making payment conditional on the payer receiving money from a third party is ineffective unless it applies in the circumstances of insolvency.

Interpreting and implementing these provisions is not straightforward and has led to many disputes. A recommended approach is to use standard forms of contract which have been specifically drafted to comply. Generally, if a construction contract is not compliant with this legislation, then the relevant provisions of the Scheme for Construction Contracts apply.

Letters of intent

In a perfect world, employers, contractors and subcontractors would finalise all aspects of their deal and record it in a written, signed contract before any work was performed or expenditure incurred.

In reality, however, the parties are often under commercial pressure to commence works or preparatory works as soon as possible and turn to a 'letter of intent' while they continue to negotiate the full contract. Typically, a letter of intent will operate in one of three ways:

- as a non-binding statement of the future intention of both parties, sometimes called a 'comfort letter';
- as an interim contract on its own terms, which will govern the relationship between the parties unless and until a formal written contract is executed and
- as a final contract which, despite there having been no formal execution, is deemed to have incorporated the terms and conditions of the formal written contract that the parties intended.

The most common argument is whether any binding contract has been reached as a result of the letter of intent. In the Supreme Court case of *RTS Flexible Systems Ltd v Molkerei Alois Müller GmbH & Co. KG* (2010), Lord Clarke noted:

> The different decisions in the courts below and the arguments in this court demonstrate the perils of beginning work without agreeing the precise basis upon which it is to be done. The moral of the story is to agree first and to start work later.

The Supreme Court held that, whilst there were some terms still to be agreed, those terms were not essential and did not prevent a contract from existing. By their conduct, the parties had affirmed the existence of the contract by carrying out the works, making payments and varying the contract.

The consequences of a poorly drafted letter of intent can be significant and costly. *Arcadis Consulting (UK) Ltd v AMEC BCS Ltd* (2018) CA concerned a pre-contractual agreement between a specialist concrete sub-contractor and a consultant designer in respect of two large construction projects. The parties commenced works and put into place a letter of intent while they negotiated a construction contract that never materialised.

The Court of Appeal found that the letter of intent actually constituted a contractual offer, which was then accepted by exchange of correspondence and by conduct when the works began. One of those terms was a liability cap of just £610,000. That proved disastrous for the party claiming damages in respect of issues with the works, as it faced a potential loss of £40 million.

If there is no contract, the contractor will at best only be paid on a *quantum meruit* basis, that is, a reasonable payment for the work done.

In conclusion, a letter of intent will always be the poor relation of a carefully negotiated and documented contract.

Discussion questions

1. Can amending a standard form of contract to change the balance of risk be justified?
2. Is the use of a letter of intent an acceptable practice?
3. Are you aware of industries, other than construction, which use standard form contracts? How do they compare, in terms of detail?
4. Why is a specialist tribunal in place to deal with construction contract breaches, at first instance?

8 Employment law

Stephen Hardy

Introduction

The law on the employment relationship in the UK has historically developed largely through statute, supplemented by case law, without any recognition of the inequality of bargaining between British employers and employees. To some extent that inequality was temporarily remedied by the growth of the power of trade unions during the 1960s–1990s. However, between 1979 and 1997, the trade unions were themselves subject to major changes which curtailed their activities on behalf of their members. Consequently, many new rights afforded to employees were usually the result of the pervasive impact of EU developments, from 1973–2020.

The Blair Labour government in 1997 sought a more relaxed approach to the role of the trade unions in the workplace and introduced various new individual employment rights, not least as part of the government's 'family friendly' and 'work–life balance' policies. Yet, alongside these UK changes, many EU developments, such as the significant initiatives in the area of discrimination on the grounds of age, sexual orientation and religion or belief, entered into UK law. More specifically, the 'New Labour' philosophy from 1997–2010 was clearly set out in the policy White Paper *Fairness at Work*, published in 1998, and enacted in its Employment Relations Acts of 1999 and 2004. The UK government, while offering new rights, demanded that employees, in return, accept their responsibilities to cooperate with employers. So far as the creation of new rights was concerned, the most radical was the provision of a new procedure for statutory recognition of unions for bargaining purposes, supplemented by an individual right to be accompanied by a trade union official or co-worker in disciplinary or grievance procedures. Also important is the introduction of a national minimum wage. The then Blair Labour Government's commitment to implement the Social Policy Agreement at Maastricht, now incorporated into the Treaty of Amsterdam, led to the promise to provide a range of 'family friendly' rights, including the right to a period of unpaid parental leave and to equality between part-time and full-time workers. On the other side of the ledger, the government committed itself to maintaining the key elements of the employment legislation introduced by the Conservative government during the 1980s, including a continuance of restrictive laws on picketing and the prohibition of the closed shop. From 2007, under the Brown premiership, new Labour sought to consolidate its reforms and even repeal some of its changes. This was evidenced most notably in its Employment Act 2008 (which repealed the statutory procedures in force since 2004) and the Equality Bill 2009 (now the Equality Act 2010), marking out the electoral banner for the forthcoming General Election in 2010.

DOI: 10.1201/9781003657101-8

Following the 2010 General Election, given the major parties' inabilities to form a majority Government for the first time post-War Britain, the Coalition (Conservative and Liberal Democrat) Government set about with zeal to reduce red tape and remove employment laws. To that end, the Equality Act 2010 was a diluted version of that originally proposed by the outgoing Labour Government. Further, the Coalition Government immediately sought to increase the qualifying period for employment protection from dismissal to a two-year qualifying period. Moreover, the minority Conservative Government that followed, first led by David Cameron and thereafter by Theresa May, sought to review employment law. The Taylor Review, *Good Work: The Taylor Review of Modern Working Practices,* of July 2017, was a review submitted to the May Government concerning employee and worker rights in UK labour law. It was chaired by Matthew Taylor, Chief Executive of the Royal Society of Arts. Its aim was to 'consider how employment practices need to change in order to keep pace with modern business models' and the report made a series of recommendations for reform. The pandemic of 2020 then beset the new Prime Minister Boris Johnson and his Government and few employment law changes were completed. Yet, the Johnson Government did enable a 'furlough scheme' which enabled the receipt of wages/remuneration throughout the COVID era. However, more importantly, with Brexit being implemented throughout the UK from 2020–2023, concluding with the Windsor Framework under the Sunak Government, the sun has truly set on some EU laws in the UK.

The 2024 General Election returned a Labour Government in the UK for the first time in 14 years. Its Employment Rights Act 2025, due to come into force in 2026, will ban exploitative zero-hours contracts (including agency workers' contracts), end fire and rehire and introduce basic employment rights from day one – like paternity and parental leave and protection from unfair dismissal. It also introduces the right to bereavement leave from day one. The UK government describes its new Act as '[t]he biggest upgrade to workers' rights in a generation' (Hansard 2025). Essentially, this new legislation will give greater protection against unfair dismissal from day one, ensuring that insecurity at work is no longer for the few. The Act's key measures will:

- grant protection against unfair dismissal from day one, whilst allowing UK employers to operate probationary periods;
- permit parental and bereavement leave from day one;
- end the usage of zero-hour contracts;
- outlaw 'fire and rehire' and 'fire and replace' practices;
- make flexible working the norm (where practical);
- enable stronger dismissal protections for pregnant women and new mothers;
- establish a new Fair Work Agency with new powers to enforce holiday pay and
- strengthen Statutory Sick Pay (SSP), meaning that all workers will either receive SSP or 80% of their weekly pay, whichever is lower, from day one of their sickness.

With a mission to end unscrupulous employment practices, the new UK Labour government's Bill aims to 'shut down the loopholes that allow bullying fire and rehire practices'. Accordingly, a new Fair Work Agency, bringing together existing enforcement bodies, to enforce rights such as holiday pay and support employers looking for guidance on how to comply with the law, will be established to ensure that the Starmer government's mission is met.

Key UK employment legislation

The primary legislation in the sphere of individual employment law is the Employment Rights Act 1996 (as amended by the Employment Rights Act 2025). This concerns the law relating to unfair dismissal; redundancy; notice rights; protection of wages; 'whistleblowing'; time off work and maternity, adoption and parental leave. At the collective level, the Trade Union and Labour Relations (Consolidation) Act 1992 concerns the law governing trade unions, their relationship with their members and employers, industrial action and collective bargaining, including the important area of the statutory recognition of trade unions contained in Schedule A1 of the Act. Certain aspects of wages have been regulated by the National Minimum Wage Act 1998, which stipulates the minimum wage for certain categories of workers in any pay reference period.

In the field of equality law, the key statute is the Equality Act 2010. This consolidating legislation encapsulates the previous Sex Discrimination Act 1975, the Race Relations Act 1976 and the Disability Discrimination Act 1995, covering sex, race and disability discrimination, respectively and extending anti-discrimination rights to other wider protected characteristics, such as age, gender reassignment, marriage and civil partnership, pregnancy and maternity, religion or belief and sexual orientation. The Equal Pay Act 1970 should be considered alongside the Equality Act 2010, as this statute concerns the elimination of sex discrimination from pay structures. The latter is supplemented by a plethora of EU jurisprudence.

Employment protection

Until recently, most of the important protections of employment law were enjoyed by persons employed under a contract of service. This type of contract can be contrasted with a contract for services, which, simply stated, is the distinction between the status of an employee and that of a self-employed person (an independent contractor). Increasingly, however, employment legislation provides rights to all workers irrespective of their employment status (e.g. rights to a national minimum wage or to paid holidays). Even so it may still be necessary to distinguish the employee from the self-employed person because:

1. Statutory rights (e.g. the right not to be unfairly dismissed and the right to be compensated for redundancy) only apply to persons employed under a contract of service.
2. The tax and National Insurance arrangements for the two groups of workers are different. Employees pay tax by means of the PAYE system, whereas the self-employed pay tax by half-yearly instalments in arrears under a different tax schedule.
3. The duties implied by law between the parties to the contract are different, tending to be more onerous between employer and employee in a contract of service.

It will be seen that many of the reasons for distinguishing between contracts of service and those for services have financial implications for the employer as well as the employee. Plainly, it may seem cheaper for the employer to encourage his or her workforce to become self-employed, and by that means avoid the increasingly strict statutory control over contracts of service. However, it is not enough for the parties simply to change the name of their relationship. It is a question of fact whether the contract is one of service or for services. In those cases where the courts think a change has been made simply to take advantage of favourable rules, they may find no alteration in the true nature of the relationship.

When the courts have to determine which relationship exists between the parties, there has been considerable difficulty in establishing a simple test by which the decision can be made. Traditionally the courts used a simple test of control. Could the employer tell the employee what to do *and* how to do it? If so, there was a sufficient measure of control to give rise to the closer contract of service. The drawback with such an unsophisticated test is that it is quite inadequate to cope with more complex employment situations. The test is only concerned with the ultimate possibility of exercising control. The fact that an employer chooses to let an experienced employee get on with the job is irrelevant. Even so, the test proved to be over-simple and by the late 1940s, the courts were searching for another test. It is only fair to say that even today, the control test still gives a reliable answer in simple employment relationships.

Employment status

In more complex situations, the judges thought an answer might lie in asking whether a person's work was done as an integral part of the employer's business. This 'integration' test was not entirely satisfactory, and at times highly artificial. More recently, British judges have come to the conclusion that there is no simple question which will provide a conclusive answer. Rather, it is a matter of looking at all the surrounding circumstances, and the modern approach is said to be a mixed or multiple test. Using this approach, the courts will balance the factors for a contract of service against those for a contract for services, considering as they do so a range of factors, including:

- What are the employer's powers of selection and dismissal?
- What measure of control is exercised by the employer?
- What agreement is there about the method and amount of remuneration?
- What arrangements exist for the payment of tax and National Insurance contributions?
- Who supplies tools and equipment?
- Does the employee bear any economic risk in the enterprise?
- Does the employee hire his or her own helpers?
- Does the employee have any responsibility for investment and management?
- Can the employee profit from sound management in the performance of a task?
- How do the parties themselves view their relationship?
- How are the employees usually engaged in the trade or industry?

This multiple test may, in reality, be a kind of 'reasonable person' approach. The difficulty in practice is that a situation may arise where the questions listed previously give equivocal answers. Consider the example of *Ready Mixed Concrete Ltd v MPNI* (1968) CA. On the one side, the lorry driver in question had to wear a uniform, obey the orders of a foreman and obey company rules. All of these features of his contract pointed towards it being a contract of service. By contrast, he was buying the lorry under HP arrangements with the company and had to bear the cost of servicing it. He also paid his own tax and National Insurance contributions. On these facts, he could be seen to be taking some risk. When all the circumstances were taken into account, the court found that he was self-employed.

It matters less what the parties call their arrangement, although, when the factors are very finely balanced, some assistance may be given by looking at what the parties have chosen to call it. This must be treated cautiously, though, as two contrasting cases will show. In *Massey v Crown Life Insurance Co.* (1978) CA, one of the branch managers approached Crown Life to suggest a change in his status from employment to self-employment. He took advice about the consequences of the change and his new status was confirmed between the parties, and in a letter sent to the tax office. Subsequently the manager had occasion to regret his decision when certain employment rights were no longer available to him. He sought to prove that, in reality, he was still engaged under a contract of service. The Court of Appeal did not agree.

Given how carefully the question of change of status had been approached, the statement by the parties that he was self-employed was of great weight and significance. Compare Massey's case with *Ferguson v Dawson and Partners Ltd* (1978) CA. Ferguson was injured on a building site and wished to sue his employers for breach of statutory duty. To succeed in his claim, he needed to show that he was engaged under a contract of service. The evidence showed that Ferguson had been brought along to the site by some of his friends and had been taken on in an extremely informal manner by the foreman, who had informed him that he would be working 'on the lump', i.e. on a self-employed basis. No doubt Ferguson was happy with that situation while everything was going well, as he had scope to keep ahead of the taxman and avoid paying appropriate National Insurance contributions. Despite the agreement of the parties to the 'lump' arrangement, the Court of Appeal held that the reality of this situation was a contract of service. It had none of the carefully worked-out and evidenced formality of Massey's arrangements, so the view of the parties could hardly be decisive.

Although there are numerous Acts of Parliament conferring rights on employees, none of them gives an exhaustive definition of a contract of service. One usual formulation is to define an employee as a person who works under a contract of employment. That, in turn, is defined as a contract of service, whether express or implied. And as that phrase is not further defined, the rules developed by the judges remain significant.

To a certain extent, it could be said that the courts follow in the wake of the drafting practices of employment lawyers as they seek to find new ways of ensuring for their client companies that workers are excluded from employment protection legislation. Thus, once it had been established by the then House of Lords (now the UK Supreme Court) in *Carmichael v National Power plc* (1999) HL that a contract of employment required as an irreducible minimum a 'mutuality of obligations' (i.e. that the employee must be ready and willing to work and that the employer had an obligation to provide work), it was no surprise to find employers inserting clauses into contracts that provided that there was no obligation on the worker to turn up or on the employer to offer work (*Stevedoring and Haulage Services Ltd v Fuller* (2001) CA). In the same way, clauses providing that personal service is not necessary, but that work can be delegated to a substitute, have been held to exclude that relationship from one of employer-employee (*Express & Echo Publications Ltd v Tanton* (1999) CA).

Another way in which employers sought to avoid responsibility for members of their workforce was by hiring via an agency. This raises the issue of the relationship between agency workers and the 'client' firms to which they are sent. The traditional view seemed to be that, whatever the relationship between the agency and the worker, the latter was not an employee of the client. There are conflicting decisions in this area, but at least two recent cases have held out the view that, at least where there is a long-standing relationship and where a worker is recruited because of specific skills, the worker is an employee of the client, thus defeating the anticipated outcome.

Indeed, in *Brooke Street Bureau (UK) Ltd v Dacas* (2004) CA and *Cable Wireless plc v Muscat* (2006) CA, the Court of Appeal confirmed that in such cases it would be surprising if the 'end-user' did not have such powers of control and direction over a worker supplied by an agency as to make that worker an employee. However, in *James v Greenwich London Borough Council* (2008), a subsequent Court of Appeal decision has cast doubt on whether a client of an agency becomes the direct employer simply because there is a long-standing relationship, particularly if the agency relationship still exists and if payment to the worker is made via the agency.

Recent cases have been notable for attempts by 'employers' to insert clauses into their contracts with their 'workers' to the effect that the work did not need to be done by the immediate contractor but might, in the alternative, be performed by a substitute. On the whole, such arrangements have been viewed by the tribunals and courts as being 'shams' seeking to hide the true nature of the relationship and thereby avoid statutory entitlements. In *Protectacoat Firthglow v Szilagyi* (2009) CA and *Autoclenz Ltd v Belcher* (2011) CA, an approach which focuses on the reality of the relationship and not on the legal form was confirmed. In the former, on commencement with Protectacoat, the claimant was required to enter a partnership agreement with another worker and to supply services to Protectacoat involving the application of exterior textured coating to buildings. Despite terms in the working agreement to the contrary, the claimant and his 'partner' were provided with a company van and the fuel to drive it and with the necessary tools. On arrival at the sites to which the claimant was sent, he was instructed to state that he was employed by Protectacoat. Each morning he had to attend the company's yard and return there on completion of each job. It is perhaps not surprising that the Court of Appeal should agree with the employment tribunal that the partnership agreement was a sham.

More recently, in *Uber v Aslam* (2021) SC, the Supreme Court ruled that Uber drivers are workers, not self-employed contractors. This decision gave Uber drivers the right to access employment rights and protections, such as paid holiday and the national minimum wage. The factors used to determine worker or self-employment status are generally independent of the label given to the relationship. Accordingly, the UK's Supreme Court preferred to look at the 'reality of the situation' rather than the contractual. In this case, the UK Supreme Court considered the five findings made by the Employment Tribunal and found that, taken together, the drivers' jobs were tightly defined and controlled by Uber, and the drivers were therefore rightly found to be workers. Although this decision was based on the facts of this case, it is potentially significant for the gig economy in the UK.

Atypical workers!

In light of the difficulties faced by so-called *atypical* workers, the casual or home worker, the worker employed under a zero-hour contract or the agency worker, the Employment Relations Act 1999 introduced a power whereby the existing statutory rights given to employees could, by statutory instrument, be extended to other groups of workers. Should this be done it would achieve for these workers the sort of protection afforded to casual building workers by the court in *Byrne Brothers (Formwork) Ltd v Baird* (2002) HL. Here the workers, clearly self-employed labour-only subcontractors, were still able to claim paid holidays when laid off over Christmas and the New Year because the relevant regulations, the Working Time Regulations 1998, applied to all workers. The court took the view that the claimants satisfied the statutory definition of the term 'worker' because they were dependent on the particular firm for work and were thus not in business on their own account. Moreover, they succeeded notwithstanding a clause that specified that a substitute worker could be provided. Under the Agency Workers Regulations 2010, temporary agency workers who have been 'employed' for at least 12 weeks are entitled to receive basic working and employment conditions on a par with those which would apply to a worker who had been recruited directly.

The rise of these atypical workers, due to the ever-increasing influence of the gig economy, has mounted a significant challenge to traditional legal notions of employment status. Notably, in the *Pimlico Plumbers v Smith* (2018) litigation SC, Gary Smith issued proceedings against Pimlico Plumbers before the Employment Tribunal (ET) alleging that he had been unfairly dismissed, that an unlawful deduction had been made from his wages, that he had not been paid for a period of statutory annual leave and that he had been discriminated against by virtue of his disability. The ET decided that Smith had not been an employee under a contract of employment and, therefore, that he was not entitled to complain of unfair dismissal, but that he (i) was a 'worker' within the meaning of s230(3) of the Employment Rights Act 1996 (i.e. a 'limb (b)' worker), (ii) was a 'worker' within the meaning of regulation 2(1) of the Working Time Regulations 1998 and (iii) had been in 'employment' for the purposes of s83(2) of the Equality Act 2010. These findings meant that Smith could legitimately proceed with his latter three complaints, and directions were made for their substantive consideration at a later date. Pimlico Plumbers appealed this decision to the Employment Appeal Tribunal (EAT) and then to the Court of Appeal, but were unsuccessful. They consequently appealed to the Supreme Court.

The UK's Supreme Court (UKSC) dismissed the appeal. The UKSC held that if Mr Smith was to qualify as a 'limb (b)' worker under s230(3)(b), it was necessary for him to have undertaken to personally perform his work or services for Pimlico Plumbers and that the company be neither his client nor his customer. The court held that the ET was entitled to hold that the dominant feature of Smith's contract with the company was an obligation of personal performance. On the second point, even though Mr Smith was allowed to reject work and accept work outside Pimlico, there were also features of the contract which strongly militated against recognition of Pimlico as a client or customer of Mr Smith. These included Pimlico's tight control over his attire and the administrative aspects of any job, the severe terms as to when and how much it was obliged to pay him and the suite of covenants restricting his working activities following termination.

More recently, the controversy in Pimlico caused further litigation in *Uber v Aslam* (as noted previously). This case concerned so-called 'gig workers', Uber drivers, who utilize Uber's smartphone app by which passengers can book rides from drivers who also have the app. The drivers own their own cars and are free to choose when they make themselves available to accept bookings. At the time of the original hearing in this case in 2016, there were about 30,000 Uber drivers operating in the London area and 40,000 in the UK as a whole. The organization has around two million passengers registered to use its services in London. The proceedings which give rise to these appeals were brought by Uber drivers against companies in the Uber group claiming holiday pay under the Working Time Regulations 1998 and under-payments of wages by reference to the National Minimum Wage Regulations 1999. One of the claimants also claims that he has been subjected to a detriment for being a 'whistleblower' contrary to Part V of the Employment Rights Act 1996. The appeal was concerned only with Uber drivers in London. Therefore, in order to bring their claims, it was necessary for the drivers to establish that they were a 'worker' within the meaning of the Regulations and the Act.

The ET held a preliminary hearing to decide that question; and also, if the drivers were workers, the period during which they were working, which is necessary for the calculation of any holiday pay due and of the national minimum wage. It held (i) that the drivers were workers, employed by Uber London Ltd and (ii) that they were to be regarded as working during any period when they were within their territory (i.e. London), had the Uber app switched on and were ready and willing to accept trips. The EAT upheld that decision. The Court of Appeal, by a majority (the Master of the Rolls, Lord Justice Etherton and Lord Justice Bean), upheld the decisions of the ET and the EAT.

The essential question as regards worker status was whether, as the drivers argued, Uber contracts with the passengers to provide driving services, which the drivers perform; or whether, as Uber argued, it acts only as an intermediary, providing booking and payment services and the drivers drive the passengers as independent contractors. The written contractual terms say the latter, but the majority held that they do not reflect the practical reality of the relationships and can therefore be disregarded in accordance with the principle established in an earlier Supreme Court decision, *Autoclenz v Belcher* (mentioned previously). They approve the reasoning of the ET, which relied on a number of features of Uber's working arrangements as being inconsistent with the driver having a direct contractual relationship with the passenger. Lord Justice Underhill, dissenting, would have held that there was no inconsistency between the written terms and the working arrangements: those arrangements were not essentially different from those commonly applying where taxi and minicab owner-drivers are booked through an intermediary. As regards the period during which drivers are to be regarded as working, drivers are free whether to switch the app on at all and when it is switched on, they have the right to choose whether to accept any particular trip offered. However, Uber has the right to disconnect drivers from the app for a period if they turn down offers too frequently. The majority of the Court of Appeal held that in those circumstances, drivers are under a positive obligation to be available for work while the app is on and that that amounts to 'work' for the purpose of the Regulations. On final appeal, the Supreme Court found that the Uber drivers were 'workers' and are not self-employed as Uber had claimed. As such, an estimated 60,000 UK Uber drivers became entitled to receive core entitlements, such as paid holiday and the national minimum wage.

Creation of a contract of employment

Although contracts of employment are governed by the ordinary rules of contract, it is important to consider the role and impact of collective bargaining. This is the name given to negotiations between trade unions and employers, usually concerning the establishment of better terms and conditions of employment for union members. Where a union (or group of unions) has established bargaining rights with an employer (or group or federation of employers), their joint negotiations may result in a collective agreement, which in practice fixes the terms of the employment of relevant employees. Such agreements, together with the increased statutory control of employment, now severely limit the true 'freedom of contract' of the actual parties, the employer and the employee. An employee may ask, 'How can an agreement (the collective agreement), to which I was not a party, affect my individual relationship with the employer?' The answer is that the collective agreement itself cannot alter the terms. But such an agreement may well become 'incorporated' into the contract, thereby altering it, and have direct force and application to the employer and employee as a result. Two simple methods by which incorporation could occur are as follows:

1 Express incorporation where, by the terms of the original contract, the parties agree to be bound by the appropriate collective agreement for the time being in force.
2 Incorporation by usage, where the employer and employees have acted on the strength of the terms of the collective agreement, e.g. the employer has paid the increased hourly rate referred to in the collective agreement and the employee has accepted the increase.

Although, since 1980, a significant number of the working population still belong to various trade unions, the impact of collective bargaining has been less significant since the early 1990s, with a move towards individual bargaining. Where collective bargaining is still prevalent, it can have an effect on the terms and conditions of everyone in a firm or industry, whether members of the union or not.

Inevitably, a collective agreement will not be an exhaustive statement of all the terms of the contract. Indeed, it has already been indicated that collective negotiation has no place in some contracts. In the higher echelons of employment, where staff have desirable skills and talents to offer, the terms of the contract may be fixed by much more individual negotiation. In either case, many of the terms will be implied into the contract by statute or by common law. There is even scope for terms to be implied by custom. These will be considered in detail later.

Form of the contract

Apart from exceptional situations, such as that of apprenticeship, there are no rules requiring a contract of employment to be made in writing. However, since 1963, many employees have enjoyed the right to receive a written statement containing the main terms of their contract. The Employment Rights Act 1996 requires an employer to issue this statement within two months of the start of employment, but from April 2020, this must be provided at the outset (per the Employment Rights (Employment Particulars and Paid Annual Leave) (Amendment) Regulations 2018), and it must contain the following:

1 Names of the parties and date of commencement of employment.
2 The job title and the place of work.
3 Details about pay, including the amount or method of calculating the amount and the intervals at which it is to be paid. (A separate right for employees to receive an itemized pay statement, setting out the purpose and amounts of deductions, is provided by ERA 1996).
4 Details about hours of work and overtime arrangements.
5 Details about holidays and holiday pay.
6 Details of rules covering sickness and sick pay.
7 Details about pensions.
8 Details about disciplinary rules and grievance procedure.
9 Details about the period of notice to be given by the employer or employee to terminate the contract.

The ERA 1996 specifically states that if there are no agreed particulars under any heading, that fact must be stated. This rather emphasizes a feature of employment which is not always appreciated. Holidays are usually a question for agreement between the parties, and if no holidays have been agreed upon, then, at common law, none are due. However, the implementation of the Working Time Directive 1993 has led to the introduction of paid annual holidays and controls over maximum working hours (see later on).

With limited exceptions, the information required in the written statement must be contained in a single document. Since the Employment Act 2002 this may in fact be a written contract of employment (or a letter of engagement), and it should be noted that there is an important distinction between a written statement and a written contract, the former providing evidence, though not conclusive evidence, of some of the terms of the contract, the latter binding the parties subject to exceptional circumstances (*Gascol Conversions Ltd v Mercer* (1974) CA and *Systems Floors (UK) Ltd v Daniel* (1981) CA).

The ERA 1996 requires the information given to the employee under the statement to be kept up to date and any changes to be notified within one month. This does not give an employer a statutory right to vary the contract but merely obliges him or her to notify employees of any lawful changes that may be made to their terms of employment. It would be extremely rare for an employer to have the right to unilaterally vary the contract; normally alterations to terms must be consensual (i.e. agreed by both parties).

If an employer fails to give a written statement, in whole or in part, as required by these provisions or to notify of agreed changes, an aggrieved employee may make a reference of the issue to an employment tribunal. If a dispute arises as to the particulars which have been given, either party may refer the issue to a Tribunal. However, it seems that tribunals have limited powers, whatever the dispute, in that they are not empowered to interpret a statement but only to rule on what should have been included in the evidence before it. Moreover, where there is no clear evidence of a relevant term, it has been suggested that tribunals must not 'invent' terms based on what would be reasonable (*Eagland v British Telecommunications plc* (1992) CA). If the means of enforcement seem ineffective, some greater protection may be afforded to employees by the introduction in the Employment Act 2002 of a right to receive between two and four weeks' pay where in specified tribunal proceedings an employee establishes a good claim and it is discovered, irrespective of the nature of the claim, that an employer is in breach of the duty to provide a statement. It should be noted that the Act does not introduce a separate free-standing right to claim compensation for failure to provide a written statement.

Unfortunately, employment tribunals have no enforcement powers and if an employer continues to default by refusing to comply with the statement issued by a tribunal, an employee will have to enforce his or her terms by a separate action in the county court or raise the matter in a subsequent relevant substantive claim before a tribunal and request the tribunal to exercise its new powers under the Employment Act 2002.

Duties of the parties

Once a contract of employment has come into existence, the very nature of that relationship imposes certain duties on both parties, whether they expressly negotiated on these points or not. These duties are summarized in the following.

Employer's duties

1 *A duty to pay the agreed wages.*

Under the National Minimum Wage Act 1998 and its attendant regulations, every worker in the UK is entitled to receive at least the minimum fixed rate of pay. This is normally reviewed annually:

Age	*Current rate (2025)*
21 and over	£12.21
18 to 20	£10.00
Under 18 and Apprentices	£7.55

In the same way, the Working Time Regulations 1998 contain detailed provisions regulating the maximum working week (generally set at 48 hours), rest breaks, night working and paid annual holidays. The latter measure adopts the same broad coverage as the 1998 Act in that it applies to workers generally (i.e. not being confined to employees but also including those employed under any other contract whereby they undertake to do or perform personally any work or services for another party to the contract not being a client or customer of any profession or business undertaking carried on by that individual).

In the UK you are allowed to work for more than 48 hours. Those hours are usually averaged over 17 weeks. However, you can opt out of the 48-hour week. Though, if you are under 18 years of age, you cannot work more than 40 hours per week. Yet, the exceptions under the Regulations are:

- where 24 hours' staffing is required by law;
- armed forces, emergency services or police;
- domestic service in a private household;
- seafaring, sea fishing or worker on vessels on inland waterways and
- where working time is not measured and/or where, as an executive, you control your own hours.

The 1998 Regulations, which have already been subject to amendment, allow for a number of derogations which significantly weaken the protection given by them. Thus, for example, the provisions on the maximum working week do not apply to those whose working hours are not predetermined or may be determined by themselves. Enforcement of the rights introduced by this measure is limited, with only the provisions dealing with rest breaks and paid holidays being directly enforceable by way of tribunal complaint. The control of maximum hours and night work is, prima facie, an issue for the Health and Safety Executive (see Chapter 10). According to the court in *Barber v RJB Mining (UK) Ltd* (1999) CA, an action for breach of contract may arise where an employee has been required to work more than the 48-hour average without his or her agreement. Indeed, the court suggested that such an employee may even work out that he or she has exceeded these working hours for a particular reference period and lawfully stop work until the next reference period commences. Even so, given the various loopholes, there is a suspicion that the 1998 Regulations have made little, if any, difference to the culture of long working hours in the UK, particularly in certain industries.

The wages paid to a worker may also need to reflect the policy of the Equal Pay Act 1970 and the related EU legislation (formerly Article 141, the Equal Pay Directive and the Equal Treatment Directive). The Act implies an 'equality clause' into all contracts of employment (broadly defined to include not only employees but those employed to personally execute any work or labour), whereby men and women who are engaged on 'like work', 'work rated as equivalent' or 'work of equal value' should receive equal treatment with regard to all terms and conditions of employment, including pay. Comparisons are limited to the treatment of members of the opposite sex employed in the same employment (i.e. employed by the same or an associated employer at the same establishment or at another establishment where common terms and conditions are observed). Even where a claimant has cleared these hurdles, it is open to the employer to raise a defence that any different treatment is genuinely based upon a material difference or factor such as age, length of service or qualifications. Claims for equal treatment are brought before employment tribunals, which may rule that a woman's contract may be modified or varied so as to be no less favourable in the future than that of her male comparator and may also award arrears of pay, which normally may extend back to six years before the claim was brought (see Equality Act 2010 later on).

English law has long concerned itself with *how* wages are paid, rather than how much is paid. The old Truck Acts 1831–1896, regulating deductions from wages and methods of payment to persons employed in manual labour have now been repealed, but in their place, the ERA 1996 covers all workers. The Act is largely concerned with deductions from wages for reasons like bad workmanship. Such deductions are permitted so long as they are made under the terms of the contract, and those terms must have been notified to the employee in writing. Of course, an employer may always deduct those amounts which are permitted or authorized by statute, e.g. tax and social security contributions. There are also court orders which permit deductions from earnings, called attachment of earnings orders. The actual method of paying the wages, e.g. cash, cheque or directly into a bank account, is a matter for agreement between the parties. Once the method is fixed, one party may not change it unilaterally without the consent of the other.

2 *A duty to indemnify the employee for expenses reasonably incurred during the employment.*

   This point will often be covered by express terms about expenses. In other cases, if an employee is out of pocket as a result of his or her work, much may depend on how the expenses arose. Where the employee is doing the work in a lawful way and acting under an employer's authority, he or she should be reimbursed. If, however, the employee chooses an unlawful method of performance, e.g. employed to drive but driving in excess of the speed limit, then the employee must bear the cost. A more specialized aspect of the duty to indemnify is the rule that if an employee commits a tort in the course of employment, the employer is usually vicariously liable. The full extent of this rule is considered in Chapter 9.

3 *A duty to take reasonable care for the safety of the employee.*

   This duty is linked with the whole question of health and safety and is considered in more detail in Chapter 10.

4 *A duty to show mutual respect to the employee.*

   This is an example of a duty which has emerged more recently in employment law, largely from suggestions made in cases being heard before employment tribunals, although it is a duty which has now been recognized by the courts. It can be said to mean that employers must treat their employees with appropriate respect and consideration. It is an example of how far the law of employment has moved in recent years from the old philosophy of the employer being able to 'hire and fire' at will. Claims of breach of the implied duty of mutual trust and respect are typically brought by employees alleging constructive unfair dismissal (see what follows). Such claims may range from a failure to respond to a complaint about a lack of adequate safety equipment or a failure properly to investigate an allegation of harassment to the imposition of a disciplinary penalty disproportionate to an offence.

It is usually argued that in most cases an employer is under no legal duty to provide work. An old case contains the comment: 'Provided I pay my cook . . . she cannot complain if I choose to take any or all of my meals out' (*Collier v Sunday Referee Publishing Co. Ltd* (1940) KBD). In a number of important ways, this rule is of practical significance:

1 It allows an employer to pay wages in lieu of notice. The employer does not usually have to provide work.
2 When there is no work, an employer may lay off an employee (N.B. this may only be done without pay if there is a term of the contract to that effect), although where this occurs over a sufficiently long period, the employee may be eligible to claim a redundancy payment.
3 If no work and no pay are available, this may amount to the employee being suspended. The usual rule is that an employee may only be suspended without pay if his or her contract expressly or impliedly provides for this. Suspension with pay normally causes no problems because the employer is under no duty to provide work.
4 If an employee has no work and no pay, the employer may have a statutory duty to make guarantee payments of fixed amounts for a specific number of days (ERA 1996).

An employer is under no duty to provide a reference or testimonial, even though an employee may suffer in consequence of this decision. This can be particularly harsh when an employee secures a new job which is dependent on satisfactory references. The employer may feel justified in refusing to give a reference because anything he or she writes could open up possibilities of legal liability, not least because under the General Data Protection Regulations 2018, any reference may be accessed by its subject from the recipient (now GDPR). For example, the employee may sue if the reference is seen to be defamatory. Employers should have little to fear on this point, as an action for defamation can be defeated by the defences of truth or qualified privilege, i.e. the employer can show that without malice or bad motive, he was simply communicating information to a person who had a genuine interest in receiving it. Nowadays, an employer is more likely to fear that statements in a reference could expose him or her to liability for negligent misstatement.

Moreover, an employer may be estopped from denying the truth of statements made in a reference. This can be relevant in cases in the employment tribunal for unfair dismissal. If the employer has written a glowing reference in the hope that a mediocre employee will move on, he or she may subsequently wish to dismiss the employee. In any later proceedings it will be difficult for the employer to deny the truth of that glowing testimonial.

Employee's duties

1 *A duty to render personal service.* In contrast to a contractor, there will obviously be no authority for an employee to substitute another to undertake their work.
2 *A duty to render faithful service.* This means that if the employee in any way damages the employer's interests, he or she is effectively in breach of duty. Where such a breach occurs, there may be justification for the employer to dismiss the employee. Sometimes a breach may cause an employer serious ongoing harm, and he or she may wish to seek an injunction to restrain a particular activity of the employee. In other cases, it may also be appropriate for the employer to seek damages from the employee. Some areas of particular importance emerge from this general duty of faithful service:

   a Employees should not put themselves in a position where their duty to their employer conflicts with their own self-interest. In *Boston Deep Sea Fishing Co. v Ansell* (1888) HL, an employee placed orders for supplies on behalf of his employers. He chose to give the orders to firms which rewarded him for putting business their way. He was dismissed for this breach of duty.
   b Employees should not make a secret profit from their employment. This will often be closely linked with the duty in (a). In Ansell's case, he had failed to disclose to his employers what he was making 'on the side'. That was quite a crude example of this problem, which crops up regularly in business and commercial life. Employees who are in a position to place large orders for supplies may be courted by hopeful companies. They should take care in accepting any gifts or favours, for they lay themselves open to being in breach of duty, even if their judgment was not affected. Many large organizations tackle the problem by imposing express rules in the contracts of employment, forbidding acceptance of gifts.
   c Employees must not compete against their employers. An example of such competition might be the employee working as a boiler service engineer who suggests to customers that they approach him personally in his off-duty hours, and he will then undertake their work on his own behalf. In a case like this the employee could be restrained by an injunction and his behaviour would almost certainly warrant dismissal. The reason here is obvious. His spare-time activities are seriously detrimental to his employer. There is nothing to prevent an employee from engaging in a second job as spare-time work so long as it does not interfere with the duties he or she owes to his employers. This rule, that an employee must not compete against the employer, is also relevant when the employment terminates. In appropriate cases, the duty of faithful service can continue to be owed *after* the employment relationship has ended. In such cases it is common to include a 'restraint of trade' clause in the contract, which will be enforceable against the employee so long as it is reasonable (see Chapter 5).

Patents, designs and copyrights

The issue of competing against the employer can also arise where an employee invents something in the course of his or her employment. It could be very unfair if he or she then leaves the job and seeks to patent the invention and profit from it. The rules covering such a situation are now contained in the Patents Act 1977 (as amended by the Patents Act 2004). Before the Act was passed, an employer who thought it likely that an employee might invent something would put an express term into the contract, requiring the employee to assign (transfer) the rights in the invention to the employer. This often produced a very unfair result, with the employee personally getting no reward even if the employer made handsome profits out of the invention. The 1977 Act aims to deal with questions of ownership of the invention and the right of the employee to be rewarded. The invention belongs to the employer if it was made by an employee in the course of his or her normal duties or in the course of duties specifically assigned to him or her (s.39). Any other inventions belong to the employee. If an employer then exploits the invention of an employee, the employee can receive compensation from the court if he or she can show that a patent has been granted which is of 'outstanding benefit' to the employer (s.40).

The Act states that the compensation must secure for the employee a 'fair share' of the benefit derived by the employer, bearing in mind the nature of the employee's duties, the effort, skill and advice of others and the contribution of the employer in terms of facilities and other assistance. Any attempt to reduce the rights of the employee by agreeing otherwise in the contract is absolutely unenforceable. In *Kelly and Chiu v GE Healthcare Ltd* (2009) Ch, the court assessed the fair share to be awarded to the inventor employees at £1.5 million, 3% of the £50 million benefit that had gone to the employer. As *Kelly* demonstrates, one problem with the Act is that an employee may have to wait many years before being able to establish an outstanding benefit. A second difficulty may arise where employees are reluctant to bring compensation claims whilst still working for the employer.

Similar problems can also arise in relation to copyright. The Copyright Act 1956 was repealed and replaced by the Copyright Designs and Patents Act 1988, which developed out of a government White Paper on intellectual property and innovation. Not surprisingly, the period since the 1956 Act had seen much change and progress, especially in the computer field, and new rules were needed to take account of these changes. Where a literary, dramatic, musical or artistic work is made by an employee in the course of employment, his or her employer will be the first owner of any copyright in the work, subject to any agreement to the contrary. The work in question could be sales and promotional literature, a users' manual or an installers' guide. It is important to bear in mind that such apparently mundane publications are governed by these rules in the same way as novels and pop songs. So complex is this area that the general public is only likely to be aware of the controversy about whether the Act should have included a levy on blank recording tapes.

Design rights are dealt with slightly differently by the 1988 Act. Designs are protected on principles similar to copyrights, but lasting only for 15 years, as opposed to the 50-year period for copyrights. The designer is the first owner of any design right in a design, but if the design has been commissioned for a fee, the person who commissioned the design is entitled to any design right in it. Generally, the rules are the same as those for copyright – when the design was created in the course of the employment, the employer will be the first owner of any design right in the design.

1 *A duty to obey the lawful and reasonable orders of the employer*. For an order to be lawful it must not contravene the general law nor must it be contrary to what was agreed by the parties. Employees could therefore refuse to obey an order which required them to contravene safety regulations or one requiring them to work in another part of the country when their contract plainly does not anticipate that they should be mobile. It may be more difficult to determine whether an order is reasonable. Each case will turn on its own individual facts. Clearly an order is not reasonable if it exposes the employee

to grave risk. Where the employee refuses to obey a lawful and reasonable order, he or she is in breach of contract, but that may not necessarily be a 'fair reason' for dismissal. In *Wilson v IDR Construction Ltd* (1975) CA, the employee refused, in breach of his contract, to move to another construction site. Given his past work record, the genuine reason the employee gave for not wishing to move, and the employer's failure to allow the employee to explain, it was held to be unfair to dismiss him for that reason. Cases like this highlight the significant difference between a common law breach of contract and the statutory protections for unfair dismissal. 'Technical' breaches have to be viewed in the overall context of the contract.

2 *A duty to carry out tasks with reasonable care and to take reasonable care of the employer's property*. Again, any behaviour short of reasonable care is technically a breach of the contract, but whether it would justify dismissing the employee depends on all the circumstances. In any well-structured employment situation, the disciplinary procedure drawn up by the employers should provide for sanctions such as written warnings, so that every act of misbehaviour does not necessarily warrant dismissal.

Statutory rights for employees

The duties previously considered arise by virtue of the employer-employee relationship. The past 40 years have seen an explosion of new rights and duties created by statute, but it should be remembered that Acts of Parliament have played an important role in some spheres of employment for much longer, e.g. health, safety, welfare and methods of paying wages.

Statutory rights are mainly granted to employees, thereby redressing to some extent the inequality of bargaining power between a single employee and a powerful employer. Some important examples include the following:

- The right to a written statement of terms.
- The right to receive an itemized pay statement.
- The right, in appropriate circumstances, to receive a guarantee payment or a medical suspension payment.
- The right to a minimum period of notice.
- The right not to be unfairly dismissed or made redundant without compensation.
- The right to request written reasons for dismissal.
- The right to maternity leave or maternity pay.
- The right not to be discriminated against.
- The right to paid annual leave.
- The right to paid paternity leave.
- The right to request flexible working.
- The right to be accompanied at disciplinary and grievance hearings.
- Rights in connection with trade union membership.

Most of these statutory protections are confined to persons who work under a contract of service, and each right has a number of qualifying conditions attached to it.

Generally, each right will lay down a period of continuous service which must have been served before the right accrues. These periods vary from one month in the case of guarantee payments to two years for redundancy payments. The concept of 'continuous service' is, therefore, vital to establishing eligibility. In any normal working life, there will be times when an employee is not at work (e.g. through holidays, leave of absence or illness). If continuity of service was broken on each occasion, none of these statutory rights would ever be earned. The law provides that where a week is covered by the contract of employment, it will not break the continuity of service.

Not only must the rules establish when there is continuous employment, but they must also determine whether the weeks in question count or not. Rules exist to govern the continuity of employment where a business is sold. Where the business is sold as a going concern, employment with the old and the new owner can count as one continuous period. Inevitably the rules about continuity are complex, but employees get the benefit of a presumption that their service is continuous; it is for the employer to prove a break in continuity. The rules are contained in the Employment Rights Act 1996 and the Transfer of Undertakings (Protection of Employment) Regulations 2006 (TUPE). TUPE has operated since 1981 to protect those employed in the service from being forced to accept inferior terms and conditions of employment or from dismissal.

Equality: Protection against discrimination

It is now nearly five decades since the first of the modern discrimination legislation aimed at protecting individuals from unlawful discrimination was passed by Parliament, and in the intervening period, the extent of the protection provided by statute and the attendant case law interpreting these measures has grown exponentially. Thus, the previous legislative framework included the following:

- Equal Pay Act 1970
- Sex Discrimination Acts 1975 and 1986
- Race Relations Act 1976
- Disability Discrimination Acts 1995 and 2005
- Part-time Workers (Prevention of Less Favourable Treatment) Regulations 2000
- Fixed Term Employees (Prevention of Less Favourable Treatment) Regulations 2002
- Disability Discrimination Act 1995 (Amendment) Regulations 2003
- Employment Equality (Sexual Orientation) Regulations 2003
- Employment Equality (Religion or Belief) Regulations 2003
- Employment Equality (Age) Regulations 2006

These provisions were largely consolidated by the Equality Act 2010, as amended. See, particularly, the Worker Protection (Amendment of Equality Act 2010) Act 2023 (the Worker Protection Act), which introduced a duty on employers to prevent the sexual harassment of their employees.

In addition, one should not forget the impact, albeit indirect, of the European Convention of Human Rights 1950 and of the Human Rights Act 1998. As well as the individual rights protected by discrimination legislation, public authorities also had legal duties relating to the main prohibited areas of gender, race and disability. Under these separate duties, public authorities had to promote equality of opportunity, good relations and positive attitudes and eliminate harassment and unlawful discrimination. Codes of Practice were drawn up, which gave practical guidance on how to meet the legal requirements of the equality duties. As with other Codes, the contents could be taken into account by tribunals and courts in determining whether a breach of duty had occurred. Responsibility for promoting equality, human rights and an effective legislative framework now falls to the Equality and Human Rights Commission (EHRC). This body also provides advice and guidance and seeks to raise awareness of a person's rights. Its remit extends in fact beyond employment to discrimination in relation to the provision of goods, facilities, services, housing and education, although here it is employment which is of concern. The EHRC has taken over the work and responsibilities previously performed by three separate bodies, the Equal Opportunities Commission, the Commission for Racial Equality and the Disability Rights Commission. However, it has a wider role in that it is charged with protecting, enforcing and promoting equality across seven areas: age, disability, gender, race, religion or belief, sexual orientation and gender reassignment. It also has a role in protecting human rights and promoting good relations in society.

One measure that the EHRC played a significant part in influencing was the Equality Act 2010, which seeks to harmonize all the existing legislation noted previously in one Act with a single approach to all the present heads of discrimination, which will in future be referred to as 'protected characteristics', specifically age, disability, gender reassignment, marriage and civil partnership, pregnancy and maternity, race, religion or belief, sex, sexual orientation.

In some cases, it may be obvious that a person has a protected characteristic (for example, that a person is a man or a woman), but in others, such as disability, the Act provides a detailed definition of what amounts to disability. In short, a person (P) has a disability if P has a physical or mental impairment which has a substantial and long-term adverse effect on P's ability to carry out normal day-to-day activities. One schedule of the Act, various statutory regulations and a set of guidance notes supplement this definition. The Act introduced both new concepts and tidied up existing provisions. It has supposedly been written in plain English so that the man or woman in the street will be better able to understand their rights.

As with earlier legislation the 2010 Act provides a definition of discrimination ('prohibited conduct') which may take different forms and then identifies the situations where acts of discrimination because of the protected characteristics are made unlawful. Thus, prohibited conduct means:

- Direct discrimination
- Indirect discrimination
- Harassment
- Victimization
- Combined discrimination – dual characteristics
- Discrimination arising from disability and failure to make reasonable adjustments

These terms are further defined by the Act. Thus, the definition of ***direct discrimination*** states, in s.13: 'A person (A) discriminates against another (B) if, because of a protected characteristic, A treats B less favourably than A treats or would treat others'.

This definition is wide enough to include less favourable treatment of a person (B) because B is associated with a third person who has a protected characteristic, for example, a mother who is the carer of a disabled child. It also covers a situation where B is perceived to have a protected characteristic (for example, a religious belief or a particular sexual orientation) but in fact does not follow that perceived belief or does not have that perceived orientation.

Indirect discrimination is, on the face of it, less obvious than direct discrimination. The latter might arise where an employer simply refuses to employ or to promote a person

(B) because the employer believes that B, a woman, is not strong enough to do the work or would not wish to do the work because it is particularly dirty. A case of indirect discrimination might involve a more subtle, less obviously discriminatory approach by an employer.

The Act defines indirect discrimination as the application by A to B of a provision, criterion or practice that is discriminatory in relation to a relevant protected characteristic of B's (s.19). A provision, criterion or practice is discriminatory if:

a A applies, or would apply, it to persons with whom B does not share the characteristic;
b it puts, or would put, persons with whom B shares the characteristic at a particular disadvantage when compared with persons with whom B does not share it;
c it puts, or would put, B at that disadvantage and
d A cannot show it to be a proportionate means of achieving a legitimate aim.

An example of indirect discrimination might arise where an employer has a policy that employees will be selected for promotion based on their prior willingness to agree to do overtime. B, a woman (a protected characteristic), is not able to do overtime because of childcare commitments. If it can be shown (which is probable) that women are more likely than men to be responsible for childcare, then B and other women are put at a particular disadvantage compared to men. It would then be up to the employer to show that the policy of selecting for promotion on this basis was a proportionate means of achieving a legitimate aim.

- ***Harassment*** is a further example of prohibited conduct and is one that attracted much attention under the earlier legislation. Three types of harassment are defined by the Act. A harasses B if A engages in unwanted conduct related to a protected characteristic with the purpose or effect of violating B's dignity, or creating an intimidating, hostile, degrading, humiliating or offensive environment for B. In determining whether the unwanted conduct has the prohibited effect, a tribunal must consider all the circumstances, including the perception of B as well as whether B's reaction was a reasonable one. The second type of harassment involves similar considerations to the first, but here the harassment involves any form of unwanted verbal, non-verbal or physical conduct of a sexual nature. Lastly, it is harassment under the Act for A to treat B less favourably because B has either rejected or submitted to harassment consisting of the other two types. This is true whether the conduct is that of A or some other person. A is aware that material of a sexual nature is displayed in the work's restroom, but does not require it to be removed when B requests that this be done. If this results in the restroom becoming an offensive environment for B, then harassment has occurred. Similarly, if B, a junior employee, rejects the advances of his female manager and does not, as a result, receive a pay rise when other employees do at the end of the year, there has been harassment. Indeed, if B had complained about his treatment and did not receive a pay rise or perhaps was not promoted when he might reasonably have expected to be, a further act of prohibited conduct would have taken place, namely *victimization*. Under the Act, A victimizes B if A subjects B to a detriment because B does a protected act, or A believes that B has done, or may do, a protected act. Protected acts include bringing proceedings under the Equality Act or alleging that A or another person (B's manager in the earlier example) has contravened the Act.

Combined discrimination – dual characteristics adds to the prohibited conduct covered by the Act by providing a potential remedy for a person who suffers discrimination on more than one ground. Thus, if B is treated less favourably by A because she is a woman and a Muslim, B will be able to bring a combined claim rather than have to proceed to establish separate claims. However, marriage and civil partnership and pregnancy and maternity, which are protected characteristics in the Act, are not potential grounds of combined discrimination.

Disability discrimination – in addition to the prohibited conduct outlined previously, there are two other claims that might arise in relation to a disabled person. Discrimination takes place where A treats B unfavourably because of something arising in consequence of B's disability and A cannot show that the treatment is a proportionate means of achieving a legitimate aim. However, A will not be liable if A did not know, and could not reasonably have been expected to know, that B had the disability. A further duty arises for an employer to make reasonable adjustments in relation to disabled workers where (i) a provision, criterion or practice, (ii) a physical feature or (iii) the lack of an auxiliary aid puts them at a substantial disadvantage.

Once an act of prohibited conduct (one of the types of discrimination as defined previously) is established, the next stage is to identify which acts of discrimination are rendered unlawful by the Equality Act. So far as employment is concerned, an employer must not discriminate against an applicant because of a protected characteristic in selection arrangements, as to the terms of a job offer or by refusing to offer a job; and once in employment, an employer must not discriminate in access to promotion, transfer or training or in receipt of other benefits, facilities or services. An employee dismissed or subjected to any other detriment because of a protected characteristic could also complain under the Act. In the same way, an employer must not victimize or harass a job applicant or an employee. An employer may also be liable for harassment of its employees by a third party if it fails to take such steps as would have been reasonably practicable to prevent that harassment. Thus, A is the main contractor on a building site where C works as a subcontractor. B, an employee of A, is gay. When C discovers this, B is subjected to abuse and 'catcalling'. If A fails to take steps to deal with C's behaviour, it is likely that A would be liable in any complaint brought by B. However, A would only be liable if B has been subjected to such behaviour, whether by C or anyone else, on at least three occasions.

Exceptionally, where an unlawful act of discrimination has occurred, the employer (A) may be able to excuse itself by establishing that, given the nature or context of the work, having a particular protected characteristic is an occupational requirement. A would also need to show that specifying the requirement was a proportionate means of achieving a legitimate aim and that the claimant (B) does not meet the requirement. Thus A, a fashion house, may legitimately specify that B, a male, does not meet the requirement to be female to model its dresses.

Where a person wishes to pursue a claim for sex discrimination, it is possible to complain to the EHRC, which may carry out an investigation into an employer's policies and may give advice and assistance to individual claimants. In exceptional cases, where a major point of principle is at stake, the EHRC may finance the bringing of a case. This could be especially useful when it is borne in mind that individual claimants are not eligible for legal aid for Employment Tribunal cases.

The more usual enforcement procedure is for an aggrieved person to commence proceedings in the Employment Tribunal, usually within three months of the date of the act of discrimination. There is no upper limit on the amount of compensation that may be awarded. At one time a serious problem facing the claimant was that he or she had to bear the burden of proving discrimination. However, this has not been the case for most employment claims since 2001 and the Equality Act extends that position to all such claims to provide that, once a claimant has proved facts indicating that an act of unlawful discrimination has occurred, it is for the employer to prove that it did not commit that act. B, a qualified surveyor of African descent, applies for a post with employer A. B is not short-listed but subsequently discovers that the only successful applicants are white and that they have less experience and fewer qualifications than he does. If B can establish these facts, A must at least respond; otherwise, a tribunal will find against it.

Where discrimination is related to the amount to be paid for the work, the issue would be resolved under the Equal Pay Act 1970. That Act provided for equal treatment regarding terms and conditions whenever men and women were doing like work, work rated as equivalent or work of equal value. An employer could pay at different rates only if it could show that it was doing so because of a 'genuine material difference' which was not just the difference in sex. The employer might have pointed to factors like length of service, previous experience or additional qualifications. Equal pay disputes were also resolved in the Employment Tribunal. The Equality Act re-enacts the 1970 Act with very little change. Comparisons under the new Act are with a colleague of the opposite sex doing like work, work rated as equivalent or work of equal value. An employer can defend differences in pay if it can establish a 'material factor' which is not the difference in sex. If the material factor is shown to be indirectly discriminatory on the grounds of sex, an employer can still justify this but only on the basis that this factor is a proportionate means of meeting a legitimate aim. Such a defence might cover a bonus payment paid to night-shift workers who turn out to be predominantly male because female employees with childcare responsibilities do not wish to work the antisocial hours. One measure designed to improve the success rate for claimants is the outlawing of secrecy clauses in employment contracts. Thus, being involved in discussions about a colleague's pay becomes a 'protected act' for victimization purposes.

Since April 2017, pay transparency has also improved by requiring employers with 250 or more employees to publish an annual pay audit. An employer must therefore comply with the regulations for any year where they have a 'headcount' of 250 or more employees on 5 April (where the private and voluntary sector regulations apply) and 31 March (where the public sector regulations apply), but employers of all sizes should consider the advantages. A wider definition of who counts as an employee is used here (from the Equality Act 2010). This means that workers are included, as well as some self-employed people. Agency workers are included, but counted by the agency providing them. There are six calculations to carry out, and the results must be published on the employer's website and a government website within 12 months. Where applicable, they must be confirmed by an appropriate person, such as a chief executive. Gender pay reporting is a different requirement to carrying out an equal pay audit. Employers have the option to provide a narrative with their calculations. This should generally explain the reasons for the results and give details about actions that are being taken to reduce or eliminate the gender pay gap. While the regulations for the public, private and voluntary sectors are nearly identical, and the calculations are directly comparable, the public sector regulations also take into account the public sector equality duty.

Termination of contract of employment

The parties to a contract of employment do not usually fix a date for its termination, unless they have negotiated a fixed-term contract. If the duration of the contract is not fixed, the law provides for either party to end the contract by giving notice to the other. Many cases are now covered by rules providing for fixed minimum periods of notice, although it is still open to the parties to agree on longer periods if they wish. The ERA 1996 provides that if an employer wishes to terminate, he or she must give a minimum of one week's notice for each year of continuous employment, up to a maximum of 12 weeks' notice. If an employee wishes to terminate, he or she must give a minimum of one week's notice, although the contract may specify a longer period. If an employee fails to give the proper period of notice, he or she is in breach of contract. Theoretically he or she could be sued for damages by the employer. This may seem a little unlikely, but it should be remembered that the employer may be holding moneys due to an employee (e.g. accrued holiday pay, bonuses or expenses or a week's pay lying on). The employer could certainly delay making those payments to the employee. What is more, the employer could also refuse to give a reference.

There are no particular legal rules about how notice must be given to be effective. This depends on the terms of the contract. It is common to require notice to be in writing, and often the contract will specify that the period of notice must be calculated by reference to a pay day. So, a weekly-paid employee who normally gets paid on a Thursday must give notice to expire on a Thursday.

Nothing in these rules prevents either party from terminating with no notice at all if the behaviour of the other party justifies such action. When an employee receives no notice, this is referred to as summary or instant dismissal. The question will often arise whether such behaviour on the part of the employer is appropriate. An employee who is the victim of such action may be able to pursue a claim for unfair dismissal.

Unfair dismissal

A statutory right to claim for unfair dismissal was introduced in 1971. Before then, an employee who had been dismissed was limited to a claim against his or her employer in a civil action in the county court for wrongful dismissal. In many cases, such an action was worthless because the county court could only award damages, and these were often limited to an amount equal to the ex-employee's wages for the period of notice he or she should have received. If proper notice had been given, there was little point in an employee bringing an action for wrongful dismissal. However, this common law remedy is still available and may sometimes be useful. This is particularly likely to be the case where the employee is highly paid, as there is no maximum limit on damages for wrongful dismissal, where the employee is outside the three-month time limit for pursuing a claim of unfair dismissal or where the employee does not yet have a sufficient period of continuous employment to be within the statutory unfair dismissal rules.

Prior to 1971, even where the proper notice had been given, an employee might still feel a sense of grievance about the dismissal, especially if there had been no good reason for it. The common law attitude at that time still reflected attitudes that the employer could 'hire and fire' at will. A long-serving and loyal employee could be dismissed for no reason at all and would have no claim so long as the employer had given the appropriate period of notice. Unfair dismissal protection was created to change this situation. An employee now has a statutory right not to be unfairly dismissed. An aggrieved employee can seek a remedy in the employment tribunal. The remedy emphasizes an idea which has developed in employment law, that an employee has a 'property right' in the job. If this is lost when the employee is not at fault (e.g. unfair dismissal or redundancy), then he or she should be protected or compensated.

The current rules on unfair dismissal are to be found in the Employment Rights Acts 1996 and 2025. Employees who may claim their protection are those working under contracts of service, except:

1 Persons employed outside Great Britain (the test for which is not entirely clear, as can be seen from *Lawson v Serco* (2006) HL) and
2 Persons employed under an illegal contract.

Definition of dismissal

To be able to proceed with a claim for unfair dismissal, an employee must show that he or she has been dismissed. There can sometimes be doubt whether a dismissal has occurred. Take the case where an employee simply walks out or an employer asks the employee for their resignation and this is reluctantly given. Section 95 of ERA 1996 defines dismissal as:

1 termination of the contract by the employer with or without notice;
2 where employment is under a limited-term contract and that term expires without being renewed under the same contract or
3 termination by the employee in circumstances justified by the employer's conduct. This situation is called constructive dismissal. The employer's behaviour must amount to a breach which is sufficiently grave so that it entitles the employee to be treated as discharged from further performance of the contract. Not every breach by an employer can be regarded as so significant. Much will depend on the factual circumstances, but a simple example is the case where the employer proposes to reduce an employee's wages without negotiation. It will always be a question of fact whether a constructive dismissal has occurred.

These cases pose a real dilemma for employees. They are faced with some kind of breach by the employer (e.g. the unilateral imposition of a new shift system) and if they leave, there is the complication of needing to prove constructive dismissal. If employees delay taking any action, the employer could argue that they had impliedly consented to the change by staying on and working under the new arrangements. To a certain extent the development of the implied duty of mutual trust and respect has assisted employees faced with unjustified actions by their employers. Tribunals have accepted that conduct which is in breach of that duty justifies an employee leaving.

Fair reasons for dismissal

The Act (ERA) has given employees who are within its scope the right not to be unfairly dismissed, and lists the reasons why a dismissal may be fair (s.98 of ERA 1996):

1 *Reasons connected with the employee's capability or qualifications*. Capability is broadly defined to include skill, aptitude and health, amongst other things. Qualifications include degrees and diplomas as well as technical or professional qualifications relevant to the particular job.

2 *Reasons connected with the conduct of the employee.*

Misconduct may range from very trivial acts to acts which can be classed as gross. Inevitably, these will vary in different employment situations. Swearing may be regarded as fairly trivial on the shop floor of a factory or construction site but may be regarded as extremely serious on a hotel reception desk or in the offices of a professional person. Generally, the more senior an employee, in terms of age and position in the firm, the more the employer can expect from him or her. The misbehaviour may take place at work, but in appropriate cases, an employer may dismiss because of conduct occurring outside work, where the conduct could have an adverse effect on the employer's business.

3 *Retirement of the employee.*

4 *Redundancy,* where the employee was fairly selected for redundancy.

The ideal situation for employers faced with inevitable redundancies is to deal with the problem by reference to an established procedure, agreed at some previous time with the appropriate unions. Otherwise, employers may be able to show they acted fairly by applying the principle of 'last in, first out'. This approach may still leave employers having to choose some from a group who all commenced work together. In that case, they could take into account experience and ability, and the hardship likely to be caused by the redundancy. Employees who think they were unfairly selected can claim unfair dismissal.

5 Where the continued employment of the employee would cause the employer to be in *breach of a statutory duty.*

6 *Some other substantial reason* justifying dismissal.

It is difficult to spell out in detail in the Act all the circumstances where dismissal can be fair, so some residual provision is inevitable. Under this heading, a number of cases have involved employees unreasonably holding out against variations to their contracts of employment. The employer cannot unilaterally change the terms of employment. Indeed, to attempt to do so would be a breach of contract. But where the commercial needs of a business dictate that it must be run in some different way, the employer may need to approach employees to seek their approval for altered manning levels or new shift patterns. In such cases, if the employer has set about negotiating the changes in a reasonable way and one employee holds out against the change, the employer may have little alternative but to dismiss that employee.

Establishing a reason capable of being fair is not sufficient in itself. The employer must also prove that it is fair in all the circumstances of the case to use that particular reason for justifying dismissal. For example, swearing at the employer is a reason capable of being fair, but whether it is fair to dismiss for swearing in any particular case would depend on the words used, the circumstances surrounding the offence, any provocation, previous treatment of employees for swearing, any work rules, any warnings received by the offending employee and any other relevant factors. The key question facing the employment tribunal is to determine whether the employer behaved as a reasonable employer would have done. In many cases this will involve the tribunal asking three questions:

- Did the employer carry out a reasonable investigation?
- Did the employer follow a reasonable procedure?
- Was the sanction of dismissal a reasonable one in all the circumstances?

As far as reasonable investigations are concerned it is a well-settled principle from the case of *British Home Stores Ltd v Burchell* (1978) CA that an employer must establish the fact of the belief that the employee had been guilty of an act of misconduct (e.g. swearing at a manager or customer), must show reasonable grounds to sustain that belief and there must have been a reasonable investigation to justify that belief.

With respect to the question of reasonable procedure, one factor which may be relevant is the extent to which the employer followed its own disciplinary practices and procedures. These should form part of the information given to employees in their written statement. A sensible employer will have disciplinary procedures suitable for the working situation which are in general conformity with the Code of Practice on Disciplinary and Grievance Procedures (2009) issued by ACAS (Advisory, Conciliation and Arbitration Service). The House of Lords has confirmed that, in unfair dismissal cases, an employer cannot ordinarily excuse a failure to follow a fair procedure by claiming that the outcome, the employee's dismissal, would have been the same even if it had done so (*Polkey v AE Dayton Services* (1987) HL). The impact of *Polkey* was later overturned by statutory changes, but has in effect been reinstated now that those changes have themselves been repealed (see later on).

As regards the reasonableness of sanctions, it has been laid down, in what has become known as the 'range of reasonable responses' test, that there is a band of reasonableness within which one employer might reasonably dismiss an employee whilst another would quite reasonably keep him or her on. It depends entirely on the circumstances of the case whether dismissal is one of the penalties which a reasonable employer would impose. If no reasonable employer would have dismissed, then the dismissal is unfair. But if a reasonable employer might reasonably have dismissed, then dismissal is fair. This approach is a long-standing one and was confirmed by the Court of Appeal in *Post Office v Foley* (2000) CA. In essence it is an attempt to ensure that tribunals do not substitute their own view of what is reasonable for that of the employer. It is seen by some as allowing too much managerial discretion to employers at the expense of employees.

Disciplinary rules

The existence of a right to claim, coupled with the possible cost of the claim to the employer if the employee succeeds, has forced employers to give more detailed consideration to their internal disciplinary and grievance procedures. These procedures must be communicated to employees who are entitled to receive a written statement.

When drafting disciplinary rules, employers can seek guidance from ACAS. The rules should aim to encourage better standards in employees and should not be concerned exclusively with imposing sanctions. Where sanctions are necessary it is better to provide a range appropriate to the different forms of misbehaviour. Not all misconduct necessarily warrants instant dismissal. It may be more appropriate for an employee to receive oral or written warnings or some other form of 'punishment', such as demotion.

The ACAS Code of Practice on Disciplinary and Grievance Procedures stresses that disciplinary rules and procedures are necessary for promoting fairness and transparency in the treatment of individuals. They also help the organization to operate effectively. Employees must know what standard of conduct is expected from them. The responsibility for formulating a policy for maintaining discipline lies with the management, who should aim to secure involvement at all levels when formulating new rules. Rules should not be so general as to be meaningless, but rather they should specify clearly and precisely what is necessary for the efficient and safe performance of work and the maintenance of satisfactory working relations. Employees should know the likely consequences of breaking the rules, particularly where such a breach is likely to lead to summary dismissal.

Disciplinary procedures should be viewed as a means of encouraging improved conduct in individuals. Great emphasis is laid on fairness in operating the procedures. This means telling an individual what the complaint is, and allowing him or her the opportunity to state their case before a decision is reached; allowing the employee the right to be accompanied by a trade union official or a fellow employee; ensuring that no decision is reached until the case has been fairly investigated and ensuring that the employee is given an explanation for any penalty imposed. An employer should keep proper records of all actions taken under the disciplinary rules. Good employers will normally provide for breaches of the disciplinary rules to be struck from an employee's record after completing a further specified satisfactory period of service. Employees should be allowed to appeal.

Under the Employment Act 2008, an Employment Tribunal may increase an award made to an employee by up to 25% if the tribunal considers that the employer has unreasonably failed to comply with the ACAS Code of Practice. An award may be reduced by a similar amount if an employee unreasonably fails to comply with the Code.

The demise of statutory procedures and Tribunal fees

In response to concerns about the increasingly expensive and time-consuming nature of employment disputes, the Employment Act 2002 and accompanying regulations introduced statutory minimum dismissal and disciplinary procedures from October 2004. The new procedures became implied terms of every contract of employment. However, dissatisfaction with these statutory procedures led to their repeal as of April 2009 by the Employment Act 2008.

Similarly, the government introduced fees of up to £1,200 in 2013, which aimed at reducing the number of malicious and weak cases. Government statistics showed 79% fewer cases were brought over three years – trade union Unison said the fees prevented workers from accessing justice. However, the Supreme Court in *R (UNISON) v Lord Chancellor* (2017) ruled that the government was acting unlawfully and unconstitutionally when it introduced the fees. The government has, since 2018, made a commitment to reimburse all fees if it was found they acted unlawfully. Fees have raised about £32 million since being introduced. Fees ranged between £390 and £1,200. Discrimination cases cost more for claimants because of the complexity and the time hearings took. The UK's Supreme Court found this was indirectly discriminatory because a higher proportion of women would bring discrimination cases.

Remedies for unfair dismissal

Claims by an employee under the unfair dismissal rules are made to an Employment Tribunal. An action should usually be commenced within three months of the effective date of termination of the employment. Claims are initially referred to a Conciliation Officer of ACAS, whose duty it is to mediate between the parties to see if a settlement can be reached. At this stage the employer may agree to take the employee back or may agree to pay a sum by way of compensation which is acceptable to the employee. A large proportion of claims are resolved at this stage. Since 2001 it has in fact been possible for the parties to agree to forego their rights to a tribunal hearing and to accept compulsory arbitration via an ACAS-appointed arbitrator.

Before a full Tribunal (ET) hearing takes place, it may be appropriate to hold a pre-trial assessment to consider whether the case has any real substance. The case is not decided at this stage, but if it is clear that one party has no real chance of success, the tribunal may warn that party that costs could be awarded against them. Unsurprisingly, a number of applications to the ET are withdrawn at this stage. If no withdrawal occurs, the case proceeds to a full hearing. An employee found to have been unfairly dismissed may be awarded:

1 *Reinstatement*. The employee's job is restored, and he or she is treated in all respects as though no dismissal ever occurred.
2 *Re-engagement*. This gives the employee a job comparable to his or her former job.
3 *Compensation*. All awards consist of two elements, the *basic* award and the *compensatory* award.

   The calculation of the *basic* award depends on a combination of factors, namely age, length of continuous service and the amount of a week's pay. An employee's working life is divided into three periods:

Age 41 and over	One and a half weeks' pay for each year of employment
Ages 22–40	One week's pay for each year of employment
Under the age of 22	Half a week's pay for each year of employment

A maximum of 20 years of employment is taken into account and a financial cut-off point on what constitutes a week's pay is fixed. This is increased from time to time to take account of rising wage levels.

The *basic award* may be reduced if the employee has caused or contributed to his or her dismissal or if the employee has been guilty of bad conduct before dismissal, which is only revealed after the dismissal takes effect.

The *compensatory* award is an amount which the tribunal considers just and equitable, bearing in mind a claimant's losses and expenses, including loss of future earnings. The calculation is often speculative, since it is impossible to know how long a claimant may remain unemployed. The limit on the maximum amount of a compensatory award is reviewed annually. This part of an employee's compensation may be reduced where his or her conduct has contributed to the dismissal or where he or she has not taken reasonable steps to minimize losses, e.g. a failure to seek new employment. The maximum figure for the compensatory award may be increased exceptionally where additional awards can be made. An additional award is relevant if an employer refuses to comply with an order to reinstate or re-engage the employee. The amount could be between 26 and 52 weeks' pay, over and above the basic and compensatory awards.

The maximum amount that you can be awarded as compensation for Unfair Dismissal is subject to a statutory cap (£118, 233 in 2025) or 52 weeks' gross salary, whichever is lower. This is in addition to the basic award, which can be ordered by the ET up to a maximum (£21,570 in 2025).

Note that both reinstatement and re-engagement are discretionary remedies. The tribunal must consider the wishes of the parties and the practicability of compliance before making such an order. Where either order is made, if an employer refuses to comply with it, they will be obliged to pay an additional amount of compensation.

Employment Tribunal procedure

Procedure at the Employment Tribunal (ET) hearing may be relatively formal, perhaps as a result of the adversarial nature of these proceedings. Cases before the ET are heard by a legally qualified chairman (an Employment Judge). In discrimination cases, an Employment Judge sits with two lay members who are intended to be representative of employers and employees, respectively. All three play an equal part in the decision-making process. Evidence is given on oath.

Although the ET may take a more informal approach than a court, it is usually better to bring witnesses who can speak directly about the facts than to rely on hearsay evidence. Applicants should bring along relevant documentary evidence, such as their written statement of terms, their written reasons for dismissal or documents relating to their financial position, which may be relevant when calculating the compensatory award. It may be difficult to persuade people to come to give evidence or to get hold of documents considered vital to the case. In these circumstances an applicant can ask the tribunal to grant a witness summons order or a discovery of documents order.

The parties will each be invited to present their case and to call any necessary witnesses, who may be cross-examined and who may also be questioned by members of the tribunal. Where an applicant is unrepresented, the members can do much to assist by the type of questions they ask. The length of each case will depend on factors such as the number of witnesses and the complexity of the evidence, but it will not normally last more than two days. The Tribunal retires to consider its decision, which is given to the parties, with full reasons.

Although large sums of money are at stake, and the law to be applied is complex and constantly changing, no legal aid is available to pay for representation in the tribunal. A claimant who is dissatisfied with the outcome of the hearing may have a right to appeal. Appeals on a point of law lie to the Employment Appeal Tribunal (EAT), then in turn to the Court of Appeal and the Supreme Court.

Redundancy payments

A dismissal is due to redundancy (see s.139 ERA 1996) (as amended by Employment Rights Act 2025) if:

1. the employer has ceased to carry on business for the purposes of which the employee was employed;
2. the employer has ceased to carry on business in the place where the employee was employed or
3. the requirements of the business for employees to carry out work of a particular kind have ceased or diminished.

Where a redundancy situation exists, there is no entitlement to a payment if the employee is dismissed for misconduct or if he or she unreasonably refuses suitable alternative employment which has been offered in accordance with the ERA 1996. Employees who have been given notice because of redundancy are entitled to a reasonable time off with pay during working hours to look for a new job or to make arrangements for retraining. What is considered to be a reasonable amount of time off will vary according to the differing circumstances of employers and employees. During such absences employees should be paid at their normal rate.

Payments are calculated by reference to rules similar to those used to calculate a basic award for unfair dismissal. Many employees are disappointed with the size of the payment they receive. This may be because they have read about especially generous redundancy schemes operating for particular industries or because they have failed to understand the basis on which the calculation is made. Vital to the calculation of both redundancy payment and unfair dismissal compensation is the concept of 'a week's pay'. Where an employee has normal working hours and the pay does not vary from week to week, a week's pay is simply the basic weekly wage. Overtime earnings are not included unless overtime is obligatory under the contract of employment. If the employee has variable earnings because of a shift pattern, piece rates or productivity bonuses, these are usually averaged over a 12-week period.

The weekly pay that will be used to work out the redundancy payment will usually be the normal weekly gross pay at the time of dismissal up to the maximum limit, which is currently £719 (2025). A week's pay does not usually include overtime pay. The maximum basic award payment is £21,570 (2025).

An employer is under a duty to consult with trade unions or other representatives if sizeable numbers are to be made redundant over a short period. This duty is quite separate from the obligation to notify the Department for Business and Trade. If employers fail to consult, a special protective award may be made against them, which will safeguard the remuneration of employees for the period for which it is made.

Hearing redundancy and unfair dismissal claims forms a significant part of the work of Employment Tribunals. However, the trend has been to increase this workload in employment-related spheres. Since 1974, the tribunals have heard appeals against Prohibition and Improvement Notices issued under the Health and Safety at Work etc. Act 1974. The equal pay and sex discrimination rules can also be enforced in the employment tribunal. Extensions to the work of the tribunals were made by the Employment Act 1980 in relation to unreasonable exclusion or expulsion from a trade union. The jurisdiction of tribunals has been extended to breach of contract claims arising or outstanding on the termination of an employee's employment, with any appeal to the Employment Appeal Tribunal. Indeed, each new statutory right adds to the jurisdiction and workload of the tribunals. However, arbitration via ACAS is a possibility in cases of unfair dismissal and requests for flexible working.

This brief account of employment law is sufficient to show it to be a complex and constantly changing area. Many firms do not have a legal or human resources department which can guide them through the maze of rules. However, the Department for Business and Trade publishes a large range of explanatory booklets and the officers of ACAS (www.acas.org.uk) are always ready to give help and advice on questions which are within their particular sphere.

Employees and trade unions

Employees governed by a contract of service may enjoy a further contractual relationship which is of major significance in employment, i.e. membership of a trade union. Several million people belong to trade unions in this country, ranging from the very large general unions, like UNITE, to small specialist unions with quite small numbers of members, like NFU and BFAWU.

The position of the trade unions under the law is unusual, which is sometimes accounted for by the history of their development. Trade unions could only begin to exist and operate as they do today once they were freed from the problems of laws like the Combination Acts and the civil and criminal rules about conspiracy. These shackles were not removed until 1871, when the growth of the movement into what it is today could get underway. After conceding the right of trade unions to exist and organize workers, the law originally did very little to interfere with or control their internal activities. The result of this early policy of non-interference was that trade unions kept control of their internal affairs, to a considerable extent, until the passing of the Trade Union Act 1984. Under provisions introduced by that Act (now to be found in the Trade Union and Labour Relations (Consolidation) Act 1992), unions are required to hold secret ballots for trade union elections, secret ballots before industrial action and secret ballots about political funding, if any.

Of the statutory rules relating to trade unions, most have been framed to regulate their external activities (e.g. political activities and industrial activities). The main reason why a union engages in industrial action is to support the claims it is making for its members in collective negotiations with employers. The process of voluntary collective bargaining is recognized in this country as one of the principal means of fixing terms of employment. If a union is to be in a position to bargain effectively with an employer, it must gain recognition for that purpose. Between 1975 and 1980 the rules of the Employment Protection Act 1975 permitted trade unions to use the services of the Advisory, Conciliation and Arbitration Service (ACAS) to gain recognition from employers who were reluctant to grant bargaining rights. These procedures proved ineffective and were repealed by the Employment Act 1980. For a period, a union wishing to be recognized for bargaining purposes had to establish that right for itself by taking industrial action if necessary. Moreover, under the provisions of the Trade Union and Labour Relations (Consolidation) Act (1992) (TULRCA) as amended by the Trade Union Reform and Employment Rights Act 1993 (TURERA), ACAS was no longer under a duty to encourage the extension and reform of collective bargaining.

However, the Employment Relations Act 1999 reintroduced a statutory recognition procedure by adding a new schedule to TULRCA 1992. The procedure, which is involved and complex, places an emphasis on securing voluntary arrangements in the first instance, with resort to the Central Arbitration Committee (CAC) only if that fails. The role of the CAC is to determine what is an appropriate 'bargaining unit' and whether there is majority support for collective bargaining by the union from those working in that unit. If necessary, the level of support may be determined by a ballot. If this reveals majority support and those voting in favour constitute at least 40% of the workers in the bargaining unit, a declaration is issued by the CAC that the union is entitled to recognition.

After a review of the operation of the Employment Relations Act 1999, the government passed the Employment Relations Act 2004, which introduced measures to modify the operation of the statutory recognition procedure. The Act also enabled regulations to be made regarding the right of employees or their representatives to be informed or consulted by their employer on a range of issues relating to their employer's undertaking. These Regulations (the Information and Consultation of Employees Regulations 2004), which apply to undertakings employing 50 or more employees, implement the Information and Consultation Directive 2002.

Consolidation of trade union legislation

Throughout the remainder of this chapter, it should be remembered that the Trade Union and Labour Relations (Consolidation) Act 1992 (TULRCA) consolidates the Employment Acts of 1980, 1982, 1988 and 1989, the Trade Union and Labour Relations Act 1974 and the Trade Union Act 1984. However, since 1992 there have been various measures that have 'undermined' this consolidation of collective employment law, notably the Employment Relations Act 1999 and the Employment Relations Act 2004. However, the Trade Union Act 2016 amended the TULRCA in order to make freedom of association much harder. Namely, in s2 of the 2016 Act, a new requirement was introduced that trade unions must have 50% of their membership vote in a ballot for strike action.

Definition of a trade union

The modern definition of a trade union was found in the Trade Union and Labour Relations Act 1974, now consolidated into the Trade Union and Labour Relations (Consolidation) Act 1992: 'An organization consisting wholly or mainly of workers . . . whose principal purposes include the regulation of relations between workers and employers'. It will be realized that trade unions are defined both by reference to their membership and their purposes. No formal act of creation is necessary. Although an organization complying with this definition qualifies for the name 'trade union', the best legal rights and protections are afforded only to 'independent trade unions' (ITUs). The status of being independent is achieved by a trade union applying to an official called the Certification Officer (CO) for a Certificate of Independence.

As a first stage towards achieving certification, a union applies to the CO to be listed. Listing is conclusive proof of the status of a trade union. The next step is to apply for a Certificate of Independence. The list of advantages conferred by independence is impressive, particularly where the union is also recognized by the employer for bargaining purposes. It includes:

1 An independent trade union has the statutory right to receive certain information for bargaining purposes.
2 ITUs have the right to be consulted about proposed redundancies.
3 ITUs have the right to be informed and consulted about transfers of undertakings.
4 ITUs can appoint safety representatives under the Health and Safety at Work, etc. Act 1974.
5 Members of ITUs may have time off work to take part in union activities.
6 Members of ITUs have the right not to be dismissed because of taking part in union activities at an appropriate time.

When the CO considers a union's application for a Certificate of Independence, he or she must take into account the criteria laid down by the 1992 Act. These provide that the union must not be under the domination or control of the employer, nor liable to interference from the employer. In other words, the advantages of independence should only be conferred on genuine unions, which can exercise a proper, vigorous and independent role on behalf of their members. Unions least likely to gain certificates are 'house unions' or staff associations, which have often emerged from a firm's social club.

The CO is also permitted to hear the views of other unions about granting the certificate. Another union may have a vested interest in the outcome, as its own position may be jeopardized by the emergence of a new independent union. This could particularly affect the recruitment of new members.

As the Certificate of Independence is the key to claiming all the statutory rights and protections, a union which is refused this status may wish to appeal against the CO's decision. A right of appeal against the refusal of a certificate lies to the Employment Appeal

Tribunal. If the application is still unsuccessful, the union may apply again, after it has put right those matters specified as reasons for refusal of the certificate.

The legal status of a trade union

Strictly speaking, a trade union is an unincorporated association, but so many special rules have been created in relation to unions that their position more closely resembles that of a corporation. The trade union is an association of individuals, bound together by their rule book, which is effectively the terms of a contract between all the members. The greatest difficulties caused by a lack of corporate status would usually be connected with holding property or making contracts. Special rules in TULRCA cover these points. A trade union may make contracts in its own name, and its property will be vested in trustees.

The Union 'Rule Book'

Generally, a trade union is free to draft its own rules subject to the following limitations:

1 The rules must not offend the discrimination laws or EU regulations.
2 The rules may not exclude the principles of natural justice, nor may they seek to oust the jurisdiction of the courts.
3 Every member has a right under TULRCA not to be subjected to unjustifiable discipline.
4 The rules of a union which has a political fund must provide for members to be able to opt out of contributing to the political fund.
5 The rules may not exclude the duty of every trade union to secure that its principal executive committee has been directly elected by the membership.

As the rule book constitutes the terms of a contract between the members, the courts have the power to enforce the rules if an aggrieved member brings an action. It must, however, be borne in mind that not all the rules may be contained in the rule book; some may exist because of custom or practice.

Membership of a trade union

The right to become and remain a trade union member has been very significant, especially when a 'closed shop' was in operation at a place of work. Basically, that meant that the worker must either be, or become, a trade union member. The closed shop (which the law calls a union membership agreement) has been gradually eroded by legislation, most recently by the Employment Act 1990. All statutory support for closed shops has now been removed, although they can still exist or be created by agreement between employers and employees.

When an applicant applies to a union for membership, the situation is controlled by the union's rule book (which also controls questions of the discipline of members, including possible expulsion from the union). It has already been seen that trade unions have great control over the contents of their rule book. If the rules are clear, unambiguous and properly applied, there should be no need for recourse to the courts for their interpretation by an applicant refused admission.

As the consequences of an exclusion or expulsion from membership of a union can be so grave, the excluded worker may refer his or her case to an employment tribunal, as the worker has a general right not to be excluded or expelled from a union (TULRCA 1992 as amended by TURERA 1993). However, in *ASLEF v UK* (2007) ECHR, the European Court of Human Rights held that a union which was opposed to racism and fascism was not obliged to admit into membership an individual who was a member of the British National Party. As a result of this decision, TULCRA 1992 was amended by the Employment Act 2008 to qualify the right of an individual not to be excluded or expelled from union membership where that individual is or was a member of a political party and such membership is contrary to the union's rules or objectives.

Freedom to join a trade union

Normally all workers are free to join a union if they choose to do so. (The exceptions to this rule are members of the police and armed forces, who cannot form themselves into unions.) What is more, workers are often free to choose between several appropriate unions. This apparent choice was once restricted to where the Bridlington Agreement operated, but at the level of the individual members, this agreement has been effectively unenforceable since TURERA 1993. However, the Bridlington Principles, an agreement made in 1939 between unions affiliated to the TUC, have continued to regulate at the collective level inter-union competition for members and the transfer of members between affiliated unions. The main object is to prevent unions from 'poaching' members from each other. Where a union wishes to complain about such activities, the dispute will be referred to a Disputes Committee of the TUC. Again, this is a body with no legal sanctions to impose. If it finds the claim of poaching is well-founded, it will usually tell the offending union to pay financial compensation to the complainant union. The Bridlington Agreement may contravene Article 11 of the European Convention on Human Rights, which guarantees freedom of association. If this is so, then the Bridlington Agreement cannot affect the rights of individuals to join the trade union of their choice. Indeed, the introduction of the general right to union membership (TULRCA 1992 as amended) did weaken the Bridlington Principles. Nowadays, changes in patterns of employment have resulted in many employers wanting to deal with only one union. In such a case it may be a question of competing unions holding a 'beauty contest' to determine which union should be recognized by the employer. Examples of this have occurred in the print and motor industries.

Recently, there has been a seven-year legal battle involving the Independent Workers Union of Great Britain (IWGB) seeking to protect Deliveroo riders' employment rights. It concluded with the UK Supreme Court unanimously determining that Deliveroo riders are self-employed and therefore, are not legally recognized as 'workers' of Deliveroo within the meaning of the Trade Union and Labour Relations (Consolidation) Act 1992.

In 2017, the Independent Workers' Union of Great Britain (IWGB) initiated legal proceedings against Deliveroo, aiming to secure improved worker rights for Deliveroo riders. The IWGB's application was brought before the Central Arbitration Committee (CAC), which ultimately rejected it. The CAC justified its decision by highlighting that the right of Deliveroo riders to engage a substitute to perform deliveries on their behalf was an 'insuperable difficulty' with the union's argument. Subsequently, the IWGB pursued an appeal in both the High Court and the Court of Appeal, but their appeals were dismissed on similar grounds.

Undeterred, the union escalated the case to the UK's Supreme Court, relying on Article 11 of the European Convention on Human Rights (ECHR), which safeguards the right to form and join a trade union.

Whilst employment status determines what employment rights individuals are entitled to and their employer's responsibilities towards them. The Article 11 right to the formal recognition of a trade union exists only within the European concept of an 'employment relationship'. Deliveroo riders, therefore, arguably have a right to use another person to cover their deliveries without Deliveroo's involvement. However, the UKSC opined that such arrangements between the two parties are 'totally inconsistent with the existence of an obligation to provide personal service which is essential to the existence of an employment relationship'. Riders do not have to comply within specific working hours, their place of work is not specified or agreed upon and they are not required to be available. 'Riders are thus free to reject offers of work, to make themselves unavailable and to undertake work for competitors'. The ruling added that 'these features are fundamentally inconsistent with any notion of an employment relationship'.

This judgment (*IWGB v CAC* (2024) SC) provides an important clarification on whether someone is a worker from domestic and European Law standpoints, providing further important guidance on who exactly has trade union rights under Article 11. This important ruling contrasts with the Supreme Court's decision in 2021 in relation to Uber drivers, when it ruled that the level of control exerted by Uber over its drivers created an imbalance of power in favour of Uber. This was a deciding factor, amongst others, in their assessment of whether the working conditions and contractual terms between Uber and its drivers met the criteria for worker employment status as defined in the Employment Rights Act 1996 (mentioned previously).

'Union only' contracts

It used to be quite common for organizations like local authorities to demand of all their suppliers that they should allow their workers to be union members. Indeed, some authorities would go so far as to refuse to invite tenders from suppliers who did not comply with this rule. Since 1982, any term in a contract for the supply of goods or services is void if it purports to require that work done under the contract shall only be done by unionized or non-unionized labour.

A new tort was also created (see TULRCA 1992). It is a breach of statutory duty if, on grounds of union membership or non-membership, a person:

1 fails to include a particular person's name on an approved list of suppliers;
2 excludes a person from tendering or
3 terminates a contract for the supply of goods or services.

Similar liabilities may arise from any actions taken to impose union recognition requirements in contracts for the supply of goods or services. It is also unlawful to refuse to deal with suppliers or customers merely because that person does not recognize or consult with a trade union.

Expulsion from membership

A union's rule book will normally provide a range of disciplinary measures to cover various acts of misbehaviour by its members. Misbehaviour can range from failure to pay union dues to breach of vague rules like 'behaviour prejudicial to the union'. The discipline imposed can range from a fine or temporary withdrawal of benefits to ultimate expulsion from the union. The remedies for a worker who wishes to contest an expulsion are:

1 To *bring an action against the union in the ordinary courts* – this can succeed if the union has acted in breach of its rules or contrary to the principles of natural justice (e.g. by failing to give the expelled worker the opportunity to put his or her side of the case).
2 To *bring an action against the union in an employment tribunal* – under TULRCA 1992 as amended, where the member has been subjected to unjustifiable discipline. The concept of unjustifiable discipline was introduced in 1988 and could include imposing a fine, suspension or expulsion on a member for refusing to take part in industrial action.

Industrial activities of trade unions

Industrial action represents the means by which employees can collectively bring pressure to bear upon an employer. The theory of this concerted action is that it produces an equality of bargaining power between workers and employers, which could never exist between one individual employee and the employer. Common reasons for using industrial action are support of claims for higher wages or demonstrating protest at unsafe working conditions.

Although the strike is the ultimate form of industrial action, a union may achieve results by using a work-to-rule, a go-slow, a boycott of customers or suppliers or a refusal to work contractual overtime. If a strike is considered to be the most appropriate action, it may not be necessary for the union to call out all its members. Results can often be achieved quickly with selective stoppages, especially if these are by key workers, e.g. those operating computers or involved in the collection of money for the employer.

All forms of industrial action can give rise to the question: 'Are the workers concerned in breach of their contract of employment?' The answer to this is not easy. For example, employees working to rule could give little cause for complaint if they were meticulously observing the rules laid down in their contract, but if they seek to give those rules an unrealistic meaning, so that they are willfully obstructing their employer's business, then they may be in breach of contract. If there were a breach, one of the remedies would be for the employer to sue the offending employees for damages for breach of contract. Moreover, on those occasions where employees take selective action and perform only part of their duties, the courts have allowed employers to deduct a 'reasonable sum' representing the duties not carried out. Indeed, where an employer makes it clear that part performance is not acceptable, it may lawfully refuse to pay anything at all, even if an employee has performed the great majority of his or her duties. However, any employer conscious of the need to preserve good industrial relations is unlikely to resort to such steps. It is far more likely that an employer who believes its business interests are being wrongly interfered with will want to take action in tort against the union and its officials. A number of specific torts (e.g. interference with contract and conspiracy, often known as the economic torts) relate almost exclusively to trade union activities. Other torts, like trespass and public nuisance, can easily be committed while participating in industrial action.

An example of a typical situation will show how these possibilities could occur. A union in discussion with an employer about new wage rates feels that negotiations are making no progress and that the employer could offer more. The employees who are union members are called out on strike after a ballot. The union members account for about 60% of the workforce. While some employees continue to work, the effect of the strike is not fully felt, so the union officials organize a picket line and attempt to persuade those still working not to cross the picket line (i.e. persuade them to break their contracts of employment). What is more, officials realize that the impact of their action can be increased by persuading lorries bringing supplies to turn back without making delivery (i.e. causing a breach of commercial contracts). The number of pickets may well get out of control, spilling onto the road and causing it to be blocked or onto the employer's premises, constituting trespass.

The possibility of the employer suing successfully in tort, either for an injunction to stop the commission of the torts or for damages, is restricted because of the immunities enjoyed by trade unions, their officials and their members. Since 1906, these immunities have existed if the person committing the tort was acting in contemplation or furtherance of a trade dispute. Because of the protection it affords against tort actions, this is frequently referred to as the golden formula, although it is subject to more recent limitations in its scope. Two elements of the formula require consideration:

1 When is there a trade dispute?
2 What constitutes acting in contemplation or furtherance of it?

Trade disputes

A definition of a trade dispute is contained in TULRCA 1992. It must be a dispute between workers and their employer, which relates wholly or mainly to one or more of the following matters:

1 The terms and conditions of employment.
2 The physical conditions in which people work.
3 The engagement or non-engagement of workers.
4 The termination of employment or suspension of workers.
5 The allocation of work between workers.
6 The question of discipline.
7 The membership or non-membership of a trade union by any worker.
8 The facilities for officials of trade unions.
9 The machinery for negotiation or consultation relating to any of these matters.
10 The question of recognition of a trade union.

The trade dispute may also relate to matters outside Great Britain if the workers involved are affected by those matters. This is of considerable significance where companies may be multinational or working on contracts abroad, engaging cheap local labour.

There are many complex cases on the full meaning of the definition, because an employer who sees business severely affected by industrial action would naturally be keen to prove that the golden formula should not apply on the ground that there was no trade dispute. An interesting example, where this argument succeeded, involved the BBC. The BBC planned to televise the 1978 Cup Final at Wembley and to 'beam' their transmission to many foreign countries. An anti-apartheid movement approached the union organizing the TV engineers and asked for support in not beaming the programme to South Africa. If the engineers were to agree to this and refuse to carry out the orders of the BBC, they would be in breach of their contracts of employment. The BBC, in turn, would inevitably be in breach of its commercial contracts with countries overseas, who would not then receive the programme. Was the proposed action of the TV engineers in contemplation or furtherance of a trade dispute? The Court of Appeal decided that no trade dispute existed here, although the engineers could easily have brought themselves within the definition. If they had asked the BBC to insert a new term in their contracts which provided that they would not be required to take part in broadcasts to South Africa, the BBC would almost certainly have refused. Then the engineers could have said that their dispute was related to 'terms and conditions of employment' (*BBC v Hearn* (1977) CA).

Sections 5–7 of the Trade Union Act 2016 require more 'information' for union members about the numbers of people voting and the results to the Certification Officer. Section 8 requires that two weeks' notice of industrial action is given to an employer (the employer can agree to one week), amending s.234A of the 1992 TULRCA. In addition, s.9 of the 2016 Trade Union Act limits the right to take industrial action, after a strike ballot, to six months or nine months, if the employer agrees.

Contemplation or furtherance of a dispute

Although the judges in the cases have often emphasized that acts can only 'further a trade dispute' if they are not too remote from the dispute, or if they are reasonably capable of furthering the dispute when viewed objectively, these approaches inevitably limit the scope of the golden formula. The matter was considered by the House of Lords twice, in 1979 and 1980, and on both occasions the court confirmed that the proper approach to interpreting the word 'furtherance' is to apply a subjective test. Did the person in question honestly and genuinely believe that these acts would further the dispute? If so, he must have the benefit of the golden formula, however unwise or damaging his actions may be.

Limitations of the immunity

The whole question of the scope of trade union immunity from actions in tort is politically very contentious, and developments in this field have continued apace. The present position can be summarized as follows:

a *Position of the trade union itself.*

The union can be liable in tort if the act in question is authorized or endorsed by the president or general secretary, the principal Executive Committee or any other committee or official, whether employed by the union or not. Where the trade union is successfully sued for damages, the damages awarded are subject to upper limits dependent on the number of members:

Number of members of a union	*Maximum award of damages*
Less than 5,000	£10,000
5,000 or more but less than 25,000	£50,000
25,000 or more but less than 100,000	£125,000
100,000 or more	£250,000

b *Need for a ballot of the members.*

Immunities will only be extended to industrial action which has been preceded by a secret ballot. The ballot must have taken place within four weeks of the action being taken and a majority of those voting must support the proposed action. Immunities will be lost if no ballot is held or if it does not conform to the statutory rules.

c *Secondary action.*

When an official or member commits a tort, he or she will be protected if it is one of the torts protected by the golden formula. Which torts are so protected has varied from time to time. With the exception of lawful picketing, secondary action does not attract immunity. The law thus protects employers far removed from the actual dispute, who may find themselves being used as unwilling pawns by a trade union experiencing difficulties in achieving its objectives by means that are more direct.

Further restrictions on the immunity from liability in tort are contained in TULRCA 1992. Thus, there is no immunity for tortious actions where the reason for the action is that the employer is employing a person who is not a trade union member or who is not a member of a particular union. Anyone who commits an industrial tort while seeking to impose or enforce a closed shop will not have the protection of the golden formula. Where the member or official concerned is picketing, limited immunity is granted as long as:

a he or she pickets in contemplation or furtherance of a trade dispute;
b he or she pickets only in the right places – this is now usually limited to the member's own place of work or
c he or she pickets for the right purposes – these are limited to peacefully obtaining or communicating information or peacefully persuading anyone to work or not to work.

One difficulty about the immunity in relation to picketing is that it is extremely restricted and bears no relation to the realities of the picket line. Although some changes in the law were made in 1980, they did not put right the worst practical problems, namely that there is no right to stop vehicles and there is no control over the number of pickets. The 1980 changes were principally aimed at putting a stop to secondary picketing, i.e. picketing on premises other than those of the employer in dispute. It was hoped that the other problems could be resolved by the issue of a Code of Practice. The most recent code was published by the Secretary of State for Employment in 1992. The rights to freedom of expression (Article 10, ECHR) and to peaceful assembly (Article 11, ECHR) under the European Convention also arguably support a basic right to picket. In reality, however, these rights may have very little impact on the legality of picketing.

By its very nature, picketing is a public activity which tends to involve the commission of criminal offences (e.g. obstruction of the highway, causing a breach of the peace). Although the immunity is so worded as to cover criminal behaviour ('it shall be lawful'), the scope of the protection is extremely narrow, as proved by the cases on the point. Thus, there was no protection for a picket charged with obstruction because he had tried to detain a lorry driver for long enough to speak to him (*DPP v Broome* (1973) HL), nor for the pickets who blocked a road by circling round continuously in it.

Picketing must also be conducted in compliance with the provisions of the Public Order Act 1986 (as amended). Section 14 of the Act empowers the police to impose conditions on 'public assemblies', i.e. assemblies of more than 20 people in a public place in the open air. The police can issue directions as to the place, the duration and the numbers involved. The power of the police exists where it appears that the assembly may result in serious public disorder, serious damage to property or serious disruption to the life of the community.

Discussion questions

1 How important is the contract of employment to employers and employees in the workplace?
2 What is the significance of the changing state of employment status to both employers, employees and trade unions?
3 Does the Worker Protection Act 2023 make a difference in equality law?
4 What are the steps for resolving employment disputes in the UK?
5 What are the pros and cons of trade unions in the workplace?

9 The law of tort

Michael Fay and Laura Healy

Introduction

The law of contract and tort forms what lawyers call the law of obligations. Together they set out the legal framework for how people, companies and other bodies (such as the UK Government) interact. Contract sets out the rules for bargains made between two or more parties. Tort – a word of French origin meaning a 'wrong' – applies when one party behaves in a way that causes or threatens to cause harm to the interests of another. The interests protected by the law of tort are as diverse as our physical and mental integrity, our property, and even our reputation, and the rules determine when one party can be compensated for, or seek an injunction against, the behaviour of another. It is not unusual for a number of torts to overlap in any one situation – for example, a person whose property is damaged might claim in negligence, nuisance and trespass – but not every wrong suffered will have a legal remedy. A defendant's behaviour must first fall into an established category of tort, and the claimant must then show that the elements of the tort are met.

Much of the law of tort has developed through judicial precedent, although some of the rules are now statutory; for example, the Law Reform (Contributory Negligence) Act 1945, the Occupiers' Liability Acts of 1957 and 1984 and the Torts (Interference with Goods) Act 1977. But case law continues to play a significant role in the development of tortious principles. Case law allows the rules to operate flexibly and to be applied to widely differing situations. The cost of that flexibility will sometimes be a lack of certainty, and it can sometimes be difficult to say, with any precision, whether liability will arise.

Compensating claimants under the law of tort is long-established in our legal system, with remedies for personal injury dating back to the Anglo-Saxons and trespass cases from the thirteenth century. The earliest forms of tort usually gave protection to the person or to property (whether land or chattels – see Chapter 11), where the injury or damage inflicted was direct, but as society became more sophisticated, the law of tort developed to include protections against less direct forms of harm. A recent example of this development is the misuse of private information, which arose following the celebrity wedding case of *Douglas v Hello Ltd* (2001) CA.

Tort distinguished from crime and contract

There are circumstances where the same act or behaviour may give rise to liability in both the law of tort and criminal law. For example, a motorist who causes death by dangerous driving may be charged with both a criminal offence and sued in the civil courts for

DOI: 10.1201/9781003657101-9

damages (compensation) by the deceased's family. Criminal law is generally concerned with preserving order in society through the punishment, deterrence and incapacitation (imprisonment) of wrong-doers, whereas the law of tort is concerned with compensating victims who have suffered harm. Although it is easy to state the differing functions of criminal law and the law of tort, the distinction can become blurred. Criminal courts have the power in certain circumstances to compensate victims alongside punishing the offender, and civil courts can award exemplary (punitive) damages when a defendant's behaviour is sufficiently egregious. However, compensation for criminal injuries may not be for the full value of the victim's losses, whereas the purpose of tortious damages is to give full compensation, returning the victim (in so far as money can) to their pre-injury position.

The same set of circumstances can also give rise to liability in both contract and tort. For example, a person who buys a drink contaminated with a dead snail (as was seen in *Donoghue v Stevenson* (1932) HL) may sue the supplier for breach of contract (under the Sale of Goods Act 1979 or the Consumer Rights Act 2015) or the manufacturer in tort (under the Consumer Protection Act 1987 or in the tort of negligence).

A significant difference between contract and tort is that contractual liability is based on consent between the parties, a form of arranged liability, whereas tort liability is imposed by law. Although this distinction may be true in essence, again, the boundaries are not so clear. In contract, many of the obligations are in fact imposed by law. For example, in a sale of goods situation, conditions are imposed by the Sale of Goods Act 1979, as amended, or the Consumer Rights Act 2015, and the parties are not free to exclude them. Another important distinction is that contractual liability arises from an exchange of promises to perform certain obligations, whereas tortious liability arises when a defendant behaves in a prohibited manner. For example, if Stevie enters into a contract to provide Donnie with a year's supply of ginger beer and then fails to do so, Donnie can sue in contract to remedy (by recission, compensation, specific performance, etc.) the broken promise. If the ginger beer arrives but is contaminated with dead snails, then Donnie can sue in negligence because Stevie failed to take reasonable care in producing the beverage.

Where tortious and contractual liability overlap, claimants can take advantage of whichever claim is more favourable to them. The choice may depend on limitation periods (the time limits for bringing legal actions) and the rules for the calculation of damages.

Although there is considerable emphasis on the compensatory function of the law of tort, an award of damages is not the only remedy when a tort has been committed. A claimant may wish to put a stop to some form of behaviour being carried on by the defendant (such as excessive noise) and ask the court to grant an injunction. An injunction is an order of the court telling the defendant to do, or refrain from doing, certain things. A defendant who fails to obey an injunction is in contempt of court, a criminal offence even when the original claim was a civil one.

As noted previously, a situation gives rise to more than one tort. When more than one tort is committed, the claimant will not recover double damages; they will simply have the protection of alternative reasons why the claim should succeed.

Tort compared with other systems of compensation

A tort action can be a complex, lengthy and costly process. While a claimant may succeed in court, there is no certainty that the defendant will have the means to satisfy the judgment made against them; i.e. they may be unable to pay the damages. As a general rule, English law does not impose an obligation to insure against the possibility of tortious liability. But there are exceptions to this rule. Statutory provisions require a motorist to insure in respect of tortious liability to third parties (s.143 Road Traffic Act 1988 and related regulations) or an employer in respect of liability to employees (Employers' Liability (Compulsory Insurance) Act 1969).

When a claim involves negligent injury, the claimant must show that the defendant was at fault. This requires that the defendant had a duty of care to the claimant, was in breach of that duty (by failing to take reasonable care), and caused damage. If fault (lack of care) cannot be established, then the claimant will be unsuccessful; for example, in *Bowen v National Trust* (2011) QBD and *Parker v National Trust* (2021) QBD, the claimants were both hit by falling tree branches, but neither could show the defendant was at fault; indeed the defendants successfully demonstrated due care. As a result of these evidential difficulties, it has sometimes been argued that victims of accidents should receive compensation from a central insurance fund on a no-fault basis, i.e. without having to prove liability against a defendant. A Royal Commission on *Civil Liability for Death and Injury* (1978, under Lord Pearson) considered this idea. It recommended that the no-fault principle should be applied to motor vehicle accidents, similar to that in operation in New Zealand. This would mean an injured person could be compensated out of a central fund, regardless of whether he or she could prove anyone was at fault. Naturally, such a system removes the vagaries and imponderables always likely to be present in a negligence action, although it raises other issues of practical implementation. The Commission did not recommend the abolition of tort actions in other cases of personal injury.

There is some experience of operating a no-fault insurance system in the UK. The Industrial Injuries Disablement Benefits scheme, a part of our social security system, allows a worker to receive disablement benefits if they suffer personal injury in an accident at work, whether or not the employer was at fault. This does not prevent an injured worker from suing their employer for damages in negligence, but in assessing tortious damages, any social security payments received will be taken into account. The same worker may also have private insurance cover, but payments made through this are unaffected by social security benefits or tortious damages.

Trespass

Trespass is commonly thought to be a tort relating to land, but the law also recognises trespass to the person and trespass to goods. Trespass to the person occurs if force or a fear of immediate force is directly and intentionally inflicted on another person. For this purpose, force means *any* physical contact, so it could be an unwanted kiss, a punch on the nose or serious sexual assault (*Lister v Hesley Hall* (2001) HL). Trespass is said to be actionable *per se* ('of itself'), meaning an action can be brought even if no damage has been suffered. Instead of physical injury or property damage, trespass arises because of an interference with a legal right; for example, the unwanted kiss infringes on our physical integrity – our right to control what happens to our bodies. Of course, a person who has suffered no tangible harm will usually receive minimal compensation.

People often consent to actions that would otherwise be a trespass to the person, such as holding out your hand for someone to shake it, taking part in a contact sport like football or opening your mouth to allow a dentist to extract your teeth. Where consent, whether express or implied, is freely given, it is a complete defence known by the Latin maxim *volenti non fit injuria*: he who consents cannot complain of the injury. However, where consent is given but the claimant was unaware of certain information (*Appleton v Garrett* (1996) QBD) or unduly influenced by another (*Re T* (1993) CA), it will be invalid.

Where there is a trespass to the person, the same behaviour could also be a crime. For example, a punch on the nose could also result in a prosecution for battery (s.39 Criminal Justice Act 1988).

Trespass to goods

Originally, there were a number of forms of tortious action that could be brought to protect goods. The law in this area was reformed by the Torts (Interference with Goods) Act 1977. Wrongful interference with goods can take the form of trespass or conversion. Trespass to goods is an intentional and direct interference by the defendant with goods in the claimant's possession. The defendant's act must be a direct and intentional interference with the claimant's rights over property, so tearing a page from someone's book, throwing a stone at cars in a warehouse or moving someone's luggage without their consent will all be trespass to goods. The claimant must be in possession of the goods, but does not have to be the owner, although the person in possession will also frequently own the goods in question.

In an action for trespass, the defendant has no intention of denying the claimant's right to possess the goods. Where a defendant commits an act that denies another's right to possession, this may be conversion. Therefore, if the defendant took the book from the claimant, rather than simply tearing some pages out of it, they have converted the book to their own use.

The principal remedy for any wrongful interference with goods is damages, but where a defendant is still in possession or control of the goods, s.3 of the 1977 Act may allow the claimant to obtain an order for delivery and payment of any consequential damages. Where a claimant seeks only damages, these are based on the value of the goods together with any consequential loss suffered.

Trespass to land

Trespass to land is a direct and unjustifiable interference with the possession of land. A defendant can intentionally trespass, such as walking over another's land, throwing things onto it or even swinging a crane jib over the land (*Wollerton and Wilson Ltd v Richard Costain Ltd* (1970) Ch). But it is not actually necessary to intend to trespass on land; in *League Against Cruel Sports v Scott* (1985) QBD, the court explained that it is enough to intend the act that leads to the trespass. Here the defendant had intended to go hunting with dogs, and it was held to be a trespass when they negligently allowed them to roam onto land owned by the claimants.

For the purposes of this tort, land means the surface of the earth itself, anything which is a fixture on it (for example, a building), the airspace above it to the extent necessary for the use and enjoyment of the land, and the subsoil below the land. Aircraft flying over land does not constitute a trespass (*Bernstein v Skyviews and General Ltd* (1978) QBD, s.76 Civil Aviation Act 1982: see Chapter 11), but boring for oil under another's land is (*Bocardo v Star Energy* (2010) SC).

Trespass protects possession of land. Where a person is rightfully in possession, they may bring an action against even the owner of the land. If a landlord lets property to a tenant but does not reserve the right of entry to those premises, entry without the tenant's permission would be considered trespass. As with trespass to the person, the tort is actionable *per se*, i.e. without proof of damage. In those cases where no harm has occurred, the claimant may be seeking an injunction, for example, to restrain a neighbour who regularly takes a shortcut through the claimant's garden. Cases such as *Anchor Brewhouse Developments Ltd v Berkley House Ltd* (1987) show the courts are prepared to grant injunctions to prevent oversailing by tower cranes, even where no damage is caused. But landowners are expected not to unreasonably refuse requests. In *Wollerton and Wilson v Richard Costain Ltd* (1970), an injunction was suspended for several months where the landowner had refused a reasonable offer of payment and insurance.

An apparent trespass may be justified because the person is on the land with express or implied permission, under statutory authority or exercising a right of way. An example of implied permission to enter a premises is the open door of a high street estate agent. Having entered the premises, however, if the visitor creates a disturbance, they may be asked to leave and would become a trespasser if refusing to do so. A number of Acts of Parliament confer powers on the police, factory inspectors, utility companies and public health officials to enter premises without permission in particular circumstances. It is not a defence to trespass to plead that the defendant did not know they were trespassing. Even if a person has lost their way, they could be liable, although a successful action is unlikely unless the trespasser is unreasonable in their delay in leaving the private property.

There is a general misconception that trespass is a crime. This is not usually true, but in a limited number of cases, the law makes trespass a criminal offence. Typical examples would be trespass on railway lines, sidings or embankments. The notice 'Trespassers will be prosecuted' is, therefore, usually meaningless. Part V of the Criminal Justice and Public Order Act 1994 introduced criminal offences for residing on land without consent, raves (although these are not always trespass) and for trespassing and disrupting lawful activities.

It is obvious that construction work could involve a constant danger of trespass. A contractor would be wise to seek permission for acts that would otherwise cause a breach of these rules. A claimant who has been dispossessed of land by a trespasser may bring a possession claim for recovery of the land and may combine it with an action for damages. If the claimant remains in possession, he or she may claim damages and, if necessary, an injunction to restrain further acts of trespass. The help of the court is not always needed in cases of trespass. The owner of the land can ask the trespasser to leave. If the trespasser does not do so, the owner may then use reasonable force to eject him or her. What is reasonable will vary with the circumstances of each case and moderation is always advised with self-help redress.

Nuisance

For the purposes of the law, nuisances fall into three categories:

- public nuisance,
- private nuisance and
- statutory nuisance.

Public nuisance

Public nuisance is primarily a crime and consists of a nuisance that 'materially affects the reasonable comfort and convenience of a class of her Majesty's subjects' (*AG v PYA*

Quarries (1957)). Examples of public nuisance include obstructing the public highway, allowing smoke from a burning field to blow across the highway, polluting a public water supply or creating some projection over the highway. An unruly march or demonstration, which causes obstruction of the highway, could also constitute a public nuisance.

It is possible for a person affected by a public nuisance to sue in tort. A claimant would need to show that they suffered special harm because of the nuisance, greater than that suffered by ordinary members of the public (see *Tate & Lyle v Greater London Council* (1983) HL). This special harm can include personal injury, which is excluded from private nuisance claims. Where this can be proved, they may succeed either in recovering damages or in obtaining an injunction or both. A right to sue for public nuisance is important in cases where a claimant cannot sue in private nuisance because, for example, there is no interest in land to protect. Sometimes, the behaviour of the defendant will constitute both a public and a private nuisance, and in such a case, the individual affected will be in an even stronger position to sue.

Private nuisance

The tort of private nuisance is committed when one person unlawfully interferes with another's use or enjoyment of land. The interference must be both indirect and unreasonable, as a direct interference would be trespass to land. A broad range of activities can amount to private nuisance, including encroachment (e.g. by tree roots or escape of sewage), physical harm to property (but not physical harm to the person), interference with quiet enjoyment (e.g. by smells, fumes, vibrations, noise, dust) and the keeping of noisy or smelly animals.

The focus of private nuisance is the balancing of incompatible uses of neighbouring properties. In short, an occupier is free to do as they please, so long as it does not spoil their neighbour's quiet enjoyment of their own land. But not every apparent nuisance will be actionable, nor will every neighbour have the ability be able to bring a claim. In the leading case of *Hunter v Canary Wharf* (1997) HL, the court explained that only those with a proprietary interest in land (i.e. a freehold or leasehold – see Chapter 11) could bring a claim in nuisance. It defined a nuisance as something emanating from the defendant's land, such as dust, noise or smells, concluding the mere presence of a building that interfered with a neighbour's television reception was not a nuisance. By contrast, in *Network Rail Infrastructure v Williams and Waistell* (2018) CA, the defendant was liable in private nuisance for Japanese knotweed emanating from its land onto residential property.

When assessing whether an interference is unreasonable, the court takes several factors into account. The duration of the interference is relevant not only in establishing the existence of a nuisance but also in determining whether damages or an injunction is the appropriate remedy. An isolated event or temporary disturbance may not be sufficient enough to constitute a nuisance, but there are always exceptions (*Midwood v Manchester Corp* (1905) CA; *Spicer v Smee* (1946) KBD). However, duration alone is not conclusive. For example, if a firm is demolishing a building and the work will be completed in a matter of weeks, it may be reasonable to expect the owners of adjoining properties to put up with the noise and dirt for that time. But if the contractor works 24 hours each day, seven days each week, the constant noise of machinery operating, vehicles coming and going, and the glare of arc lamps at night may well constitute a nuisance. Certainly, in *Andreae v Selfridge and Co. Ltd* (1938), a claimant hotel owner, Mrs Andreae, was able to recover damages from Selfridges, who had created excessive amounts of noise and dust during demolition work in a developing part of the same London block as occupied by the hotel.

The nature of the locality where the alleged nuisance is occurring will be relevant. In *Sturges v Bridgman* (1879), a doctor had built a new consulting room on his property adjacent to a confectioner. Concluding the confectioner's use of land caused a nuisance, the court explained that 'What would be a nuisance in Belgrave Square would not necessarily be so in Bermondsey'. In *Laws v Florinplace* (1981) QBD, the court, likewise, explained that the defendant's 'private' shop would be a nuisance in Pimlico (where they intended to open it) but not by Victoria Station, where there were already several such shops. However, though what constitutes a nuisance will vary according to location, it does not mean certain activities will always be acceptable in a town or country setting. While keeping pigs, for example, would certainly be a nuisance in a town, in *Wheeler v JJ Saunders Ltd* (1995) CA, it was also a nuisance in a rural area.

Although locality may be an important factor, the courts will be more inclined to find a nuisance if the interference has caused actual harm to property (in which case locality is generally irrelevant) as opposed to mere interference with the occupier's comfort and enjoyment of their land. A claimant is far more likely to succeed where actual physical damage can be shown. If they are complaining about behaviour that causes personal discomfort in the use of the property, they must show that the interference is substantial, and one that ordinary people of ordinary sensitivity would complain about. The essence of nuisance is to achieve a balance. If the defendant has behaved reasonably, the claimant cannot complain if they have suffered damage only because their own property is extraordinarily sensitive or their use of it is in some way unusual or special. If a defendant's behaviour would cause no harm to ordinary plants, there is no cause of action for a claimant who has suffered harm only because they grow sensitive orchids (*McKinnon Industries v Walker* (1951) PC). If, however, the defendant's behaviour would have damaged ordinary property, the claimant can recover for the additional damage to their sensitive property. In short, you cannot make your neighbour liable by using your own property in some special or extraordinary way (*Robinson v Kilvert* (1889) CA).

When the court is assessing the reasonableness of the interference, it may take into account that the defendant is doing something of value to the public – for example, making early morning deliveries of milk. A greater degree of noise and disturbance may be acceptable before such acts become a nuisance, but this does not mean the behaviour never can. Public utility is more likely to be relevant in determining whether an injunction should be granted. This issue arose in two sports-related cases, *Miller v Jackson* (1977) CA and *Kennaway v Thompson* (1981) CA.

In *Miller v Jackson*, cricket was played on a village pitch that had been in use since 1905. Adjacent land was developed for private housing and the claimants purchased a house directly in line with the wicket. They soon discovered they had to live under very trying conditions during the cricket season, when they felt it was impossible to use their garden if play or practice was taking place. Although balls had come into the garden and bounced off the house, no one had yet been struck. On the facts, there was indeed a nuisance, but partly because of the public interest, an injunction was refused.

In *Kennaway v Thompson*, the nuisance arose out of water sports activities on a lake adjacent to the claimant's house. The noise from motorboats was excessive, and the court agreed that the activities did indeed constitute a nuisance. Considering the public interest in taking part in such recreations, the court decided that some balance could be achieved between the conflicting interests of the claimant and the club users. An injunction was granted, not in terms of a total restraint on water sports, but limiting the size of boats that could be used and the number of times each year competitions could be staged.

The UK Supreme Court recently broadened the scope of private nuisance to include certain cases of visual intrusion. In *Fearn and others v Board of Trustees of the Tate Gallery* (2023) SC, the Tate Modern's viewing platform, which allowed visitors to overlook neighbouring residential flats, was a substantial interference with the residents' use and enjoyment of their property. Landowners must therefore consider how their property's use may visually impact neighbours, as excessive overlooking (such as with CCTV cameras) may be an actionable nuisance.

If the defendant can be shown to have acted out of malice, that will usually be conclusive evidence of a nuisance. In *Christie v Davey* (1893) Ch, the defendant lived next door to a music teacher, and whenever a music lesson took place, he would bang tin trays, whistle and shout. His acts were shown to have been done only for the purpose of annoying and interrupting the claimant and were held to amount to a nuisance. In *Hollywood Silver Fox Farm v Emmett* (1936) KBD, the defendant stood on their own land close to the claimant's breeding pens and repeatedly fired a gun. The court held the defendant was acting with malice and liable in nuisance, even though silver foxes were unusually sensitive to miscarriages when startled.

There is a popular perception that if the alleged nuisance was taking place before the claimant came to their property, there can be no claim in nuisance. The UK Supreme Court clearly rejected this idea in *Coventry v Lawrence* (2014) SC, so long as the claimant uses their property as their predecessors did. If the claimant changed the use of the property, making the defendant's activities a nuisance, then the outcome may be different. The court stressed that each claim would turn on its facts. Although 'coming to the nuisance' is no defence, the right to commit a nuisance by noise could be acquired by prescription (long usage).

Liability for nuisance

Where a potential nuisance exists, problems arise as to who can bring an action and against whom that action must be brought. Private nuisance is designed to protect a person's use or enjoyment of land. A claimant must, therefore, be able to show some legal interest in the land, whether as owner or tenant (*Hunter v Canary Wharf* (1997) HL). No action in private nuisance is available to a mere licensee (a person who only has permission to be on the land but no legal rights to the land itself). Thus, a lodger or guest cannot sue, although if either has suffered personal injury, they may have a right of action in another tort, for example, negligence. Actions in public nuisance are not restricted by this 'interest in land' requirement.

The Human Rights Act 1998 may, in the future, have an impact on the rule that the claimant must have a legal interest in the land to sue for private nuisance, in view of the provisions relating to respect for private and family life and home, although the point has not yet been taken.

Who can be sued

The occupier of the land from which the nuisance arose is the obvious person on whom to impose liability. But there may be a difficulty if the occupier did not actually cause the nuisance. It seems that any of the following persons may be liable:

1 The actual creator of the nuisance, who remains liable even if they have disposed of the land and is no longer able to put the matter right. The reasoning is that they committed the wrong in the first place and should remain liable.

2 The occupier of the land, if they or their employees (in which case, the employer is vicariously liable – see later on) created the nuisance. Although an employer is not normally vicariously liable for the acts of independent contractors, an occupier who engages a contractor for construction work on their land may well be primarily liable for any nuisance that the contractor creates, if the occupier should have realised that the work was likely to cause a nuisance.
3 A trespasser who creates the nuisance is liable, but may not be easily found or identified. The occupier will be liable for a nuisance created by a trespasser or resulting from an act of nature if, once they know about it, they adopt or continue it. In effect, this means that if the occupier fails to take reasonably prompt and sensible steps to stop the nuisance, they will be liable (*Sedleigh-Denfield v O'Callaghan* (1940) HL).
4 The tenant occupier. Where premises are leased by a landlord to a tenant, the tenant, as the occupier, seems the obvious person to be liable for a nuisance arising from the premises. However, the landlord is also made liable in circumstances where they authorised the nuisance, for example, by leasing the property to a go-karting club (*Tetley v Chitty* (1986) QBD or where it is caused by disrepair of the premises for which he is responsible.

An example that shows how the occupier can become liable for an act of nature is *Goldman v Hargrave* (1967) PC, involving a tree that was struck by lightning and then caught fire. The occupier called someone in to chop it down, but took no steps to put the fire out, leaving it to burn itself out. When the fire spread to the claimant's property, the defendant was liable in nuisance for the damage caused.

In *Leakey v The National Trust* (1980) CA, the Trust was found liable in nuisance for damage caused by a naturally occurring hazard on their land due to their knowledge of the matter over some time and failing to take reasonable precautions to protect neighbouring property, even refusing to act on offers by neighbours to share remedial costs. By contrast, in *Holbeck Hall Hotel v Scarborough BC* (2000) CA, land on a cliff top that was owned by the council fell away, taking the hotel along with it. It was alleged that the local council had known about the risk of this happening but had done nothing to prevent it. The court decided that the council was not liable because it could not have known about the risk to land near the cliff top without undertaking an expensive survey of the area.

A statutory enforcement mechanism also does not preclude liability in private nuisance. In *Manchester Ship Canal Co Ltd v United Utilities Water Ltd* (2024), the Supreme Court rejected an argument that the owner of a watercourse was prevented from bringing a claim in nuisance because of enforcement provisions in the Water Industry Act 1991. The defendant had discharged foul water into a canal in breach of its statutory duties, but argued that allowing an action in nuisance would conflict with the Act's enforcement mechanisms. The Supreme Court held the pollution of a watercourse constituted an actionable nuisance at common law and, as such, remedies would be available to the owner, unless a polluter could show they were acting within their statutory powers or had been granted statutory immunity (see Statutory Authority later on). As the defendants had breached their legal duties and did not have any immunity, the canal owners were not prevented from bringing a claim in nuisance.

Unlike trespass, where a claimant does not need to prove damage, an action in nuisance is only possible where damage can be proved. The damage may be to the property itself (e.g. some nuisance that causes windows to break or masonry to crack) or may consist of interference with the claimant's quiet enjoyment of their property.

Defences

Even where all the various elements of private nuisance can be proved by the claimant, defendants may escape liability if they have a valid defence.

Statutory Authority: A defendant may be able to argue that the conduct complained of is authorised by an Act of Parliament. This defence can often be used by local authorities or public authorities who have been authorised by statute to conduct certain operations. Whether the defence of statutory authority will succeed depends partly on whether the statute lays down a duty or merely a power. Much will depend upon the precise wording of the legislation. However, in *Allen v Gulf Oil Refining Ltd* (1981) HL, the courts were prepared to give a wide interpretation to the words of a statute. The Gulf Oil Refining Act 1965 granted the defendant authority to acquire land and build an oil refinery. It was held that the defendant's powers must be taken to include the power to operate the refinery on the land and such operations would not, therefore, establish any liability in nuisance unless negligence could be proved. Statutory authority can only be derived from an Act of Parliament. Planning permission, although granted under a statutory authority, does not grant immunity to an action in nuisance (*Wheeler v JJ Saunders* (1995) Ltd CA).

Prescription: Defendants may also be able to show that they had a prescriptive right to do the act complained of. The right to commit a nuisance may be acquired under the Prescription Act 1832 if defendants can show that they have committed the nuisance openly and continuously for more than 20 years. This is limited to those cases where the right is capable of existing as an easement, which is a question to be determined in accordance with land law principles (see Chapter 11).

Consent: Claimants may have consented to the state of affairs (*volenti non fit injuria*), but this defence would not operate where claimants acquired the land with knowledge of an existing nuisance. A defendant's behaviour will still be tested by the stated principles and could be found to amount to nuisance.

Remedies

Abatement (Self Help): It is true that the law allows a person affected by a nuisance to take appropriate steps to put a stop to it, but this form of self-help should not be undertaken lightly. The courts have said that this is not a remedy the law favours and is usually inadvisable. The sort of problem a claimant could face is shown where branches from fruit trees encroach over another's land. If a claimant cuts off the branches, strips off the apples and then sells the fruit, although within his or her rights to lop the branches, the claimant has no right to keep them or the fruit, and must return them to the tree owner.

Injunction: It will usually be much safer for a claimant to seek an injunction when wanting to put a stop to a nuisance. The granting of an injunction is entirely at the discretion of the court and may be refused. That may happen where the nuisance occurs only very infrequently or in circumstances where the injunction would cause undue hardship to the defendant.

Damages: Damages are always available as a remedy to compensate the claimant for harm suffered as a result of nuisance.

Statutory nuisance

Most claimants simply want to put an end to the state of affairs causing a nuisance, but they may be put off by the thought of bringing a costly and time-consuming court action, which may not improve relations with their neighbours. In some cases, the state of affairs they wish to complain about may also constitute a statutory nuisance under the Environmental Protection Act 1990, which came into force on 1 January 1991.

Examples of nuisance covered by the 1990 Act include:

- premises in such a state as to be prejudicial to health or a nuisance;
- accumulations or deposits or animals which are prejudicial to health or a nuisance;
- emissions of smoke, gas, fumes, dust, steam or other effluvia and
- noise or vibration.

The Clean Neighbourhoods and Environment Act 2005 introduced two new statutory nuisances: insects emanating from premises and artificial light. For a state of affairs to be a statutory nuisance, it must either be prejudicial to health or a nuisance. A state of affairs is prejudicial to health if it is injurious to health or likely to cause injury to health. The nuisance element of the phrase can only be satisfied by showing what amounts to a nuisance at common law. A statutory nuisance cannot exist where there is merely an interference with the personal comfort of the occupiers, although noise that has an irritating quality has been held to constitute a statutory nuisance.

A local authority is under a positive duty to make inspections within its area to establish whether any statutory nuisance exists. Additionally, the local authority must take such steps as are reasonably practicable to investigate complaints of statutory nuisance made by persons living within the local authority's area. The inspections are made by Environmental Health Officers, who have the right of entry to property for detection purposes.

Where a statutory nuisance exists or is likely to occur or recur, the local authority must serve an abatement notice on the person creating the nuisance or, in some cases, on the owner of the property from which the nuisance emanates. The notice indicates what must be done and a time limit for action. Non-compliance with such an abatement notice is a criminal offence, for which the offender can be fined on a continuing daily basis until compliance. Alternatively, the local authority may undertake the work necessary and recover its expenses.

The 1990 Act includes a right of appeal against an abatement notice. The appeal must be lodged within 21 days and will be heard in the magistrates' court.

Where there is evidence of a statutory nuisance and the local authority declines to act or is itself responsible for the nuisance, any person aggrieved may bring proceedings in the magistrates' court. In such instances, the aggrieved person must give at least 21 days' notice in writing to the defendant, stating the intention to bring proceedings and outlining the matters complained of. When the case comes before the magistrates' court, the court has the power to make a nuisance abatement order, as well as impose a fine of up to £5,000 and order reimbursement of necessary expenses. Failure to comply with a nuisance abatement order is a criminal offence. The court may order the local authority to perform the necessary work under the order where a defendant has defaulted.

The rules under the Environmental Protection Act do not give rise to any liability in tort. A person suffering harm from a nuisance would still need to pursue a claim in private nuisance to claim significant damages. However, in smaller problems, the possibility of an award of criminal compensation under the 1990 Act may suffice. The advantage of the statutory rules is that, where the local authority is prepared to act, the state of affairs constituting the nuisance can be put right without any need for individual action.

The rule in *Rylands v Fletcher*

This is a tort of strict liability. This means that if a set of given facts fit the requirements of the rules, then the defendant will be liable, regardless of whether they displayed reasonable care. This tort is known by the name of the case that first gave rise to it. An occupier who engages in potentially hazardous activities on their property will be liable where those activities cause injury or damage to neighbouring property. The same set of facts may also, but not necessarily, give rise to an action in negligence. The importance of the *Rylands v Fletcher* rule was that it pre-dated the general development of negligence principles. Nowadays, it is often suggested that the rule is of diminished significance, and certainly, its technical requirements mean that it is difficult to prove. In addition, many of the hazardous activities it may have covered are now governed by statutory frameworks (e.g. the Reservoirs Act 1975, as amended; the Animals Act 1971 and the Nuclear Installations Act 1965, as amended).

The rule in *Rylands* was stated by Lord Blackburn as:

> Where a person, for his or her own purposes, brings on to his or her land, and collects and keeps there anything likely to do mischief if it escapes, he or she must keep it at their peril, and if not doing so, is liable for all the damage which is a natural consequence of the escape.

The rule involves several elements:

Accumulation: The dangerous 'thing' must be accumulated on the land by the defendant. No liability can arise from an escape of something that is naturally on the land, e.g. weeds or rocks.

Mischief: The 'thing' accumulated must be likely to cause harm if it escapes. The rule has been applied to a wide range of substances and items, from oil, gas and explosives (*Read v J Lyons & Co. Ltd* 1947 HL) to flagpoles and fairground equipment. There is no requirement that the 'thing' must be intrinsically dangerous.

Escape: There must be an escape. The claimant must show that they were not on the defendant's land at the time when the harm occurred.

Damage: The escape must cause damage. This may be damage to property, but it does not include personal injury (*Hunter v Canary Wharf* (1997) HL). The damage must also be foreseeable (*Cambridge Water Co. v Eastern Counties Leather plc* (1994) HL).

Non-Natural Use: The defendant must be making a non-natural use of the land. This point is probably the most difficult and contentious. The defendant must be using their land in some extraordinary and unusual way. The idea of non-natural use is closely bound up with public benefit (or lack of it) and will tend to change with changing social needs. In the case of *Rylands v Fletcher* (1868) itself, the non-natural use was bulk storage of water in a reservoir, but in *Transco plc v Stockport Metropolitan Borough Council* (2003) HL, a large water main supplying a block of flats was held not to be a non-natural use of land. According to Lord Bingham, a cause of action under *Rylands v Fletcher* should involve establishing that the use of the land was extraordinary and unusual rather than non-natural. In the same case, Lord Hoffman thought that a useful guide to determine whether there has been a 'non-natural' use of land is to consider whether the damage caused was 'something against which the occupier could reasonably be expected to have insured himself'.

Although liability is said to be strict, i.e. not dependent on whether the defendant took reasonable care, several defences may be available to the defendant. These include:

Fault of the claimant: If the damage is caused by the claimant's own wrongful act, they cannot recover.

Consent: If the claimant consented to the presence of the 'thing' that escaped, then the defendant may not be liable. Consent could be express or implied. For example, tenants on different floors of a building may impliedly consent to the presence of a water supply running through the building.

Statutory Authority: A defendant may be able to point to an Act of Parliament that excuses their behaviour, which would otherwise be tortious.

Act of a stranger: If the escape was caused by an act of a stranger that was unforeseeable, then the defendant will not be liable. In *Box v Jubb* (1879) ExD, a reservoir overflowed onto the claimant's land, but the flood resulted from the actions of another reservoir's owner, over which the defendant had no control, so was not liable. In *Rickards v Lothian* (1913) PC, the claimant's premises were flooded due to a continuous overflow of water from a sink on the top floor of the building. A third party had deliberately plugged the wastepipe and turned on the taps. The defendant was not liable as they could not reasonably have known of the act and so could not do anything to prevent the harm. However, if the act of a stranger should have been foreseen, the defence will fail. In *Perry* v *Kendricks Transport Ltd* (1956) CA, it was held to be reasonable to expect children to meddle with a disused motorcoach on the defendant's land. When it was set alight by a group of young boys and injured another child, the defendant was liable.

Act of God: An act of God is an extraordinary natural occurrence that could not have been foreseen. In *Nichols* v *Marsland* (1876) CA, the defendant had several ornamental pools on their land. These had been created by damming up a stream that ran through the property. Extraordinarily high rainfall caused the banks of the pools to collapse, and the escaping water destroyed four bridges on the claimant's land. The court held the defendant was not liable for the extraordinary weather, which they could not have reasonably foreseen. However, extraordinary rainfall was not a defence in *Greenock Corporation* v *Caledonian Railway* (1917) HL. The local authority had constructed a concrete paddling pool in the bed of a stream, altering its natural course. An extraordinary level of rainfall caused the pool to overflow and, due to its construction, water that would have ordinarily flowed downstream spilled onto a public street. The House of Lords held that such an event did not qualify as an 'act of God'.

Finally, it must be remembered that the same set of facts could variously give rise to liability in negligence, nuisance and *Rylands v Fletcher*.

Negligence

Negligence is a relatively young tort, with its modern form only emerging in the 1930s. Unlike private nuisance and trespass to land or goods, it is not solely concerned with protecting interests in property. Instead, it can compensate for harm caused to people and property in a wide range of circumstances.

More people are injured by careless acts than by intentional acts. As the law of tort developed, several situations were recognised where liability would be imposed on a person for negligent behaviour. Early examples included where employers failed to take reasonable care to ensure the safety of employees, and where occupiers failed to take reasonable care for the safety of visitors. But initially, there was no general principle established that could be applied to any and every set of circumstances.

Such a general principle emerged in 1932 from the case *Donoghue v Stevenson*. Here, a manufacturer of ginger beer distributed the product in opaque bottles. A bottle was sold by a café owner in Paisley, Scotland, to a customer who bought it for a friend, Mrs Donoghue. A decomposed snail plopped out from the bottle when Mrs Donoghue poured it over her ice cream float, causing her to become seriously ill. As she had not bought the ginger beer, Mrs Donoghue had no claim in contract (note that this would be different today under the Consumer Rights Act 2015). Undeterred, she brought a case against the manufacturer, arguing they were liable in negligence. The case never actually dealt with whether or not there was a snail in the bottle (some suspect there was not) or whether the manufacturer had failed to properly quality check the ginger beer. Instead, it focused on a single question: Could a manufacturer be liable in negligence to someone who had not purchased their product?

The House of Lords decided a manufacturer owed a duty to take reasonable care to ensure its products were not contaminated and the duty was owed to all consumers of the product, regardless of whether they purchased it. Lord Atkin explained that a person owed a duty of care to their 'neighbours', who in law were people 'so closely and directly affected by my acts or omissions that I ought reasonably to have them in contemplation as being so affected'. The case was then sent back to the Court of Session (Outer House) for Mrs Donoghue to bring her claim, but Stevenson died in the interim, and the claim was eventually settled for £200 (around £12,000 in 2025).

The importance of *Donoghue v Stevenson* is that it created the 'neighbour principle', a test that linked together all the previously disparate claims of negligence and provided a touchstone for new claims. Over the years, negligence has overtaken all other torts in importance: first, in the number of claims that are brought and second, in the way that negligence has begun to take over some of the liability that would previously have been covered by torts like trespass or nuisance.

The tort of negligence is made up of three elements:

- Duty of care
- Breach of duty
- Causation

Duty of care

Liability in negligence can only arise if a duty to take care is owed to the claimant. When a court is faced with a situation that has been considered before, such as a car accident, it is usually able to apply the precedent from an earlier case when determining if a duty exists. An Act of Parliament may also impose a duty, e.g. the Occupiers' Liability Act 1957. In either case, establishing the existence of the duty is relatively straightforward.

Where the court is presented with an entirely novel situation to consider, it must decide whether or not a duty to take care exists, and Lord Atkin's 'neighbour principle' was the first method for answering this question. His Lordship explained:

> The rule that you are to love your neighbour becomes in law, you must not injure your neighbour . . . . You must take reasonable care to avoid acts or omissions which you can reasonably foresee would be likely to injure your neighbour. Who, then, in law, is my neighbour? The answer seems to be – persons who are so closely and directly affected by my act that I ought reasonably to have them in contemplation as being so affected when I am directing my mind to the acts or omissions which are called in question.

In short, a person or company had to take reasonable care to avoid causing a foreseeable type(s) of harm. A 'neighbour' for this purpose is any person who is 'closely and directly affected' by the defendant's acts or, in limited circumstances, omissions. The 'neighbour principle' was originally thought to rest exclusively on reasonable foresight of harm. For example, if you are driving along a road next to a school at home time, you can reasonably foresee that a schoolchild will be injured if you do not take reasonable care (such as driving to the speed limit).

However, as negligence has developed over the past 93 years, foresight of harm has become merely one of the ingredients necessary to establish a duty of care. A second ingredient is 'proximity', a word that immediately conjures an image of physical closeness. This is partially true – proximity between the claimant and defendant can exist because they are both in the same place, at the same time. In the earlier example of driving by a school, you and the schoolchildren share a physical location, so you would have sufficient proximity for a duty of care. Of course, this is only part of the story and proximity can exist because of a pre-existing relationship between the parties (such as a doctor-patient) or because the claimant was present at a particular time (such as a spouse or civil partner arriving at the scene of an accident). A close look at the neighbour principle suggests proximity was always relevant, as a claimant had to be 'closely and directly affected' by a defendant's acts, but it is now treated as a distinct question. The advantage of including a test of proximity is said to be that it can be more stringent where the circumstances require it.

The modern approach to the duty of care was originally set out in *Caparo v Dickman* (1990) HL and was recently restated in *Robinson v Chief Constable of West Yorkshire Police* (2018) SC. In both cases, the court explained that there is not actually a single test for proving a duty of care exists and, in most situations, the law should develop by analogy with decided cases. So, if a case was brought about a segway rider knocking over a pedestrian, the court would look at decided cases involving pedestrians, motorists and other road users. As a segway is a motor vehicle, it is analogous to a car or motorbike, so the rider would owe the same duty of care as any other motorist.

But this incremental approach to developing duties of care does not work when the court is presented with a novel situation. In both *Caparo* and *Robinson*, the court explained that when faced with a wholly new situation, the starting point is a three-part test.

Foresight of harm: the defendant must actually foresee, or a reasonable person in their position would foresee, that some type of harm will occur to a category of people (e.g. consumers) if they do not take reasonable care.

Proximity or 'Neighbourhood': There must be a proximate relationship between the parties, either because they are in the same place, at the same time (e.g. a road accident: *Nettleship v Weston* (1971) CA) or there is some other kind of relationship (e.g. a doctor-patient relationship: *Cassidy v Ministry of Health* (1951) KBD).

Fair, Just and Reasonable: It must be fair, just and reasonable to impose a duty of care on the defendant. This is essentially a policy decision and asks if there is enough of a reason to impose a duty of care. In novel situations, this may be a significant barrier for the claimant. An example of a case where what was fair, just and reasonable was considered is *Commissioners of Customs and Excise v Barclays Bank plc* (2006) HL. In this case, Barclays had failed to comply with orders freezing the bank accounts of two of its customers. The order had been obtained by Customs and Excise for unpaid VAT. Customs and Excise sued Barclays for the money it was unable to recover since the bank had allowed these customers to withdraw large sums of money from the supposedly frozen accounts. Although it was foreseeable that Customs might be unable to recover sums owed to it if the accounts were not frozen, the House of Lords decided that it was not fair, just and reasonable to impose liability, potentially of over £2 million, on Barclays for not complying with a court order that it had no opportunity to resist.

In essence, the purpose of requiring a claimant to establish a duty of care is simply to determine the extent of a defendant's legal culpability. Where it is not considered appropriate for a defendant to be responsible for certain harms (e.g. the birth of an undesired child: *McFarlane v Tayside Health Board* (2000) HL), they can be insulated from liability by restricting the scope of their duty of care. The duty concept is inherently flexible, capable of expanding and contracting as needed, although this flexibility can lead to decisions based on policy reasons that are difficult to justify in the modern world (*Michael v Chief Constable of South Wales* (2015) SC).

There are certain situations where the courts have encountered difficulties when establishing whether a duty should exist because of policy considerations. These include:

1 Pure economic loss
2 Negligent misstatements
3 Judicial immunity
4 Psychiatric injury
5 Negligent exercise of statutory powers
6 Pure omissions

Pure economic loss

A significant proportion of the losses compensated by a negligence claim are economic in nature (e.g. a loss of earnings), but as a general rule, a defendant does not owe a duty of care when a claimant has suffered only 'pure' economic loss. These are essentially financial losses that are not consequences of the harm to a person or property. In *Weller v Foot and Mouth Disease Research Institute* (1966) QBD, a research institute negligently allowed a virus causing foot and mouth disease to escape. Cattle needed to be slaughtered over a wide area and livestock markets were closed to stop the spread of the infection. The claimant was a cattle auctioneer who lost profits because of the closures. Though their losses resulted from the foot and mouth outbreak, they were not a consequence of any actual harm (i.e. the loss of cattle to infection or slaughter). As their only loss was pure economic loss, they could not recover damages for negligence. The court was likely conscious that if they allowed the claim, every cattle transporter, supplier of cattle feed and dairy would come forward to claim. Liability would have been excessive in terms of the number of claims and the possible amounts involved.

The courts are particularly anxious not to create 'indeterminate liability', where a negligent act gives rise to a large number of (potentially small) claims out of proportion to the defendant's wrong. They also keep in mind that a person who suffers pure economic loss can frequently correct the situation by, for example, increasing prices. This may justify treating economic loss differently from personal injury, which can be irreparable. The courts may also be mindful of which party could more easily have insured against the loss.

The rule can undoubtedly be very harsh and sometimes almost illogical in application. Take, for example, the case of *Spartan Steel and Alloys Ltd v Martin and Co. (Contractors) Ltd* (1972) CA. The defendants had negligently cut the electric power cable under a road. This resulted in power being cut off in the claimant's metal foundry. Work in progress (a 'melt') had to be aborted, resulting in damage to the metal and the furnaces. As this was physical harm, the claimants could receive compensation for the harm to the furnaces, the metal they had to discard and the consequential loss of profit on that 'melt'. Since there was no power, they lost the possibility to carry out other 'melts' and so lost further profits, but this was classed as pure economic loss and the claimant could not recover damages.

One of the reasons the claimant in *Spartan Steel* could not recover for pure economic loss was that the length of the power outage was not the defendant's fault, as it was up to the utility company to repair the cable. This meant the loss of profits was independent of the injury caused by the defendant's negligence.

There have been cases where the courts have allowed claims for pure economic loss, but these have become increasingly restricted over time. One such example is *Dutton v Bognor Regis Urban District Council* (1972) CA. Here the claimant sought compensation from the local council, which had certified her house was sound, when it emerged the foundations were defective because the house was built on a rubbish tip. The defendant would have discovered this had the proper checks been made. The claimant had bought the house from the initial owner, who had bought it from the builder, so she had no direct contract with either builder or council. The question for the court was whether the claimant, as a subsequent purchaser, had suffered physical damage or pure economic loss. Since the house was already defective at the moment of purchase, she had not suffered damage to the property, but only to its value. However, the court accepted that the damage was damage to property.

At the time *Dutton* was decided, judges had adopted an expansive approach to the duty of care concept. The high-water mark for claims of pure economic loss was *Anns v Merton LBC* (1977) HL, another defective foundation case, where Lord Wilberforce set out a two-stage test for the duty of care (requiring the defendant to show a reason why a duty should not exist) that was swiftly backed away from over the following decade. A detailed analysis of the *Anns* test is beyond the scope of this chapter, but it is interesting to note that this more liberal test continues to be used in other countries, such as Canada and Singapore. It is quite possible that the conservative political climate in the UK during the 1980s motivated what has become known as the 'retreat from *Anns*'.

That 'retreat' was completed in *Murphy v Brentwood District Council* (1990) HL. Here, a builder carried out defective construction work, resulting in an unstable structure. The purchaser of the property sold it at a loss due to the unstable foundations and sued the council for their failure to exercise reasonable care in carrying out their regulatory inspection. It was held that there can be no claim for pure economic loss and that claims for defective building must be in contract, through the Defective Premises Act 1972 or through the National House Building Council (NHBC) ten-year warranty.

In a number of cases involving building works, the courts have emphasised the role of contractual agreements and that claims should be pursued in contract where possible. Within contracts, the parties can establish their responsibilities, subject to statutory limitations. Once they have done that, it was said to be wrong to impose any further liability ((see *Simaan General Contracting Co. v Pilkington Glass Ltd (No. 2)* (1988) CA; *Greater Nottingham Co-op Society Ltd v Cementation Piling* (1988) CA; *Foundations Ltd* and *D & F Estates Ltd v Church Commissioners for England* (1989) HL)).

The difficulty of extending negligence to cover pure economic loss lies in the need for a connection between the defendant's act and the claimant's loss. For this reason, negligence is simplest when presented with a collision of two parties that ends in a tangle of broken limbs or another physical malady. In *Donoghue v Stevenson*, for example, the defendant's negligence caused harm to the claimant by making her ill. But when a builder constructs a house with defective foundations, the argument goes that the owner who discovers that defect is not physically harmed, nor has the house damaged any other property. The house suffers from a defect, which damages the property itself, reducing its value. The decisions in *Dutton* and *Anns* are at odds with this point of view, favouring the claimant.

In *Robinson v PE Jones (Contractors) Ltd* (2011) CA, it was again stressed that remedies against builders should be pursued through contract law and that there would only be liability in tort if a special relationship could be established. The mere existence of a contract did *not*, of itself, indicate a special relationship and assumption of responsibility. This was echoed in *Thomas v Taylor Wimpey Developments Ltd* (2019) TCC. Here, a householder sued builders for structural damage. They relied on an *obiter* statement by Lord Bridge in the *Murphy* case, such that, although no claim could be made in the absence of damage to person or property, if a structural fault was near a boundary and presented a danger to members of the public, then there could be a claim. This apparent exception to what has become the general rule around claims for economic loss was rejected

Negligent misstatements

Excluding pure economic loss claims caused problems in those cases where the negligence consisted of a negligent statement rather than a negligent act. In such cases, pure economic loss is the most likely form of harm to result. For example, if a firm wishes to do business with a new and unknown customer, it may have reservations about extending credit to that customer. To protect itself, the firm may ask permission to approach the customer's bank for a credit reference. If the reference is negligently given by the bank and the firm relies on it, extends credit and sustains financial loss as a result, the exclusion of pure economic loss would prevent the firm from suing the bank for negligence.

When presented with a similar set of facts in the case of *Hedley Byrne and Co. Ltd v Heller and Partners Ltd* (1964) HL, the courts chose to distinguish between negligent acts and negligent misstatements. If a negligent statement causes physical damage, the normal rules of negligence apply – thus, in *Clay v Crump* (1964) CA, an architect who negligently stated a wall was safe to be left standing was liable in negligence when the wall collapsed and injured a worker.

Where a negligent misstatement causes pure economic loss, liability may now arise either under the principles set out in *Hedley Byrne* (discussed later) or the Misrepresentation Act 1967 (see *Patarkatsishvili and another v Woodward-Fisher* (2025) Ch). Liability for false statements in tort had previously been covered only by the tort of deceit. To succeed in an action for deceit, it is necessary to prove that the false statement has been made knowingly (without belief in its truth) or recklessly (careless whether it be true or false). There must be dishonesty; mere carelessness is not enough. Relief could also be obtained in a contract where the false statement induced the claimant to enter into the agreement and, in such circumstances, the claimant may be able to rescind the contract (see Misrepresentation in Chapter 5). Where a contract exists between the parties, using the rules of the Misrepresentation Act 1967 is preferable, as the burden of proof is placed upon the defendant, who must prove they had reasonable grounds to believe the facts represented were true. Where the false statement does not induce a contract, however, then a claim for negligent misstatement will be the only option.

The general effect of *Hedley Byrne* is that certain negligent misstatements are now actionable, even if there is only pure economic loss. For liability to arise, the defendant must present themselves as having a special skill (they do not need to be a professional: *Chaudry v Prabhakar* (1989) CA), reasonably know about the recipient of their advice (this could be a specific individual or a finite group) and it is reasonable for the recipient to rely on the advice. Thus, if a stockbroker gave a friend negligent advice on the housing market while in the pub, they lack the requisite special skill and it would also likely be unreasonable to rely on advice given in a social environment. If the negligent advice was about investments and it was delivered to a client in the office, then it would be a completely different outcome.

The claim in *Hedley Byrne* ultimately failed because the defendants had expressly disclaimed responsibility for their reference. The Unfair Contract Terms Act 1977 now limits the possibility of using such a disclaimer successfully. Under s.2, where a person seeks by a notice to exclude or restrict his or her liability in negligence that results in loss or damage other than personal injury or death, they may only rely on that disclaimer if it satisfies the requirements of reasonableness. The Unfair Contract Terms Act 1977 applies only to a business liability situation. For consumer contracts, s.62 and Schedule 2 of the Consumer Rights Act 2015 state that exclusion clauses are only applicable where fair.

The possible scope of the *Hedley Byrne* principle is very wide. It applies, for example, to professional advice given by architects, surveyors, engineers, accountants or lawyers, amongst others. For this reason, there is inevitably an overlap between the rules of tort and contract. Most professional people are likely to have a contract with their client, so it is more likely that they will be sued for breach of contract in respect of any negligent service. However, two cases show how reliance on a negligent misstatement made outside of any contract can cause extensive loss. In *Smith v Eric S Bush* (1989) HL, a prospective purchaser of a house sought a mortgage. The building society instructed a surveyor to carry out a house valuation for them and, on the strength of that survey, loaned money to the purchaser. Also relying on the survey, the purchaser decided to go ahead and buy the house. The valuation said no essential repairs were needed, but this was inaccurate, as bricks from the chimney collapsed through both the roof and loft space. There was no contract between the purchaser and the surveyor, who had been engaged by the building society. However, the court held that the surveyor was liable to the purchaser, as the situation was akin to a contract. The surveyor knew the payment they received (£36.89) came from the purchaser and, furthermore, knew the purchaser would rely on the report when deciding whether to buy the house. The surveyor sought to rely on a disclaimer of liability, but the court held it was not reasonable under s.11 of the Unfair Contract Terms Act 1977, as the surveyor had to know the purchaser would be supplied with a copy of the valuation report and would be likely to rely on it.

In *Caparo Industries plc v Dickman* (1990) HL, the auditors of a company issued inaccurate and misleading accounts for a company in which Caparo held shares. Caparo bought further shares on the strength of the audited accounts, which were circulated to all shareholders. Later the same year, they mounted a takeover bid, but once they acquired the company, Caparo discovered the true state of its finances. They alleged the auditors owed them a duty of care when making statements, both as a shareholder and as a potential bidder to take over the company. In the House of Lords, it was held that no duty was owed to them in either capacity. As a member of the public, looking at the audited accounts with a view to takeover, there was no proximate relationship between the auditors and Caparo. There was no duty as a shareholder purchasing more shares in the company, either; the court stated an individual shareholder should not occupy a different position from a member of the public. The accounts were prepared and circulated to members of the company to enable better management, not to permit them to make a personal profit.

Judicial immunity

As a result of *Hedley Byrne*, the courts have recognised there are times when there will be liability in tort for negligent misstatements. Nevertheless, no liability attaches if the statement is made by a judge or arbitrator during legal proceedings. The reason for this is public policy. It is regarded as being against the public interest to have the constant possibility of litigation being reopened. The remedy, where it is believed that an error has been made, is to use the appeal system.

The immunity of arbitrators was considered in the case of *Sutcliffe v Thackrah* (1974) HL, where an architect who had negligently certified work claimed to be acting as an arbitrator in the certification process. That claim was not accepted by the court, but had it been correct, no action in negligence would have lain against the architect. The immunity of arbitrators is dealt with by s.29 of the Arbitration Act 1996.

In *Mazhar v Lord Chancellor* (2017) CA, it was held that nothing in the Human Rights Act 1998 overruled the common law principle of judicial immunity. However, the long-standing immunity of other professionals regarding legal proceedings has been removed in the courts over the past two decades. For barristers and solicitors in *Arthur J S Hall & Co. v Simons* (2000) HL and for expert witnesses in *Jones v Kaney* (2012) SC.

Psychiatric injury

Psychiatric injury refers to where the claimant has suffered a recognised psychiatric disorder because of the defendant's negligence. It was previously referred to as 'nervous shock', but such a label is reminiscent of a Victorian-era approach to alienism (as psychiatry was then called), in that it assumes mental harm is caused by a 'sudden assault on the nervous system' (*Alcock v Chief Constable of South Yorkshire* (1990) HL). Though the courts have struggled to adapt to claims for psychiatric harm, accepting them begrudgingly, injury to the mind (as opposed to the physical brain) was recognised as far back as the case of *Wilkinson v Downton* (1897) QBD. This case allowed recovery for intentionally inflicted mental injury, caused by means of a practical joke. A claim in negligence finally arose in *Dulieu v White* (1901) KBD, where the claimant was at risk of physical injury when a horse and cart crashed through a pub wall. Since the mid-1980s, the courts have accepted that 'pure' psychiatric injury (i.e. unconnected with physical harm) is foreseeable in certain classes of persons, but the House of Lords severely restricted who could access compensation for mental harm. Various (debatable) reasons have been given for this strict approach, including fear of fraudulent claims (even though the same problem exists with physical injuries, for example, whiplash), indeterminate liability, the cost of expert witnesses (which is just as much a problem in claims for physical harm) and a widely criticised theory that permitting claims would act as a disincentive for recovery. In truth, the leading cases in this area are sorely outdated, and the rules – created in response to extraordinary events rather than more common accidents – are far too inflexible to accommodate a nuanced understanding of mental injury. To complicate matters, the courts have found a significant number of exceptions to allow deserving claims to succeed. It is fair to say the law here is a 'patchwork quilt'.

How negligence treats a claimant who suffers psychiatric injury depends upon which of two categories they fall into: primary victims or secondary victims.

A primary victim is someone who has either sustained physical injury *and* psychiatric injury or was at risk of suffering physical injury but only sustained psychiatric injury. In either circumstance, the usual rules of negligence apply and the defendant will be liable for any psychiatric harm so long as a risk of 'personal injury' is foreseeable (*Page v Smith* (1995) HL). So, a victim seriously injured when a building collapsed would have a claim for both physical *and* psychiatric injury. A victim of the same collapse who narrowly escaped injury but suffers a recognised psychiatric disorder in the aftermath would only claim for psychiatric injury.

A secondary victim is someone who was never in any physical peril but has suffered psychiatric injury because of what they witnessed. For example, if a parent watches an unattended lorry roll downhill towards where they know their child is playing, they may suffer mental injury despite not being in danger (as happened in *Hambrook v Stokes Bros.* (1925) CA). The problem, in so far as the courts see it, is that a significant number of people could (in theory) claim as secondary victims without robust criteria. Where the circumstances are sufficiently horrific, it is conceivable that people who did not witness anything of the incident might suffer psychiatric harm after learning about it via the news or even from a third party. This is where the concern of indeterminate liability inevitably arises.

The House of Lords answered some of these questions in the case of *McLoughlin v O'Brian* (1983). The claimant's husband and two of her children were badly injured in a horrendous road accident, and one child was killed. Her fourth child was unharmed. A friend came to tell the claimant the news of the accident and took her to the hospital. There, the claimant learned the full extent of the tragedy and saw her family in harrowing and distressing circumstances. It was held that her psychiatric injury was reasonably foreseeable and she should be permitted to recover damages.

Lord Wilberforce, giving the leading judgment, set out a number of principles for claims involving secondary victims. First, a close family relationship had to exist between the victims of an event and the claimant; second, the claimant had to be close to the event 'in both time and place', which meant the claimant was either present for the accident or its immediate aftermath; third, the claimant's injury had to 'come through sight or hearing of the event, or of its immediate aftermath'. Lord Wilberforce's principles meant claims by the unrelated bystanders (which had already been rejected by the House of Lords in *Bourhill v Young* (1943) HL), those who were not present for the event or its aftermath and those who learned of the event from a third party were all excluded. Lord Bridge was less enthusiastic about setting fixed rules for certain types of claims, believing the ordinary rules of duty were sufficiently robust. Indeed, the approach in *McLoughlin* meant that certain types of deserving claims could fail for lack of a close family relationship. For example, a person who suffered psychiatric harm as a result of assisting in an accident or disaster, where they saw and experienced dreadful injuries and suffering, would be unable to sue in negligence.

If damages for psychiatric harm were dependent on reasonable foresight of harm to the claimant in question, then that, in itself, would act as a significant filter to ensure that only a proportion of cases were likely to succeed. However, *McLoughlin* narrowed the scope of liability further by stating the need for a close family relationship.

The rules on psychiatric injury were tightened even further in *Alcock v Chief Constable of South Yorkshire Police* (1991) HL. This was a test case brought by ten relatives of victims of the Hillsborough football stadium disaster, where negligent management of the match-day crowd resulted in a crush that killed 97 people and injured over 700. They claimed psychiatric injury suffered from watching the events unfold from either elsewhere in the stadium or on television. The House of Lords rejected their claim for damages. Watching edited pictures on television where victims could not be identified did not satisfy the test of proximity, namely that the psychiatric disorder must have been the product of what the claimant saw or heard of the event or of its immediate aftermath. The court went further, though, introducing a presumption that only a spousal or parent-child relationship was a sufficiently 'close ties of love and affection'. Siblings, grandparents, aunts, uncles and so on all had to prove the closeness of the relationship. As this was not a requirement when the claimants brought their case, they had not shown sufficiently close bonds. The case had long been criticised for being harsh and not corresponding with medical knowledge and the Law Commission proposed reforms in 1998, but the '*Alcock* rules' for secondary victims remain entrenched.

In *White v Chief Constable of South Yorkshire* (1999) HL, the police officers on duty at Hillsborough brought a separate action for psychiatric injury. The court set out its reasons for treating claims for mental harm differently (see the beginning of this section), adding that the claim also had to fail on policy grounds – as the relatives' claim was rejected in *Alcock*, it would not be just for the police officers to succeed.

Thus, the position remains that unless someone is either a primary or secondary victim, they have no access to compensation via a claim in negligence.

Negligent exercise of statutory powers

Dutton v Bognor Regis UDC (1972) CA opened up a significant development in the law of negligence, when an owner of a building sued the local authority, alleging negligence regarding the inspection of the foundations of the building. In many of these cases, the builders will not be worth suing, for example, if they have gone into liquidation or there are doubts about their solvency. Local authorities, therefore, look like richer pickings in these circumstances.

Where a local authority acts under statutory powers, as, for example, in the building inspection processes, there may be liability for negligence if, once having decided to inspect, the inspector fails to carry out the inspection properly. Of course, where there is merely a statutory *power* to act, as opposed to a statutory *duty*, the local authority has some discretion about whether to do anything at all. Its resources may be so scarce that it must determine priorities about which inspections it will make. In these circumstances, if it decides to make no inspection, the question then arises whether the authority could be negligent for failing to inspect. It seems that there would be no liability in negligence, so long as the local authority has given proper consideration to the question of whether or not to inspect.

These issues were all considered in *Anns v Merton London Borough Council* (1977) HL. Like *Dutton*, this was a case involving defective foundations. Although the decision has now been overruled (in *Murphy v Brentwood District Council* (1991) HL, discussed later), some consideration of the approach taken by the House of Lords in *Anns* is vital to an understanding of the later developments. So far as local authorities are concerned, *Anns* appeared to decide that a local authority may be liable in negligence to a building's owner regarding their statutory control over building operations. The extent of liability would be the cost of remedying a dangerous defect in the building resulting from a negligent failure by the authority to ensure that it was constructed according to the appropriate standards in building regulations.

The difficulty created by this broad-ranging approach can best be illustrated by comparing a defective product with a defective building. If you make a product that contains some latent defect that causes damage or injury to persons or property, you may be liable in negligence if you made the product negligently (e.g. *Donoghue v Stevenson*). However, if the product is merely defective in quality, then your liability in that regard lies only in contract (under the Sales of Goods Act 1979 or the Consumer Rights Act 2015). Anyone acquiring the product who does not have a contract with you may suffer economic loss, either because they need to repair the product or have to discard it, but you would owe no liability in tort in respect of mere economic loss in the absence of a special relationship (see Negligent Misstatements mentioned previously).

If a builder constructs a house with a latent defect (e.g. defective foundations), the builder may be liable in tort for injury to the person or damage to property arising from the dangerous defect. But if the owner of the house gets to know about the defect before it causes any injury or damage, then the owner would have rights in contract if the house had been built for them. They would have no rights in tort, because their loss would be mere economic loss, i.e. the cost of the repair or the cost of abandoning the building if it cannot be repaired. To hold otherwise would be to impose on the builder a warranty as to the quality of their work that would be owed not only to the original owner, but also to all who subsequently acquired an interest in the property.

There is a distinction between defects that cause external harm and defects that merely harm the product itself. This raises the question of the liability of local authorities. If the builder's duty should be limited as suggested previously, then it seems altogether wrong to make the local authority liable to a greater extent than the builder. The function of the local authority in inspecting building work is to ensure that there is compliance with building regulations. If the local authority negligently failed to ensure compliance with those regulations, it seems absurd that it would then incur a liability greater than that of the builder himself. In the *Murphy* case, Lord Mackay made the point that it is not a proper exercise of judicial power for the courts to create large new areas of responsibility on local authorities in respect of defective buildings. That, properly, is the role of Parliament. He was strengthened in his view by the consideration he gave to the Defective Premises Act 1972 (see what follows).

In *Murphy v Brentwood District Council* (1991) HL, the case arose out of defective foundations. A pair of semi-detached houses had been built atop an infilled site on a concrete raft foundation to prevent damage from settlement. The local authority approved the plans. Some years after the house was completed, the owner noticed serious cracks and then discovered that the raft foundation was defective and had become distorted. Necessary repairs would have cost £45,000, which the claimant could not afford. Instead, they sold the house and obtained £35,000 less than the market value, had the house been in sound condition. The claimant sued the local authority to recover that sum. The House of Lords held that the claimant was unable to recover from the local authority, as their loss was pure economic loss.

Inevitably, these complex cases are often the result of an aggrieved person looking for someone who is financially worth suing. It has already been suggested that the erring builder may be insolvent, but the other difficulty a claimant faces is the problem of limitation periods. In all tort actions, a claim must be commenced within the appropriate time limit laid down by the Limitation Act 1980. An action relating to damage to property must be commenced within six years of the cause of action accruing. Usually, this is the moment when damage occurs. That does not necessarily mean when the defective building was constructed. Since the decision of the House of Lords in *Pirelli General Cable Works Ltd v Oscar Faber and Partners* (1983) HL, the cause of action arises when physical damage actually occurs to the building, whether or not the claimant knew, or could have known, about it. This rule is undeniably harsh, especially as it fixes the date for the cause of action accruing for subsequent as well as present owners. It runs counter to an earlier test proposed by the courts, that time should only run from a point when the claimant could have discovered the defects. Some help with this problem has been given under the Latent Damage Act 1986. In those instances where damage cannot be discovered immediately (i.e. defective foundations), the claimant must bring an action within three years from the time when damage was discovered or when it should have been discovered, if using reasonable care, subject to an overriding rule that the action must be commenced within 15 years of the building being completed.

Omissions

As a general rule, there can be no liability in negligence for a pure omission to act. The law distinguishes 'between making matters worse (or harming) and failing to confer a benefit (or to protect from harm)' (*Tindall and another v Chief Constable of Thames Valley Police* (2024) SC). Thus, if one or more victims of a car accident are trapped in their vehicle, a passerby commits no legal wrong if they make no effort to rescue those trapped. This is described as a 'pure omission', in the sense that there was no existing duty to act. If a person creates the danger or there is an existing duty, then an omission can be as wrongful as a negligent act. A parent of a toddler, for example, has a duty to care for the child. If they omit to feed the toddler, such behaviour would be actionable.

In *Tindall v Chief Constable of Thames Valley Police*, the Supreme Court explained that the distinction between negligent acts and omissions now has added significance because the liability of public authorities 'is governed by the same principles as private individuals'. Public authorities (e.g. the police, fire brigade, etc.) can have statutory powers and duties to protect the public from harm, but an authority will only be liable for failing to do so if a private individual would also have been liable in the circumstances.

To recover damages, a claimant must show a public authority did not just fail to protect them but actually caused the harm they suffered.

If there is no duty to act, but someone decides to act – such as the passerby deciding to try to release the crash victims from the car – the law generally takes the view that there will be no liability for negligence unless the passerby unreasonably makes matters worse, i.e. the law operates in favour of rescuers. As there is no duty to act, if you do act, there is no duty to affect an improvement. This principle is subject to an important exception in relation to the exercise of statutory powers. In such a case, although there is no duty to act, the defendant has a duty, if they choose to exercise the power, to affect an improvement.

Breach of the duty of care

If a duty to take reasonable care is owed to the claimant, the next stage is for the claimant to prove that the defendant was in breach of duty – in short, that the defendant failed to take reasonable care. This raises the question of what would amount to reasonable care in the circumstances of the case. If the defendant's behaviour falls below what is reasonable in the circumstances, then they are in breach of duty.

Inevitably, there is no absolute precision in measuring what amounts to reasonable care in any given situation. A court will need to weigh up the degree of risk involved in the defendant's behaviour (*Bolton v Stone* (1951) HL); how serious was the harm that was likely to be caused (*Paris v Stepney Borough Council* (1951) HL); whether there was any particular social value or utility in the defendant's activities (*Tomlinson v Congleton Borough Council* (2003) HL) and how costly it would be to take precautions to eliminate or reduce the risk (*Latimer v AEC* (1953) HL).

The standard of care expected is that of the fictional 'reasonable person', who once rode on the Clapham Omnibus (*Blyth v Birmingham Waterworks* (1856) Exch) and now takes the London Underground (*Rees v Darlington Memorial Hospital* (2003) HL). Essentially, the defendant's behaviour is assessed against an objective standard, and whether they tried hard or did their best is irrelevant. Of course, where a defendant holds out as possessing a particular skill or capability, they will be expected to display a level of competency with the skill in question (*Bolam v Friern Hospital Management Committee* (1957) QBD). The cases setting out the principles of 'professional' negligence mainly concern healthcare practitioners, such as *Bolam v Friern*, but the test applies equally to other professions. A builder must build with the care that a reasonable builder would take, a surveyor must carry out their work with the care a reasonable surveyor would take and so on.

In *Large v Hart* (2021) CA, a surveyor completed a Homebuyer's report on a refurbished cottage with a value of £1.2 million. Such a report grades items using a traffic light system – 'red' for urgent repairs, 'amber' for non-urgent repairs and 'green' for no repairs needed. The report on the cottage noted one 'red light' (drainage) and one 'amber light' (pipes and gutters). The purchasers relied on the report and bought the property, but soon discovered serious water ingress, which required extensive rectification. The purchaser brought an action against the surveyor for negligence, and the court held they were liable because their conduct had fallen below the reasonable standards of the surveying profession. Rendering on the property made it impossible to check for damp-proof membranes, but the surveyor had assumed these were present and had not informed the purchaser that further investigation was needed.

The courts do not generally consider the age or experience of a person relevant to the question of breach. An 18-year-old learner driver, for example, is held to the same standard as a fully qualified driver (*Nettleship v Weston* (1971) CA) and a graduate at the beginning of their career is held to the same standard as a more experienced colleague (*Wilsher v Essex Area Health Authority* (1988) HL). However, when the court is trying to determine if a child has been negligent, it considers whether the child in question took the care that could be expected of a reasonable child of the same age.

When formulating a standard of reasonable care, everyone is assumed to have knowledge about common facts, e.g. fire burns, water can drown and knives cut. A person with expert knowledge will be expected to display a reasonable amount of such knowledge. When judging whether a defendant has taken reasonable care, some guidance may be given by looking at what is common practice in the activity, industry or operation in question. But common practice is not an absolutely reliable guide or protection. An employer who failed to provide protective clothing to their employees may argue they acted reasonably because their practices were in step with other employers in the industry, but it could also be said employers in the sector are lacking care (*Izzard v Field Palmer* (1999) CA). As the court in *Bolam* put it, a professional cannot 'obstinately and pig-headedly carry on with some old technique if it has been proved to be contrary to . . . the whole of informed [professional] opinion'.

In assessing how much risk is involved in the defendant's activities, one judge has said that 'people must guard against reasonable probabilities, but they are not bound to guard against fantastic possibilities' (Lord Dunedin in *Farndon v Harcourt-Rivington* (1932) HL). If it is only possible to conceive of harm occurring in an obscure or bizarre way, then the risk may be so small a defendant was justified in ignoring it. The law only requires reasonable care against foreseeable risks, not an absolute guarantee of safety in all circumstances.

In practice, it may be very difficult to separate questions like the degree of risk from the seriousness of potential harm. Even in cases where a very small risk is involved, it may be necessary to take appropriate precautions where the harm, if caused, would be very serious. If an employer operates a factory with huge tanks of liquid, it is foreseeable that employees or objects may fall in and cause the liquid to splash. If the liquid is cold water, a barrier to prevent employees from falling in may be all that is necessary to show reasonable care. If the liquid is a caustic solution, likely to burn at the slightest touch, then the employer may have to consider a protective cover for the tanks. What precautions are reasonably necessary may also be determined by asking whether employees are meant to be in the vicinity of the tanks or not and whether those employees have been issued with protective clothing.

A defendant engaging in some activity designed to save life and limb may well escape liability for negligence even though his or her behaviour may amount to a failure to take reasonable care in other circumstances. In *Watt v Hertfordshire County Council* (1951) CA, firefighters rushed to the scene of an accident where a jack was needed to free a woman who was trapped. Because of the emergency nature of the call, the jack was transported on an unsuitable vehicle and was not secured during the journey. It shifted on the lorry, and a firefighter was injured. Though the defendant was found not liable, it was made clear that if the jack had been transported in a similar way in a commercial enterprise, there would have been a failure to take reasonable care.

Even in those cases where some minute risk is foreseen, it may be possible to eliminate that risk, but only by disproportionate expenditure. The law does not impose such an impossible standard. If the risk could be eliminated or reduced by reasonable expense or practicable precautions, then the defendant should have done what they could. Again, it must be emphasised that they are only required to act reasonably. In *Latimer v AEC Ltd* (1952) HL, oil was spread over a factory floor when the premises were flooded. The factory occupier scattered some sawdust, mopped up as much as possible and gave a warning to employees to take care, as the floor was slippery. The claimant slipped and was injured, but the defendants were not liable as they had taken reasonable care. Any harm they could foresee was not likely to be serious and, short of closing down the factory (a disproportionately costly measure), there was nothing more they could have done.

The discussion of reasonable foresight of harm may raise the idea that what the courts are concerned with is hindsight, but breach is measured against the knowledge available at the time of the alleged breach. So, if a defendant only learns of a risk after the fact, they will not have breached their duty of care if it was not something they could have reasonably known about at the time. In *Roe v Minister of Health* (1954) CA, the claimant was paralysed when contaminated anaesthetic was injected into their spine. The defendant, however, was found to have exercised reasonable care because the mode of contamination (hairline cracks in glass ampoules) was not known at the time.

It has already been indicated that it is up to the claimant to show the defendant was in breach of duty. The claimant may be assisted in proving negligence by s.11 of the Civil Evidence Act 1968. If the defendant has been found guilty of a criminal offence involving negligence, that fact may be used as evidence of negligence in subsequent civil proceedings. This is obviously a useful rule where a person has already been convicted of certain driving offences, but it could also be relevant where an employer has been found guilty of offences under health and safety legislation.

A claimant seeking to prove negligence may also rely on the maxim *res ipsa loquitur*: 'the facts speak for themselves'. This could be pleaded in cases where the facts indicate no explanation other than negligence. In *Scott v London and St Katharine Docks Co.* (1865) ExCh, a man was walking past a warehouse when sacks of sugar fell on his head. What explanation could there possibly be other than that someone was being negligent? It is only appropriate to plead *res ipsa loquitur* if the defendant had management or control of the thing that caused the harm. This tends to depend on whether there was any possibility of some external interference. Where a claimant relies on the maxim, it is open for the defendant to rebut the finding of negligence if they can do so.

Causation

Once a claimant has proven a defendant was in breach of duty, they must go on to prove causation. This, in effect, means that they must show that the defendant's negligence caused the harm they have suffered. To establish causation, it must be shown that the breach of duty is both factually and legally the cause of the harm. Factual causation refers to the chain of events linking (or not, as may be the case) the defendant's breach to the claimant's injury. Legal causation (sometimes called 'remoteness') assesses whether that injury was a reasonably foreseeable consequence of the defendant's actions.

Factual Causation (the 'but for' test)

A defendant may behave negligently towards a claimant and yet still not be the cause of the injuries. In *Barnett v Chelsea and Kensington Hospital Management Committee*

(1968) QBD, a doctor was negligent in failing to examine a patient who had been brought into casualty, complaining of feeling sick after drinking tea. In fact, the patient was dying from arsenic poisoning, but as nothing could have saved him, the doctor's lack of care did not *cause* his death. The claim by the patient's widow failed as a result.

Factual causation is assessed using the 'but for' test. Essentially, the question is: 'but for' the defendant's negligence, would the claimant have been injured? In *Barnett*, the answer was 'yes' – there was no possibility of the treatment for arsenic poisoning being administered in time to save his life.

The 'but for' test is by no means conclusive, and it may be particularly ineffective in giving an answer where the harm caused to the claimant was inflicted by two or more people, in quite separate incidents. Claims involving asbestos-related disease often pose such evidential problems, as the claimant may have been exposed to asbestos by multiple employers. As it is impossible to pinpoint which employer caused the harmful exposure, the 'but for' test imposes too high a burden on claimants. In these cases, as it was considered unjust if the claimants could not succeed, the courts were prepared to accept a 'material increase of risk' as proof of factual causation (*Fairchild v Glenhaven Funeral Services Ltd* (2003) HL; *Sienkiewicz v Greif* (2011) SC).

In *Chester v Afshar* (2004) HL, the House of Lords stretched the principles of factual causation to the limit (and perhaps beyond). The claim related to a surgeon's failure to advise a patient on the risk of paralysis associated with an operation, which she later suffered. Had the patient known of the risk, she would not have undergone the operation at that time. The defendant argued the risk associated with the operation was the same, whether or not they gave advice about potential side-effects – that is, it was not the lack of advice that caused the damage, it was the operation, which had not been carried out negligently and which the patient may have chosen to have at some later date, even knowing the risks. The majority of the court held that the test of causation was met on grounds of policy. The risk of suffering side effects as a result of the operation came within the scope of the surgeon's duty to warn the patient, and therefore the side effects suffered could be regarded as having been caused by the surgeon's breach of that duty. In *Montgomery v Lanarkshire Health Board* (2015), the Supreme Court stated that doctors have a duty to take reasonable care to ensure that patients are aware of any material risks involved in any recommended treatment, and of any reasonable alternative or variant treatments.

Legal Causation (Remoteness)

Assuming there is no difficulty in proving factual causation, the claimant must then show that the harm they suffered was a reasonably foreseeable consequence of the defendant's negligent act. This is legal causation, or 'remoteness of damage'. Any act can give rise to endless consequences, and it is often possible to say with hindsight that had a particular event in the past not happened, none of the later events would have occurred.

As a feature of not imposing excessive liability, the law has created the rule that a claimant can only recover for harm that is a reasonably foreseeable consequence of the negligent act. An objective test is used to determine whether the defendant should have reasonably foreseen a certain outcome, but it is important to note that the question is whether or not the *type* of harm – physical, economic and so on – was foreseeable, not the exact way it occurred. For example, if you light a bonfire, it is reasonably foreseeable that people may be burned or that a spark from the fire may set alight nearby property.

But it is likely unforeseeable that the fumes from the bonfire would suffocate the battery chickens being reared on adjoining land.

In a case that established the approach to legal causation, a ship called *The Wagon Mound* negligently discharged oil into Sydney Harbour as it was sailing out. The wind and tide carried the oil all around the wharf area, where scraps of cotton wool were mixed with it by the wind, and then onto where the claimants were carrying out welding. The welding was suspended until the claimants checked whether the oil would ignite. They discovered that this was unlikely, as it was furnace oil with a high ignition point. Welding recommenced and an extensive fire was caused when the oil did, in fact, ignite. The defendants, the owners of *The Wagon Mound*, were held not liable. Although they owed a duty to take care in relation to the claimants and were in breach by negligently discharging the oil, the damage they had caused was not a reasonably foreseeable consequence of their negligence (*The Wagon Mound, No. 1* (1961) PC).

In *The Wagon Mound*, no doubt, the ship owner would have been liable if the claimants had been complaining about oil fouling their slipways, but as fire damage was an unforeseeable type of harm, the claimants could not recover. This case, of course, related to property damage. When the harm suffered is physical, it is much more difficult to say an outcome was not reasonably foreseeable; if the type of harm is foreseeable in any form, then the claimant is able to show it is reasonably foreseeable. In *Bradford v Robinson Rentals Ltd* (1967) CA, a van driver was sent on a long journey in the depths of winter in a defective and unheated van. When he suffered frostbite, he sued the employers, who indisputably owed him a duty of care. Frostbite is relatively unusual in the UK; nevertheless, *some* harm to the driver's health was reasonably foreseeable and frostbite was therefore in the same class of risk. Oddly, the court discussed how cold sores were foreseeable, but since these are caused by bacteria rather than extreme cold, it is a rather peculiar point.

Doughty v Turner Manufacturing Co. Ltd (1964) CA is at odds with the general approach to cases of physical injury. Here the employer had a vat of acid maintained at a very high temperature. It was reasonably foreseeable that employees might be injured by falling in or by being splashed if any object fell in. The employer would certainly be liable if an accident had occurred in either of those ways. But when the lid of the vat, made from asbestos, slid into the acid, the chemical reaction resulted in an explosion and employees were injured by erupting acid. The employers were held not liable as the damage was too remote. The court accepted it had occurred in a manner that was not reasonably foreseeable. But as the question is where the type of harm (burning) was reasonably foreseeable, not the way it occurred, the court appeared to misapply the test. In both *Hughes v Lord Advocate* (1963) PC and *Jolley v London Borough of Sutton* (2000) HL, the courts looked at whether the injury was of a type that was foreseeable and answered in the affirmative.

Eggshell Skulls, Intervening Acts and Defences

When assessing causation, a court will bear in mind the following points:

The defendant must take the claimant as he finds him. If a negligent driver hits a claimant who suffers from a weak heart and dies as a result, they may argue that, whereas the claimant's injury has proved fatal, any normal victim would only have been shocked and bruised. Again, they may crash into the back of another vehicle, which may be old and rusty or an expensive Rolls-Royce. But in each example the defendant will be liable to the full extent of the injury or damage caused, if injury or damage of that type could have been foreseen. This is sometimes referred to as the 'eggshell skull' principle.

(*Smith v Leech Brain & Co Ltd* (1962) QBD)

The consequences of the defendant's breach of duty may have been 'overtaken' by some act which intervenes. This principle is known as *novus actus interveniens*. Essentially, another act has 'intervened' to break the 'but for' chain of causation. For example, in *Knightley v Johns* (1982) CA, the claimant was a police officer who attended a car accident blocking one carriageway of a busy tunnel. The police sergeant overseeing the scene sent the claimant to close the opposite end of the tunnel, but this required him to ride his motorcycle against the flow of traffic. When the claimant was hit by a car, the sergeant's acts were said to break the chain of causation, meaning the driver who caused the original accident was not to blame. The 'intervening act' will usually be the sole cause of the claimant's harm. Note here that medical intervention very rarely breaks the chain of causation.

(*Webb v Barclays Bank* (2001) CA)

If a person is responsible for the claimant's harm more than once, the court will then assess contribution. So, if an employee is injured at work and then has their foot wrongly amputated during treatment, both the employer and the healthcare provider will be liable (*Webb v Barclays*: the contribution was assessed at 75% on Barclays and 25% on the healthcare provider). The situation is governed by the Civil Liability (Contribution) Act 1978. The Act provides that where two or more persons are liable in respect of the same damage, each may recover from the other, or others, a contribution of an amount regarded as 'just and equitable', considering their share of responsibility for the damage.

Any applicable defences. Even if the claimant has satisfied all the mentioned aspects of proof, the claim may still fail if the defendant has a complete defence (e.g. *volenti*). More usually, if it could be shown that the claimant was partly to blame for their own injuries, the award of damages will be reduced. This is called contributorily negligence. The present rules are to be found in the Law Reform (Contributory Negligence) Act 1945. If a claimant suffers damage partly through their own negligence and partly through the negligence of the defendant, the compensation will be reduced to the extent that the court thinks just and equitable, considering the claimant's share of the responsibility. In *Ayres v Odedra* (2013) QBD, the claimant drunkenly dropped their trousers and flashed his derriere at motorists waiting at a pedestrian crossing. When the lights changed, the claimant dove for the island in the middle of the road, but their trousers became tangled around their ankles and they landed in the road instead. When they were hit by a car, the High Court said the claimant was 20% responsible for his own injuries. The reason the driver was still 80% liable was that, once they lost sight of the claimant, they could have stopped their vehicle.

To use the defence, the defendant must show that the claimant failed to take reasonable care for his or her own safety. As with negligence itself, the standard of care varies according to the circumstances. Thus, an experienced, mature worker can be expected to take more care when carrying out his or her work than a new apprentice. This defence is used in motoring cases when an injured claimant has failed to wear a seatbelt or crash helmet. In such cases, the claimant may well find themselves regarded as contributorily negligent up to a maximum of 25% (*Froom v Butcher* (1975) CA). Similarly, not wearing

PPE (personal protective equipment, such as a hard hat or steel-toe cap boots) where appropriate on construction sites will likely lead to a reduction of damages in a negligence claim.

If the claimant consented to the risk, they should have no claim if injured, but the scope of the *volenti* defence is very limited, and a defendant is more likely to plead contributory negligence. The defence can be relevant, however, when a defendant alleges that the claimant impliedly consented to a lower standard of care, for example, in relation to sporting events, both for spectators voluntarily sitting there to watch and participants in the sport.

Occupiers' liability

Although the law can now be said to have general principles relating to negligence, there are some categories that pose their own special problems and for which specific rules have been created. An important example is the liability of occupiers of premises, whose duties are now laid out in the Occupiers' Liability Acts 1957 and 1984. The main duty under the 1957 Act is a common duty of care owed by occupiers to their lawful visitors. The important questions, therefore, are:

- Who is an occupier?
- What are premises?
- Who are visitors?
- What is a common duty of care?

The occupier

This word is not defined by the Act but is generally understood to mean the person who has control of the premises. They are the person in the best position to know who is likely to be using the premises, and to know about and be able to put right, anything wrong with it. The occupier does not need to be exclusively in occupation; it is possible for premises to have more than one occupier at a time (*Wheat v Lacon* (1966) HL). For example, in *AMF International Ltd v Magnet Bowling Ltd* (1968) QBD, a building contractor was held to be a joint occupier with the landowner because they were undertaking substantial building work.

The premises

This word is wide enough to include not only buildings but also land itself. What is more, by s.1(3) the Act extends its meaning to cover vessels, vehicles or aircraft. Premises will also include sporting stadia.

The visitors

The range of people 'using' premises can be very wide, from a canvasser, the mail person, an invited guest, a gas or electricity meter reader or even a burglar. The most important point is to distinguish between people lawfully on premises and trespassers. A person's presence on premises could be lawful because they have express or implied permission from the occupier (such as friends invited to a house party); they have statutory authority to be on the premises (a police officer with a search warrant); or because they have a contract (i.e. a cinema ticket).

Even when entry onto premises is initially lawful, a visitor can become a trespasser if permission is subsequently withdrawn (*Snook v Mannion* (1982) QBD). For example, a shopper might enter a department store in a drunken state, create a disturbance and then be asked to leave. If they refuse to go, they become a trespasser. A visitor can only use the premises for the purposes for which they are invited or permitted to be there, so the house party guests cannot go for a nap in your bedroom (unless you give them permission).

It should be noted that persons who enter under a public or private right of way are not visitors for the purposes of the 1957 Act (*McGeown v Northern Ireland Housing Executive* (1994) HL), but in certain circumstances, the common duty of care is still owed, such as in respect of cattle (*McCaskie v Cameron* (2009) Blackpool CC). This also does not prevent liability for general negligence.

The common duty of care

Under s.1 of the 1957 Act, the occupier's duty is owed in respect of dangers due to the state of the premises themselves *and* due to things done or omitted to be done on the premises. Section 2(2) states the duty is to take all reasonable care in the circumstances to see that the visitor is reasonably safe in using the premises for the purposes they are invited or permitted to be there.

The Act also sets out several factors to be considered when determining whether or not there has been a breach of duty. In s.2(3)(a), it states that an occupier, when considering how careful a visitor is expected to be, must be prepared for children to be less careful than adults, although young children can also be expected to be in the guardianship of a parent or carer (*Phipps v Rochester Corporation* (1955) QBD).

An occupier can expect visitors who use the premises in the exercise of their profession to appreciate and guard against the associated risks (s.2(3)(b)). So, there was no liability to the deceased (and their widows) in *Roles v Nathan* (1963) CA, where two industrial chimney sweeps, who had been warned not to work when the fire was alight, died from fume inhalation. If the sweeps had been hurt by, for example, a broken pane of glass in the front door, i.e. something unrelated to their profession, then the occupier *would* be liable.

If the occupier has given a warning of a danger (e.g. by a notice stating 'Cliff edges crumbling – danger of collapse'), this is not enough to protect them from liability (s.2(4)). It may, in the circumstances, be sufficient to keep the visitor reasonably safe, and so the occupier may have discharged their duty. But a sign is rarely enough when the visitors are younger children.

Where the occupier knows of a state of affairs on the premises that they cannot personally remedy, and will involve asking a contractor to undertake work of construction, maintenance or repair, s.2(4) provides that a visitor injured as a result of the faulty execution of such work by a contractor will not succeed in a claim against the occupier so long as:

- the occupier acted reasonably in entrusting the work to a contractor *and*
- the occupier had taken reasonable steps to ascertain the contractor's competence *and*
- the occupier has taken reasonable steps to believe that the work had been properly done.

Note that this could require the occupier to employ someone to supervise or check the contractor's work, e.g. where building work is being carried out by the employment of an architect.

This section was extensively considered by the House of Lords in *Ferguson v Walsh* (1987). The claimant had been injured while working for the first defendant on a demolition job. The claimant had been asked to undertake the work by X, who in turn had been given the job by the site owners, the district council. By the terms of the contract between X and the district council, X was not allowed to subcontract the work without the council's permission. Nevertheless, that was exactly what X had done. The first issue in the case was to decide if the claimant was a visitor for the purpose of the Act. It was decided that they were, because they were there by X's authority and X was a joint occupier. Whether the council, as joint occupiers, was liable to them depended on s.2(4). The council had acted reasonably in entrusting the work to X and was held not to be liable. On its wording, s.2(4) appears to be limited to 'construction, maintenance or repair', but the House of Lords said that those words should be construed broadly and purposively, by which interpretation they were sure that 'construction' would embrace demolition.

The liability of the occupier may be limited or excluded, to the extent permitted by law. Section 2 provides that the occupier owes a common duty of care 'except in so far as he is free to and does . . ., restrict . . . or exclude his duty to any visitor or visitors by agreement or otherwise'. There are two principal limitations to an occupier's right to exclude liability:

a Even where liability can be successfully limited, it can only be reduced to the level of responsibility owed to a trespasser. Otherwise, a person entering lawfully could have less legal protection than a trespasser.
b More importantly, any limitation of liability must be read subject to the Unfair Contract Terms Act 1977 and the Consumer Rights Act 2015. In a 'business liability' situation, an occupier *cannot* exclude or limit liability for death or personal injury caused by negligence. Nor can they limit liability for other damage caused by negligence unless the restriction satisfies the requirement of reasonableness. When a person enters or uses premises by virtue of a contractual right, then, unless the contract states otherwise, it is implied in it that the occupier owes a common duty of care. If the contract does contain express terms about the occupier's liability, these may be subject to the Unfair Contract Terms Act 1977.

In considering whether an occupier has taken reasonable care, in *Debell v Dean and Chapter of Rochester Cathedral* (2016) CA, the court stressed that not all foreseeable risks create a duty on the occupier to take remedial action and that occupiers should take a 'practical and realistic approach' to the kinds of dangers they are obliged to remedy.

Damage and defences

A visitor to premises may suffer personal injury, damage to property or both. When they claim against the occupier, the defences that may be pleaded are either consent (*volenti*) under s.2(5) or contributory negligence (see earlier).

Occupiers' liability to trespassers

The 1957 Act extends protection to lawful visitors; thus, it has no relevance to trespassers injured on premises. The rules developed by the courts, culminating in the House of Lords' decision in *British Railways Board v Herrington* (1972) (where, for the first time, it used its powers to overturn its own previous decisions (*Robert Addie & Sons (Colliery) Ltd v Dumbreck* (1929) HL), have now been overtaken by the Occupiers' Liability Act 1984. Cases like *Herrington* had determined that an occupier owed a lower standard of care to a trespasser than to lawful visitors, a standard described as 'a duty of common humanity'. The lower standard took account of the fact that a trespasser was uninvited. Nevertheless, common humanity was a variable standard, depending on factors such as whether the occupier knew of the likely presence of the trespasser; the kind of trespassers they were likely to be, e.g. children, burglars, poachers or squatters; the seriousness of the risk or danger; and whether or not the occupier knew of the risk.

The interplay of these factors can be seen in a case like *Herrington*. A child trespassed on the railway line abutting open access parkland and was electrocuted. The fence guarding the line was in a state of disrepair. The stationmaster knew that children were trespassing and had alerted the police. The line was something of an enticement, as the broken fence allowed children to take a shortcut by that route to the nearby park. It was held that the British Railways Board had failed to show common humanity. In view of the high level of danger, the likelihood of trespass and the relative ease and cheapness of eliminating the danger for an organisation with such resources, the child succeeded in an action for damages.

Under the 1984 Act, 'premises' and 'occupiers' are defined in the same way as for the 1957 Act. The duty under the Act is owed to persons 'other than visitors' and the duty consists of taking such care as is reasonable in all the circumstances of the case to see that the non-visitor does not suffer injury on the premises by reason of the danger concerned. For these purposes, 'injury' is not to include loss or damage to property; it covers personal injury only. In this respect, the scope of the 1984 Act is narrower than the scope of the 1957 Act.

The occupier of premises only owes a duty under s.1 if they are aware of a danger or have reasonable grounds to believe it exists; knows, or has reasonable grounds to believe a trespasser is, or may come into, the vicinity of that danger; and the risk is one that, in the circumstances of the case, the occupier may reasonably be expected to offer the trespasser some protection.

As the occupier owes a duty to take such care as is reasonable in all the circumstances, factors relevant under the 1957 Act – like the age of the trespasser, the nature of the premises, the character of the entry and the extent of the risk – all still play a part in determining whether there is a breach of duty. What is reasonable in all the circumstances may include a consideration of the relative burden of available precautions, as is the case for breach of duty under general principles of negligence.

Section 1(5) allows the occupier to discharge their obligations in appropriate cases merely by giving warning notices. It is doubtful if a notice saying 'Keep out' would suffice, as it would merely alert persons entering the premises to the fact that they were trespassers and not to the fact that there was some danger. Warning notices may be ineffective if the trespassers are children. Moreover, as the Act is concerned largely with trespassers, they may choose to enter premises at some unconventional place where no notice is displayed. The Act also retains the defence of *volenti* (consent).

The House of Lords assessed an occupier's responsibilities under the 1984 Act in *Tomlinson v Congleton Borough Council* (2003) HL. In this case, the claimant was injured whilst diving into a lake. The lake was used for boating, but signs warning that the water was dangerous and that swimming was forbidden were placed around its edge. The defendant was aware that, despite these notices, the lake was used for swimming and had considered altering the edges of the lake to make it less accessible to swimmers. In the Court of Appeal, it was held that the danger posed by swimming was sufficient to have required the defendant to make the lake totally inaccessible. In failing to do so, the council was therefore liable to compensate the claimant for the injuries suffered (albeit reduced by two-thirds for contributory negligence). When the case was heard in the House of Lords, the court held that the Occupiers' Liability Act 1984 required that the relevant danger was 'due to the state of the premises or to things done or omitted to be done on them'. The court decided that the lake was not unusually dangerous and that the claimant's injuries were due to their own actions (diving into the lake despite the warning signs) rather than to 'the state of the premises'. The council was therefore not liable for the claimant's injuries.

In fact, the 1984 Act has made little difference to the sort of trespasser claimant who can succeed against an occupier. In cases like *Pannett v McGuiness & Co.* (1972) CA, decided on the old 'common humanity' rules, demolition contractors were occupiers of a site in a busy urban area where it was reasonable to expect that there would be large numbers of children. It became necessary to burn large quantities of rubbish and, realising that this would attract children, three workmen were employed to keep them away. Children, including the claimant (aged five), were chased away from the site on several occasions. While the fire was still burning, the three men left the site and the claimant re-entered and was injured by the fire. The nature of the activity was hazardous, the likelihood of trespass was great and the attraction to children was obvious. Applying the principles of the 1984 Act, the defendants would be aware of the danger, would know that trespassers were in the vicinity of the danger and in all the circumstances, the risk would surely be one against which the contractor could reasonably be expected to offer some protection. The defendants were held liable under the pre-Act rules and the same outcome would be likely to occur today.

Liability of non-occupiers and the Defective Premises Act 1972

For complicated reasons connected with the overlap of duties in contract and tort, the law has been slow to develop protection in tort for the person who suffers damage because of a defective building, where the defects are due to a non-occupier. In defective foundations cases like *Anns v Merton London Borough Council* (1972) HL and *Murphy v Brentwood District Council* (1991) HL, builders sold houses to the initial purchasers, who later sold them on to the claimants. The builders had been negligent, but the claimants had no contract with them, and nowadays have no action in negligence because the harm is pure economic loss.

The Defective Premises Act 1972 imposes a duty on persons who take on work for the provision of dwellings. That duty is owed to the person ordering the work and to any person who subsequently acquires an interest in the dwelling. The extent of the duty is to see that the work is done in a competent and professional manner with proper materials so that the dwelling will be fit for habitation when completed.

Some very significant limits operate, however, to cut down the apparent effectiveness of the protection given by this Act:

- The protection does not apply to dwellings already protected by an approved scheme of the National House Building Council (such as NHBC Buildmark).
- The Act only applies to dwellings, not to all types of buildings.

- Any action based on the Act must be speedily brought, as there is a limited period of only six years, which runs from the date of completion of the dwelling, not the date of the defect.

Although the protection afforded by this Act is useful, the very limited extent of it has encouraged claimants to seek alternative ways of obtaining redress.

In *Murphy v Brentwood DC*, Lord Mackay, the Lord Chancellor, said,

> I am of the opinion that it is relevant to take into account that Parliament has made provisions in the Defective Premises Act 1972 imposing on builders and others undertaking work in the provision of dwellings obligations relating to the quality of their work and the fitness for habitation of the dwelling. For this House in its judicial capacity to create a large new area of responsibility on local authorities in respect of defective buildings would in my opinion not be a proper exercise of judicial power.

Lord Mackay was supported in this view by Lord Jauncey, who added,

> Parliament imposed a liability on builders by the Defective Premises Act 1972, a liability which falls far short of that which would be imposed on them by *Anns*. There can therefore be no policy reason for imposing a higher common law duty on builders, from which it follows that there is equally no policy reason for imposing such a high duty on local authorities.

It seems now that a first purchaser of a defective building is, therefore, limited to contractual remedies and/or the protection provided by the 1972 Act. A subsequent purchaser of the property will have no contract with the original builder and may therefore have to rely exclusively on the 1972 Act, subject to any rights acquired under the Contracts (Rights of Third Parties) Act 1999.

A practical application of the Act can be seen in *Andrews v Schooling* (1991) CA. An Edwardian house had been converted into flats, and the ground-floor flat was bought by the claimant. Their flat also incorporated a basement. When the builders undertook the conversion work, no damp-proof course was installed. Once the claimant took up occupation, they found that the flat was affected by penetrating damp in the basement. They sued the builders under s.1 of the Act. The builders sought to argue that they would indeed be liable if they had put in a damp-proof course and done it badly (misfeasance) but that they could not be liable if they had done nothing (non-feasance). The court rejected that argument, indicating that it made no difference to the claimant whether it was misfeasance or nonfeasance, as in either case the builders had not provided a house fit for habitation. Of course, in awarding damages to the claimant, we have to accept that this was to compensate her in respect of economic loss and it is arguable whether that was ever the intention of the Act.

To date, there is no sign of Parliament tackling the issues raised by the *Murphy* case and the limitations indicated in the 1972 Act.

Defective products

Where a person is injured by using a defective product and is looking for some legal redress, there can be considerable overlap between the law of contract and the law of tort.

In contract

If the injured person purchased the product, they may sue the seller (under the Sale of Goods Act 1979 and Sale and Supply of Goods Act 1994 or the Consumer Rights Act 2015) if there has been a breach of any of the implied terms relating to reasonable fitness for purpose and satisfactory quality. Moreover, in a consumer transaction, the seller cannot exclude or limit liability for breach of these implied terms, because of the provisions of the Consumer Rights Act 2015. In exceptional cases, a buyer of defective goods may also have limited contractual rights against the manufacturer, where there is a guarantee in operation. This contract route to legal redress is limited to persons who can prove that they bought the goods, so it cannot assist a person who receives the goods as a present or who is injured when using borrowed goods. It is rather inefficient, in the sense that it may set in motion a chain of legal actions: buyer v seller, seller v wholesaler, wholesaler v importer, importer v manufacturer. Several costly actions may thus be necessary to lay liability at the door of the manufacturer. The contract route relies on suing a retailer who owes a strict liability, who will be liable whether at fault or not. This does nothing to ensnare the manufacturer, who is best placed to remedy such faults for the future and who could redistribute any losses much more widely if made liable for the defective product.

In tort

The tort route for damages in respect of defective products finds its origins in *Donoghue v Stevenson*, where the defective product was contaminated ginger beer. It was consumed by a friend of the purchaser. The friend had no contractual rights. The purchaser had contractual rights but had suffered no loss or damage. The House of Lords recognised in that case that a manufacturer owed a duty to take reasonable care to the ultimate consumer of the product where injury could be foreseen and it was unlikely that there would be any intermediate examination of the goods. Of course, under this rule, the range of people who can sue is potentially much wider, and the manufacturer's liability is not strict. The manufacturer owes a duty to take reasonable care. The claimant will bear the burden of proving negligence against the manufacturer. That may not prove easy, although the claimant may be able to rely on the *res ipsa loquitur* principle. A claim may be defeated or reduced if the manufacturer proves that the claimant was contributorily negligent, e.g. by failing to follow the instructions issued with the product. And of course, a claimant needs to show that they suffered harm of a foreseeable type. In an action for negligence, harm is normally personal injury or damage to property. A claimant will not usually succeed if they suffered pure economic loss unless a special relationship can be established.

Thus, the tort and contract routes can both be seen to be problematic. This area is now also governed by legislative controls. This is particularly important as there are inevitable policy decisions to be taken into account. English law had to come into line with the European Community, which issued a Directive in 1985 requiring member states to harmonise their rules on product liability. The current law, giving rise to both civil and criminal liability, is largely in the Consumer Protection Act 1987 (CPA 1987) and the General Product Safety Regulations 2005, which impose strict liabilities that cannot be excluded or limited, contractually or otherwise (s.7 CPA 1987). The CPA provides that a producer of defective products is liable for any damage caused by the defect (s.2). The terms 'producer', 'product' and 'defect' are all defined by the Act.

The producer is defined by s.1 as the person who manufactured the product, the person who won or extracted it (where the product is a substance won or extracted) or, where essential characteristics are due to an industrial or other process, the person carrying out the process.

A product, defined by s.1(2), means any goods or electricity and includes a product that is comprised in another product as a component part.

The Act recognises that it may be easy to identify the person who sold goods, but not so easy to establish the producer. This problem is met by s.2(3), under which a supplier of a product is liable if the claimant asks the supplier within a reasonable time of damage occurring to identify the producer, in circumstances where it is not reasonably practicable for the claimant to identify the producer and the supplier fails to satisfy the request within a reasonable time.

A defect in a product is defined by s.3. There is a defect if the safety of the product is not such as persons generally are entitled to expect. In determining what persons generally are entitled to expect, a court could take into account:

1 The way and purposes for which the product has been marketed and any instructions for use or warnings in relation to the use of the product.
2 What might reasonably be expected to be done with, or in relation to, the product.
3 The time when the product was supplied by its producer to another.

Nothing in this section is to be taken to infer that a product is defective merely because a later issued product adheres to an even higher safety standard. The factors specifically listed in s.3 are not necessarily exhaustive.

In *Tesco Stores Ltd v Pollard* (2006) CA, a childproof cap that could be opened with far less force than that required by the relevant British Standard was deemed not defective under the terms of the Act. The Court of Appeal held that the expectation with this type of cap was only that it was more difficult to open, not that it met the relevant British Standard.

Section 4 provides that it is a defence for a producer to show (amongst other things) that at the relevant time (usually when the producer supplied the product) the state of scientific and technical knowledge was such that producers could not have been expected to discover the defect. This is often described as 'the state of the art' defence. It is thought to be vital that technical and scientific innovation is not discouraged. It was one aspect of the Directive where member states had some choice about whether to incorporate such a defence.

To sustain a claim under the Act, a claimant would need to prove damage. This is defined by s.5 as including death or personal injury or any loss or damage to any property, including land. It expressly excludes loss or damage to the product itself. The property lost or damaged must be property intended for private use or consumption. Where the harm to property is valued at less than £275, no claim arises, but this restriction does not operate where the damage is personal injury.

The Act provides rights that are in addition to existing common law rights. Section 2 specifically states that it is without prejudice to any liability arising by means other than the Act. This could be relevant in cases involving pure economic loss. The Act places an embargo on recovery for such loss, but cases on the common law rules of negligence may develop differently. An action under the 1987 Act is obviously advantageous to a claimant because the burden of proof is primarily on the manufacturer.

Vicarious liability

The word 'vicarious' means 'in place of another person' or 'on behalf of the proper person'. A rule has emerged in tort that an employer is vicariously liable for the torts of employees committed during the course of their employment. The rule extends beyond employees to other relationships described as 'akin to employment', such as members of a Christian religious group (*The Catholic Child Welfare Society v The Institute of the Brothers of the Christian Schools* (2012) HL), but not *normally* to contractors. The rule is one of policy. It ensures that an injured person can sue a more substantial organisation rather than an individual. The employer is likely to have greater financial resources and insurance. There is no other particularly satisfactory justification for the existence of the rule, although it can be said that an employer profits from an employee's work and should, therefore, bear the risk of it.

The two main criteria for the operation of the rule are that the person who committed the tort must be an employee or akin to an employee (meaning it is possible for liability to extend to sub-contractors) and there must be a close connection between the wrongdoer's act and their employment or other relationship.

A Close Connection

The courts have developed a 'close connection' test (*Lister v Hesley Hall Ltd* (2001) HL) for assessing whether vicarious liability arises. This requires the court to investigate how close a connection there is between the wrongdoer's act and the employment or other relationship. For example, in *Mohamud v WM Morrison Supermarkets Plc* (2016) SC, the Supreme Court found there was a close connection between an assault and the employee's role as a petrol station kiosk employee, because the employee had told the claimant not to return to their employer's property. However, in *Trustees of the Barry Congregation of Jehovah's Witnesses v BXB* (2023), the Supreme Court held that the Jehovah's Witness organisation was not vicariously liable for a rape committed by one of its 'elders'. Although the relationship between the organisation and the elder was 'akin to employment', the rape had been committed while the victim was in the elder's home offering emotional support (which the other elders had encouraged and was important in the High Court), and not when the elder was carrying out his authorised activities. As such, there was no close connection between the perpetrator's position as elder and the rape was not committed in the course of his authorised activities.

Despite the development of the close connection test, decisions on vicarious liability tend to turn on the individual facts of the case. However, certain situations are certain to meet the requirement, such as:

a Torts committed while the employee is doing what they are employed to do. For example, a bus driver injures a pedestrian by negligently driving a bus (*Limpus v London General Omnibus Co* (1862) ExCh).

b Torts committed while the employee is doing what they are employed to do, but in a way forbidden by the employer. For example, a petrol tanker driver is forbidden by his employers to smoke while making deliveries. He disobeys this order and causes an explosion which injures a passerby (*Century Insurance v Northern Ireland Road Transport Board* (1943) HL).

c Torts committed while the employee is doing what they are employed to do, but in a criminal manner. For example, a solicitor's clerk, employed to advise clients, fraudulently advises a client to transfer property to them (*Lloyd v Grace, Smith & Co.* (1912) HL).

d Torts committed while the employee is doing acts which are reasonably incidental to their work. For example, an employee opens the office window for ventilation and, in doing so, negligently knocks a potted plant out of the window onto a passerby below.

e Torts committed while the employee is acting in an emergency for the protection of the employer's person or property, when their actions would normally be outside their employment. For example, an employee negligently moves a vehicle which they have no authority to drive, but which they reasonably fear is about to be damaged in a fire.

Where the employee's behaviour has amounted to a 'frolic of his own' (*Joel v Morison* (1834) KB), the employer will not be vicariously liable. Cases include the conductor who decides to drive the bus (*Beard v London General Omnibus Company* (1900) CA) and the delivery driver who varies his proper route to visit a friend (*Storey v Ashton* (1869) QB). In these cases, there was no course of employment, but it must be remembered that the question is now one of close connection. Thus, a nightclub owner was vicariously liable for the acts of a bouncer who assaulted a customer, after returning home for a knife, because there was a close connection with the bouncer's employment – the nightclub owner had encouraged the bouncer to act aggressively (*Mattis v Pollock (t/a Flamingos Nightclub)* (2003) CA). In *Bellman v Northampton Recruitment Ltd* (2018) CA, the employer was also liable when an employee was assaulted at an office Christmas party by the managing director. Although there was said to be no close connection at first instance, on appeal, due to the managing director very clearly and vociferously asserting their role during the party, there was found to be a close enough link between the act and their employment.

The rule of vicarious liability is often said to be one of the distinguishing features of an employer–employee relationship. Generally, an employer is not vicariously liable for the torts of an independent contractor. If a builder is engaged to build a house, they are liable for any acts of trespass or negligence they may commit; the engaging client is not. However, in some cases, either the liability imposed is strict, or the duties owed by a person are so onerous that there can be no delegation to someone else (e.g. *Woodland v The Swimming Teachers' Association* (2014) SC, where a child was left severely brain-damaged after a school swimming lesson where the teaching was sub-contracted). Then the employer is liable in addition to the contractor.

Although these may look like examples of vicarious liability, it is more appropriate to regard them as examples of an employer continuing to be directly (or personally) liable. This can arise in the following situations:

1 When the employer is negligent in choosing the contractor, they remain directly liable. For example, an occupier might need repair work of a technical nature undertaken on their premises. They know that the work involves electrical rewiring and yet appoint a friend with no training or expertise. The job is not properly done and causes a lawful visitor to the premises to receive a severe electric shock. The occupier would be in breach of their duty under the Occupiers' Liability Act 1957 and could not apply s.2(4) of the Act, because they have not taken reasonable steps to ensure the contractor was competent, i.e. the choice of contractors is, in itself, demonstrating a lack of care.

2 When the liability of the employer is strict (e.g. as in the tort of *Rylands v Fletcher*) and can occur even if reasonable care has been taken, the employer is liable whether the tort has been committed personally or through an independent contractor.
3 When the employer undertakes a particularly hazardous activity, they are liable for torts committed personally or by an independent contractor. In *Tarry v Ashton* (1876) QBD, a heavy lamp suspended over a footpath fell into disrepair. The employer was liable when a passerby was injured by the activities of the contractor called in to repair it. Similarly, in *Balfour v Barty-King* (1957) CA, a contractor was engaged to thaw out frozen pipes and chose to use a blowlamp. The employer was vicariously liable when a fire broke out as a result of the contractor's negligence.
4 When the employer authorises or instructs the contractor to do something involving the commission of a tort, they remain liable. In *Ellis v Sheffield Gas Consumers Co.* (1853) QB, a gas company engaged a contractor to dig up part of a street over which they had no such authority. The gas company was held liable when a passerby fell over a heap of earth left by the contractor.

Dual vicarious liability

It had long been established law (dating at least from *Laugher v Pointer* (1826) KB), that only one party can be vicariously liable for the wrongs of another. This was overturned (or perhaps, more correctly, dismissed) in *Viasystems (Tyneside) Ltd v Thermal Transfer (Northern) Ltd* (2005) CA where it was found, through a consideration of leading authorities such as *Mersey Docks and Harbour Board v Coggins & Griffith (Liverpool) Ltd* (1947) HL, that there was no legal principle against dual vicarious liability and that each case should be decided on its facts. Thermal Transfer was engaged to install air conditioning in a factory. They contracted the work to S and P Darwell Ltd, who, in turn, sub-contracted to CAT Metal Services. CAT engaged fitters and fitter's mates. A young fitter's mate negligently caused extensive damage. It was held that S and P Darwell Ltd and CAT were vicariously liable in equal part.

Discussion questions

1 A claimant 'coming to the nuisance' is not a defence to a claim for private nuisance. Why might the courts not wish to accept such a defence? What would it mean for subsequent purchasers of land if it were allowed?
2 In *Caparo v Dickman* (1990) HL, the courts set out two methods for expanding the duty of care: an incremental approach and a three-part test. What are the elements of these tests, and why might the courts wish to limit the expansion of the duty of care?
3 A breach of duty is when the defendant fails to meet the behaviour expected of them. The test is objective and does not usually consider the characteristics of the defendant, but the 'reasonable' person. When are the characteristics of a defendant relevant?
4 Causation is about proving that a breach of duty is both the factual and legal cause of the harm. Why might problems arise when applying the 'but for' test, for example, in a case involving negligent exposure to asbestos? What is 'remoteness' and how do the courts approach this issue in cases involving (1) personal injury and (2) property damage?

10 Health and safety

Carrie de Silva

The background to legislation

After agriculture, construction has the highest rate of fatalities of any industry in the UK. Although the construction industry has been subject to significant review and improvement, particularly since the 1980s, to quote the title of Rita (now Baroness) Donaghy's report to the government in 2009. 'One death is too many' (*Inquiry into the underlying causes of construction fatal accidents*, Great Britain, 2009a).

Construction management and practice have to contend with problems which vary enormously from site to site and during the life of the site and can include climate, regional attitudes, geology, the type of contract, the scale of the job, the methods of payment, the type of employment and the rapid turnover of labour. There is also an often-complex relationship between a main contractor and a changing group of subcontractors. All these factors affect management in all its aspects, including the management of safety. This chapter provides a general overview of health and safety law and practice. Refer also to the Building Safety Act 2022 with reference to provisions applicable in the construction and management of certain buildings, the scope of which is beyond this book.

For a long time, the law has played a part in controlling how employers must manage safety. Acts of Parliament laying down duties and standards in specific working places date back to the Act for the Better Regulation of Chimney Sweepers and their Apprentices 1788 and the more widely applicable Health and Morals of Apprentices Act 1802 (sometimes known as the Factories Act). The earliest rules related to the employment of the very young in chimneys and mills, where the particular evil was the importing of large numbers of pauper children and engaging them in appalling working conditions. The procedure of making rules to control specific workplaces was established, and this, ironically, became one of the worst features of the safety legislation.

With the pattern set, whenever a new loophole, crisis or new hazardous work area was discovered, special rules would be made to cover the situation. The end result was piecemeal legislation covering mines, factories, offices and shops, agriculture and so on, culminating in a series of specific workplaces each governed by its own Act of Parliament. The Acts were not sufficiently detailed in themselves to cover every conceivable situation, so they all had various sets of regulations (delegated legislation) made under them. The consolidating Factories Act 1961 (now largely, but not wholly, superseded by the Health and Safety at Work etc. Act 1974) alone had more than 200 sets of regulations made by virtue of its authority.

The main aim of the safety legislation was to establish standards and duties for the workplaces covered. Failure to abide by the prescribed standards and duties would lead

DOI: 10.1201/9781003657101-10

to employers being prosecuted and fined. By operating in this way, the rules could, therefore, be seen to fulfil an accident prevention role. When an employer's breach of duty led to a workman (and older legislation was almost exclusively couched in the masculine) being injured, the various Acts were interpreted by the courts as giving the injured employee the right to sue for compensation.

The health and safety question can be seen to have three dimensions:

1 *The point of view of an injured employee*
Should the accident be reported? Will compensation be available? Is the employer insured? Will state benefits be available? Does it matter that the employee was partly to blame?
2 *The point of view of the employer*
Will they be prosecuted? Will insurance cover any compensation and other related costs? Will insurance premiums be increased? Will the employee claim damages? Importantly, should existing safety arrangements be amended?
3 *The point of view of the state*
Here the issues will include the cost of health care for the injured person, lost productivity, the cost of state benefits and whether there are any lessons to be learnt for future safety policy.

The number of fatal injuries to workers reported in 2023–24 (1138) was slightly higher than the previous year, but there is a continuing trend that has seen fatalities at work reduced to a level which is less than one-third of that recorded in the 1980s. On the other hand, there is little room for complacency as the number of reported major injuries to employees has only decreased marginally in recent years. And fatalities are, of course, and thankfully, a tiny proportion of the issue. Many millions of days are lost each year due to work-related injury and ill health. In part the improvement in fatal and non-fatal injuries is attributable to shifts from manufacturing and heavy industries to service sectors, where risks are generally lower.

Dissatisfaction with the existing system of piecemeal legislation led to a review by the Robens Committee on Safety and Health at Work (1972). Robens recommended a major revision of the approach to health and safety with a move away from legislation which applied to specific types of workplace to a single Act which would provide comprehensive coverage to workplaces in general. It also stated that the promotion of safety and health at work was the joint responsibility of both management and employees. The implementation of this new thinking on health and safety took the shape of the Health and Safety at Work etc. Act 1974 (HSWA). This was a piece of framework legislation which established certain general duties on employers, workers and others, introduced new requirements for safety representatives and committees, and amended the existing regulatory and enforcement machinery with the establishment of the Health and Safety Commission and the Health and Safety Executive. It was never intended that the existing legislation be substituted overnight; rather, the aim was that over time the Acts which related to specific places of work would be replaced by regulations and codes of practice of more general application. However, although some important changes were made, overall progress on this process of substitution proved to be slower than anticipated and towards the end of the 1980s much of the old legislation was still in place. But a new impetus was brought to the process of reform in the shape of the Single European Act of 1987, which introduced Article 118A, thereby enabling improvements to the work environment and to health and safety to be adopted by qualified majority voting rather than by the unanimous assent of member states. Since that time a programme of safety, hygiene and health reforms has been put in place and, starting with the European Framework Directive in 1989, a series of directives has been passed which has had a profound effect on the working environment in the UK and across Europe. With the leaving of the EU, any relevant provisions not already in UK legislation were adopted through Health and Safety (Amendment) (EU Exit) Regulations 2018.

On 1 January 1993 new regulations were introduced, intended to implement the UK's obligations under the Framework Directive and its subsequent measures. These became known as the 'six pack', but before turning to these, we must consider in more detail the structures and bodies established by the Health and Safety at Work etc. Act itself.

The Health and Safety Executive

It is worth remembering that safety is not just an issue affecting those in employment. The general public is awakened to safety issues from time to time when a major disaster occurs, such as the Piper Alpha oil platform explosion in 1988 (167 fatalities), the King's Cross tube station fire in 1987 or the Paddington rail crash in 1999 (each resulting in 31 fatalities). At such a time, there are often demands for safety standards to be improved, and a public inquiry (held by virtue of s.14 of HSWA 1974) may highlight changes that are necessary. The government has overall responsibility for introducing legislation to promote health and safety, and several of its departments, such as the Department of Transport, will see this as part of their work. Other organisations (e.g. the Royal Society for the Prevention of Accidents, RoSPA) exist to promote health and safety, and although their work may be more general, it can have useful implications for health and safety in employment.

The Health and Safety Commission (HSC) was established by the HSWA in 1974. It had up to nine members, who were required by the Act to be representative of employers, trade unions, local authorities and other bodies, such as RoSPA, concerned with health and safety. The role of the Commission was to undertake, by whatever means it considered appropriate, to secure the health, safety and welfare of people at work and to protect other people from risk to their health and safety created by work activities. The Commission carried out research, provided training, disseminated information and advice and submitted proposals to the Minister for new regulations.

The Commission's accident investigation or inquiry role was significant, as it tended to have a high public profile and created significant pressure for change or some government action. An investigation would be less formal than an inquiry, so whereas the latter was the proper method to adopt in the King's Cross situation, an investigation would be more appropriate for, say, a serious accident at a fairground. The person appointed to conduct an inquiry would be given formal powers to compel the attendance of those required to be present. Inquiries would usually be held in public and might well involve visits to the site.

One major initiative of the Health and Safety Commission was to establish advisory committees on issues such as major hazards, nuclear safety and toxic substances, as well as industry-based advisory committees designed to promote the health and safety of workers in a particular industry. One example is the Construction Industry Advisory Committee (CONIAC, now hosted within the HSE web communities). This kind of specialist committee has played an important role in assisting the Commission in tasks such as drawing up guidance and promoting awareness of health and safety issues. CONIAC also plays a part in identifying training and education needs and promoting improvements in competence through continuing professional development.

The most significant development which resulted from the establishment of the Health and Safety Commission was the new unified focus for safety issues and policy-making, which had previously been lacking. The Commission operated on a day-to-day basis through the Health and Safety Executive (HSE), which was the enforcing authority. Some enforcement duties were allocated to local authorities, and to avoid duplication between HSE inspection and local authority inspection, some types of premises, e.g. shops, catering establishments and offices, were specifically allocated as the responsibility of the local authorities. However, local authorities did *not* inspect their own offices! That task was performed by health and safety inspectors.

In April 2008, after consultation on the roles of the two bodies, the HSC and the HSE merged to become a single regulatory body under the title of the Health and Safety Executive. In its first annual report the objectives of this merged body were stated to be to:

- influence people and organisations – duty holders and stakeholders – to embrace high standards of health and safety;
- promote the benefits of employers and workers working together to manage health and safety sensibly and
- investigate incidents, enquire into complaints about health and safety practices and enforce the law.

More immediately an employer, landowner or occupier with contractors on his land must pay particular regard to the following legislation:

- the Health and Safety at Work etc. Act 1974 and
- the 'six pack' and other regulations stimulated by EU Directives.

The Health and Safety at Work etc. Act 1974

The 1974 Act adopted the recommendations of the Robens Report, placing emphasis on greater involvement and self-regulation rather than reliance on state regulation. The Act provided the framework of a new system, and under this 'umbrella', new codes of practice have been created. As this happened, the intention was that old Acts and Regulations would gradually be repealed.

The main objectives of the Act are set out in s.1:

1 To secure the health, safety and welfare of persons at work.
2 To protect persons other than persons at work against risks to health and safety arising out of work activities.
3 To control the keeping and use of dangerous substances.
4 To control the emission of noxious and offensive substances.

To achieve these objectives, the Act lays general duties on employers (s.2), employees (s.7), the self-employed (s.3), occupiers of workplaces (s.4) and manufacturers and suppliers (s.6).

The Offshore Safety Act 1992 now extends the 1974 Act to cover the health, safety and welfare of persons on offshore installations.

Duties owed by the employer – sections 2 and 3

So far as is reasonably practicable, an employer must ensure the health, safety and welfare of employees at work. This broad generalist approach is certainly a remarkable contrast with the detailed format previously adopted by safety Acts and Regulations. The employer must (so far as is reasonably practicable):

1 Provide and maintain plant and systems which are safe and without risk to health.
2 Make proper arrangements for the safe handling, storage, use and transport of articles and substances.
3 Provide necessary information, training, instruction and supervision.
4 Maintain the place of work and access to it so that it is safe and without risk to health.
5 Provide and maintain a safe and adequate working environment.

The Act does not define the terms 'health', 'safety' or 'welfare', nor does it define the phrase 'so far as is reasonably practicable'. The latter phrase had long been familiar to the lawyer, because it was also used to qualify certain duties created by the Factories Act 1961. For example, s.28 of that Act provided that floors and stairs had, so far as was reasonably practicable, to be kept free from obstruction. The meaning of this phrase has been judicially considered and has a narrower meaning than 'physically possible'. It implies that: 'A computation must be made in which the quantum of risk is placed on one scale and the sacrifice involved in the measures necessary in averting the risk (whether money, time or trouble) is placed on the other. If the risk is insignificant in relation to the sacrifice, the defendants discharge the onus on them' (*Edwards v The National Coal Board* (1949) CA). The following case provides an illustration.

As a result of an accident at work, the Health and Safety Inspector issued an Improvement Notice, requiring a dairy to provide protective footwear free of charge to all employees involved in operating hydraulic trolley-jacks. The employer successfully appealed against the notice. The expense of providing the footwear free of charge was disproportionate to the risk to the employees. The dairy's existing arrangements, whereby employees could buy the footwear at cost price and pay by instalments, were adequate to satisfy the firm's duty to secure its employees' safety so far as was reasonably practicable (*Associated Dairies v Hartley* (1979) IT). This principle stands today, but be aware that expectations regarding specific provisions may have changed.

Nothing in the Act allows these duties to be used as the basis for a claim for compensation for the victim. The sanctions for health and safety enforcement are criminal. However, an injured person would be able to proceed against the employer under ordinary negligence principles.

Section 2(3) provides that an employer must prepare and, when necessary, revise a written statement of general policy with regard to health and safety, and must bring the statement to the notice of employees. The introduction of such written policy statements was seen as an important application of the ideas about greater involvement and self-regulation expressed in the Robens Report. If employees are to be involved, they must know what the safety policy is. The Act gives no guidance on the form or content of the safety policy, although the HSE gives a template on their website. What is required will naturally vary with the size and type of organisation and the specific hazards involved. The lack of a model form is sensible, as it compels employers to think positively about safety problems. This approach to employer and employee awareness is also encouraged by some of the 'six pack' regulations. Failure to abide by the rule about issuing a safety policy is an offence usually punishable by a fine. Note that those with fewer than five employees are exempted (Employers Health and Safety Policy Statements (Exception) Regulations 1975) from the requirement of a written policy, although small businesses still need to comply with health and safety law.

Note that from October 2015, self-employed persons not employing anyone else, not involved in any of a prescribed list of activities and not posing harm to others are now excluded from health and safety provisions (Health and Safety at Work etc. Act 1974 (General Duties of Self-Employed Persons) (Prescribed Undertakings) Regulations 2015). This light touch followed recommendations of the 2011 Löfstedt report, 'Reclaiming health and safety for all: an independent review of health and safety legislation'. Undertakings where the self-employed are *not* exempted include agriculture, construction, those working with asbestos, genetically modified organisms or on the railways.

Section 3 imposes on employers and self-employed persons a duty to conduct their undertakings in such a way as to ensure, as far as is reasonably practicable, that persons *not* in their employment are not exposed to risks to their health and safety. This section is a valuable protection for members of the public, but an unusual application of it can be seen in *R v Swan Hunter Shipbuilders Ltd* (1981) CA. In that case a fire broke out on board a ship being built by Swan Hunter. In one part of the ship the fire became particularly intense because a subcontractor had left an oxygen hose there. Swan Hunter was aware of the dangers of using oxygen equipment, particularly because oxygen enrichment could occur in confined spaces. In order to alert its own employees to this danger, Swan's had distributed a booklet to them, setting out practical safety rules. These booklets were not given to all subcontractors as a matter of course. As a result of the fire, eight men died. Swan Hunter was convicted of a breach of its duties to provide a safe system of work and to provide information and instruction (s.2). It was *also* convicted of a breach of its duty under s.3. By not providing information to non-employees, Swan Hunter had failed to ensure that persons not in its employment were not exposed to risk to their health.

Duties laid on controllers of premises (s.4)

Section 4 imposes duties on controllers of premises, where the premises are used by persons who are not employees and where those persons may use plant or substances provided there for their use. A frequently quoted example of such premises is a coin-operated dry cleaner. Here, the controller must take such measures as are reasonable to ensure that the premises, plant and substances are safe and without risk to health so far as is reasonably practicable. (This is the criminal law provision corresponding with an occupier's duties at civil law under the Occupiers' Liability Act 1957.)

Section 4 can also be significant where a site is in the control of the main contractor. That site will be used by subcontractors and their employees, i.e. people who are not employees of the main contractor.

Duties laid on manufacturers and suppliers (s.6)

These duties are regarded as among the most important methods of securing an improvement in safety standards at work. Generally, manufacturers or suppliers of any article or substance for use at work must, so far as is reasonably practicable:

1 Ensure that the article or substance is safe and without risk to health when properly used.
2 Carry out such tests and examinations as are necessary to achieve (1).

3 Take such steps as are necessary to ensure that adequate information is available about how to use the article or substance, so that when properly used, it is safe and without risk to health.

These duties allow faults and defects to be traced back to their source, and prosecutions can be brought against persons with the ability to remedy them. The effect of s.6 is not to absolve the employer from its duties, but it is recognised that the power of the employer to do much is often limited. These duties do not create any new civil liability. Any civil proceedings must be brought under ordinary negligence rules. However, the possibility to prosecute under s.6 is a significant increase in the scope of the inspectorate's power. Note also that similar duties are imposed on designers, importers, erectors and installers.

Duties laid on employees (s.7)

Under the old Acts and Regulations, duties were placed directly on employees, and this section merely extends an old principle. The duties which an employee owes while at work are:

1 To take reasonable care for his or her own health and safety.
2 To take reasonable care for the health and safety of others affected by his or her acts or omissions at work.
3 To cooperate with the employer where necessary to enable the employer to fulfil its duties.

As with the other duties under the Act, s.7 cannot give rise to a civil claim for compensation, but an offending employee can be prosecuted. Employees are most likely to be prosecuted in those cases where the employer has fully complied with its duties. The most notable feature about prosecutions brought against employees is that fines tend to be smaller than those imposed on employers, although the maximum fine in the Magistrates' Court is now unlimited and/or 12 months' imprisonment. In *R v Sidebottom and Golding* (2014) Southwark CrC, a health and safety officer received a custodial sentence when his inadequate site monitoring led to the death of a construction worker.

Although s.7 relates specifically to employees, they could also, with others, be prosecuted under s.8, which provides that no person shall intentionally or recklessly interfere with or misuse anything provided under the Act for health and safety purposes. Those guilty of an offence under s.8 could face unlimited fines or 12 months' imprisonment, or both, on summary conviction.

Under the Management of Health and Safety at Work Regulations 1999, employees are under a duty to use machinery and equipment as they have been trained and instructed to do so, and they must inform the employer of any situation which is an imminent danger or where there is noted to be a gap in the health and safety arrangements.

Safety representatives

In safety matters, the Robens Report considered that management and employees had an 'identity of interest' which should lead to 'participation in working out solutions'. Indeed, they found evidence in some working situations of existing participation by employees, e.g. coalminer representatives appointed *by* coalminers to inspect mines. Many firms and organisations had also established voluntary safety committees before 1974. This voluntary effort needed statutory support, and the 1974 Act permitted regulations to be made providing for the appointment of safety representatives and for the establishment of safety committees. Such regulations have been in force since 1978 and provide an accompanying code of practice for recognised trade unions to appoint safety representatives from amongst the employees. Generally, representatives must have been employed for two years by the employer or have two years' experience in similar employment. The main tasks of the representative are:

1 To be consulted on safety matters.
2 To inspect the plant and premises, usually at three-monthly intervals but more often if there has been a change in work methods or an accident.
3 To investigate complaints from employees with regard to health, safety and welfare.
4 To investigate potential hazards and dangerous occurrences.

To perform these functions effectively, the representative needs time and training. The Regulations provide for time off with pay for carrying out their responsibilities.

The appointment of safety representatives was a step designed to overcome the apathy on safety matters which had been identified by Robens. This can be regarded as enforcement through cooperation. Two points may be noted, however:

1 The representatives have no obvious 'teeth' for enforcement. Given that they are appointed by recognised trade unions, it could be argued that any necessary 'teeth' would be the usual forms of industrial pressure or action. If this is what is envisaged, it seems to contradict the idea of a 'common interest' in safety matters.
2 By limiting the power to appoint safety representatives to recognised trade unions, no statutory machinery exists to appoint representatives in non-unionised areas of employment. Presumably, employees in these areas were to be left to 'flourish' in the state of apathy which the Robens Report identified. The initiative in these cases had to come from management.

With the decline in union recognition in the 1980s and 1990s, the part played by safety representatives decreased correspondingly. With many workplaces not being unionised, the Health and Safety (Consultation with Employees) Regulations 1996 introduced just such a requirement and an obligation on employers to give elected employee representatives such information as is 'necessary to enable them to participate fully and effectively in the consultation'. Employee representatives receive many, but not all, of the rights and protections given to their trade union-appointed counterparts. Thus, they are protected against detrimental treatment or unfair dismissal based on their role as safety representatives and have the right to paid time off, but have no right to inspect the plant and premises.

Safety committees

Regulations made under the Act provide for the establishment of safety committees if the employer receives a request in writing from two safety representatives. However, employee-elected representatives have no right to request that a safety committee be set up. Once a committee is established, s.2 lays down that it should keep health and safety measures under review. The recommendation about health and safety committees is that they should not be given any other functions or responsibilities. There are no prescribed rules about membership, but the make-up of the committee, and the seniority of the management representatives chosen to sit on it, may well give a good indication of a firm's general attitude to safety questions!

One useful function which a safety committee can fulfil is to monitor the advice available which is disseminated by bodies like the Health and Safety Executive. That advice can then be pursued and applied more specifically to the particular work situation. For example, in demolition work, an area notorious for the number and seriousness of accidents, the committee might consider the Executive's guidance notes and then recommend to the employer that it should adopt a policy of drawing up a 'written methods statement' for each demolition job. This would focus the employer's attention on the hazards of the particular job to be undertaken (e.g. the presence of asbestos, unusual structures, toxic chemicals, safe access for work to be carried out at height, dust levels and the need for protective clothing). If this is prepared after a site visit and at the tender stage, it will allow the contractor to tender at a price level which takes into account doing the job safely.

By monitoring all the reports and advice available, the safety committee can also play a useful role in alerting employers to wider health and safety issues. Again, taking demolition as the example, it may be thought that the hazards are all too obvious. But this type of work may be even more hazardous than conventional construction work because of hidden dangers. It may seemingly have been fungal spores which caused severe pulmonary illness in Lord Caernarvon and some of his team who had excavated the tomb of Tutankhamun in Egypt in the 1920s. Indeed, the illness was referred to as the 'curse of the Pharaohs'! Some old buildings could present similar hazards, calling for breathing equipment or protective clothing.

It should be remembered that the inspectors employed by the Health and Safety Executive have always seen advice giving as one of their key functions and a safety committee can act as an impetus to the employer to take the benefit of their advice.

Enforcement of the Act

A system of inspectors for the enforcement of safety legislation has existed since the appointment of the first inspectors under the 1833 Factories Act. Indeed, it is the oldest established inspectorate in the world. With the establishment of the Health and Safety Executive, all inspectors were brought together under one umbrella.

Training for new inspectors is partly 'on the job' and partly by formal training courses and there is some industry specialism, e.g. construction or agriculture, although this is less than in the past. A warrant is issued immediately to new inspectors. This is their authority to act. It specifies the powers conferred on them and they must produce it on request when seeking to exercise those powers.

Although inspectors have always had the power to prosecute, there has been internal conflict over their role, which they have seen as the improvement of standards of safety, rather than the mere enforcement of the law. Indeed, the time and effort needed to mount a prosecution were often felt to be wasted because of the low level of fines imposed. The inspectorate was fragmented with separate inspectors for factories, mines and quarries, etc. Valuable time was often wasted making routine inspections of places where there was no great danger, leaving insufficient time to concentrate on places where there were real hazards.

Section 20 of the HSWA sets out inspectors' powers to enable them to perform their duties effectively. They may:

1 Enter premises (if necessary with a constable), taking necessary equipment with them.
2 Make investigations and examinations.
3 If necessary, order the premises to be left undisturbed.
4 Take photographs and measurements or recordings and, if necessary, take away samples.
5 Question persons whom they believe can give them information relating to their investigation.
6 Inspect records, accident books, test certificates, etc.

When establishing how their time can best be spent, inspectors involved in the construction industry might prioritise as follows:

1 Investigating fatal accidents and large incidents.
2 Visiting demolition sites and sites where steel erection is taking place. This is seen as particularly important, not only because of the high level of accidents, but also because these may involve examples of employers using untrained employees.
3 Investigating any site involving asbestos. This takes a high priority because of the high incidence of asbestos-related deaths.
4 Investigating non-fatal accidents, where there has been a breach of law.
5 Investigating complaints from the general public. For example, groups of parents who are worried about an adjacent building site or possibly rival contractors who are disgruntled that they failed to get the job!

Inspectors also lay considerable stress on their instruction role and point to guidance notes issued, e.g. on the use of ladders or demolition practice. In 2008, for example, the HSE issued a pamphlet entitled 'The absolutely essential health and safety toolkit for the smaller construction contractor'.

Guidance notes are less formal than codes of practice, of which there are very few. Ironically, codes and guidance are not always popular with 'customers', who actually seem to prefer the clarity of strict regulations.

Most inspectors emphasise that, by and large, they get a good reception from employers, who accept how wide the powers of the inspector are. This may be due in part to their training in the skills necessary to present new ideas and persuade people to make changes in their working practices. The inspectors deal with people from the shop floor and people at the director level, with back-street garages and multinational corporations. In all these dealings, they hope to advise and persuade. Often the threat of action by an inspector is the only 'enforcement' necessary.

If that fails, or the situation otherwise merits it, an inspector may use the powers granted by ss.21 and 22 to serve either an improvement notice or a prohibition notice. The advantage of both of these notices is that they can be issued by the inspector immediately.

Improvement notices (s.21)

The inspector may discover a breach of HSWA 1974 or the Regulations and decide to serve an improvement notice on the person in breach. This notice will set out the breach of duty which the inspector thinks is occurring, specify the Act or Regulation in question and give the inspector's reasons for this opinion. The notice will also set out the time limit for correcting the breach, the matters to be put right and the way in which this is to be done.

The earliest date which an inspector can set for compliance is 21 days after service of the notice. This is to allow time for an appeal to be lodged, but it creates the biggest drawback about improvement notices – it means that it is impossible to get something done immediately. This can sometimes undermine the inspector's authority. Take the example of a site with no welfare hut. This is certainly not likely to be a situation warranting the use of a prohibition notice, but if the inspector serves an improvement notice, the employer can continue to ignore the notice for the next 21 days.

There is a right of appeal against the notice:

- Against the time limit imposed.
- Against the substance of the notice.

The effect of lodging an appeal is to suspend the operation of the improvement notice. On appeal, which is heard by the Employment Tribunal, the notice may be affirmed, cancelled or modified. The most frequent modification is an extension of the time limit. There must be proper grounds for an appeal. Mere lack of finance to carry out the improvements is doomed to failure as an excuse. Similarly, arguments that there has never been an accident, that the employees have not complained or that the breach is trivial, have not found favour with tribunals.

Prohibition notices (s.22)

This type of notice can be served when activities being carried on involve a risk of serious personal injury. Unlike an improvement notice, no specific breach of the Acts or Regulations need be specified, but the inspector must give reasons to support the decision. The notice directs that the activities in question must stop either immediately or within a specified period until matters have been remedied. One significant difference about a prohibition notice is that lodging an appeal does not suspend its operation. The repercussions of a prohibition notice are serious for an employer, as business may be brought to a standstill. In consequence, these appeals are usually heard as a matter of urgency.

Anyone served with either variety of notice who fails to comply with it commits an offence for which he or she may be prosecuted. As well as imposing a fine, the court may order him or her to attend to the matters specified in the notice.

Prosecutions (ss.33–42)

The two types of notices already mentioned may not be considered appropriate by the inspector. Knowledge of an employer's past record may convince the inspector that immediate prosecution is necessary. Prior to 1974, proceedings were always summary in the magistrates' courts, and the levels of fines tended to be quite low. The Act has now created the possibility of summary proceedings in the magistrates' court or proceedings on indictment in the Crown Court. The Health and Safety (Offences) Act 2008 has modified the penalties available to the courts.

Prosecutions in the Magistrates' Courts must be brought within six months of the date of the offence, but there is no time limit if a case is tried on indictment. In magistrates' courts, inspectors will conduct prosecutions themselves. Commonly, the charge may be failing to comply with s.2 of the 1974 Act. Breach of ss.2–6 could lead to a fine in either the magistrates' court (unlimited fine and/or 12 months' imprisonment) or the Crown Court (unlimited fine and/or two years' imprisonment). Failing to comply with an improvement or prohibition notice or a court remedy could result in a fine within the same limits and/or a period of imprisonment (maximum of 12 months or two years, respectively).

Breaches of all health and safety regulations are subject to similar sanctions, as well as giving rise to civil liability and for larger organisations, at least, fines have increased in recent years. In 2018 Bupa were fined £3 million by Ipswich Crown Court, one of the highest health and safety fines ever, for failings in care home management leading to Legionnaires' disease. In 2019 DB Cargo (UK) Ltd was fined £2.7 million by Newcastle Crown Court after the serious injury of several children on railway premises. And in construction, Balfour Beatty was fined £2.6 million in 2016 for multiple breaches of health and safety legislation (s.3(1) Health and Safety at Work etc. Act 1974, Reg. 31(1) Construction (Design and Management) Regulations 2007 and Reg. 3(1)(a) Management of Health and Safety at Work Regulations 1999) leading to the death of a sub-contractor working in a trench whilst laying ducting for an off-short wind farm.

In any proceedings for an offence under the Act or its regulations, the onus of proving that everything reasonably practicable has been done to satisfy a particular duty rests upon the accused. Despite the imposition of criminal liability by HSWA the Court of Appeal has held in *Davies v Health and Safety Executive* (2003) that the reverse burden of proof in health and safety cases is justified, necessary and proportionate, and does not infringe the principle in Article 6(2) of the European Convention of Human Rights that anyone charged with a criminal offence must be presumed innocent until proven guilty.

Prosecutions are only brought in extreme cases. In 2018/19 there were 394 offences prosecuted, leading to 364 convictions and 11,040 prohibition and improvement notices issued by the HSE and local authorities. In the construction industry, with its relatively high rate of fatalities and other injuries, weapons such as notices and prosecution form a useful part of the inspector's tools.

Supporting regulations

On 1 January 1993 the government introduced six new sets of regulations under the Health and Safety at Work etc. Act 1974. These were intended to implement the EU Directive on Health and Safety. The regulations are commonly referred to as 'the six pack' and cover the following areas:

- Management of Health and Safety at Work Regulations 1992
- Provision and Use of Work Equipment Regulations 1992
- Personal Protective Equipment (EC Directive) Regulations 1992
- Manual Handling Operations Regulations 1992
- Workplace (Health Safety and Welfare) Regulations 1992
- Health and Safety (Display Screen Equipment) Regulations 1992

In fact, all of the 'six pack' have subsequently been amended to a greater or lesser extent. There are extensive guidance notes attached to the regulations. They are cross-industrial and apply to all places of work. Originating from the EU, they go beyond the Health and Safety at Work etc. Act 1974 and are designed to harmonise health and safety legislation.

The central theme is, however, that there must be a continuing proactive approach to health and safety. An example is that where traditionally British legislation refers to employers doing as much as 'reasonably practicable', the new regulations are, in a sense, more prescriptive, and terms such as 'suitable and sufficient' or 'effective and suitable' are used. These imply a direct application to the job or site in question and require an employer to prepare individual risk assessments by competent persons.

The Manual Handling Operations Regulations provide a good example. Previous legislation applied to specific occupations, whereas the new regulations apply to employees, whatever their occupation, including the self-employed, and apply to the movement of loads by hand or other bodily force. Their application is therefore extremely wide, involving the preparation of individual risk assessments for any task involving an element of manual handling and the reduction of risk wherever possible.

Management of Health and Safety at Work Regulations 1999

These regulations added detail as to how the 1974 Act should be complied with and, in particular, introduced the requirement that employers carry out and document risk assessments.

Provision and Use of Work Equipment Regulations 1998

The 1998 Regulations impose on employers various obligations with respect to the provision, use, maintenance and inspection of 'work equipment', namely 'any machinery, appliance, apparatus, tool or installation for use at work'. The breadth of this definition was illustrated in *Whitehead v Trustees of the Chatsworth Settlement Trust* (2012) CA, where a gun, used by a gamekeeper, was found to be workplace equipment within the meaning of the regulations. And in *Spencer-Franks v Kellogg Brown and Root Ltd* (2008) HL, where a door closer was held to be work equipment even though at the time of the incident it was being repaired, rather than being used as a door closer. In the same way a ramp giving access for carers to a wheelchair user of care facilities has been held in *Smith v Northamptonshire County Council* (2009) CA to be work equipment, although on the facts, the employers were held not to be liable to a care assistant who was injured when the ramp collapsed as her employer had not constructed or installed the ramp. An employer may also escape liability under these Regulations where an employee improvises equipment and is injured as a result. In *Couzens v T McGee & Co Ltd* (2009) CA, a lorry driver's claim failed when a piece of metal, which he had stored in the cab of his lorry, had caused a crash and resulted in injury to the driver. His employer had neither expressly nor impliedly consented to the use of the piece of iron by the driver for scraping out the lorry and it was thus not 'work equipment' for the purposes of the Regulations. A similar outcome was reached in *Ammah v Kuehne & Nagel Logistics Ltd* (2009) CA, where an employee had used a plastic box to stand on to reach a high shelf despite the provision of suitable equipment by his employer and warnings and instructions on proper working methods. The employee's claim for compensation for breach of duty at common law and under statute was rejected.

Control of Substances Hazardous to Health Regulations 2002

These rules, generally known as COSHH, were originally enacted in 1988 in response to an EU Directive on hazardous agents. Substances are defined as 'any natural or artificial substance whether solid, liquid, gas or vapour'. Harm may be caused by inhaling, ingesting or absorbing the substance.

Employers are required to assess the risk presented by the substances, both to employees and others. They must prevent or control exposure to substances, monitor it and carry out health surveillance of employees exposed to risk. The regulations impose obligations on employers regarding training and information for employees.

Many, if not most, businesses will use substances to which COSHH Regulations 2002 and the subsequent amendments apply, although COSHH does *not* apply to asbestos or lead, which have their own regulations.

Employers may even be liable under COSHH when they have provided protective equipment, such as latex gloves, where those gloves themselves cause a harmful reaction to an employee. In *Dugmore v Swansea NHS Trust* (2003), the Court of Appeal held that employers are subject to an absolute duty which requires them to seek out risks and prevent them. There was no reason to limit liability to situations where a risk was reasonably foreseeable or could only have been revealed by a risk assessment.

In June 2007, a European Union regulation operating alongside COSHH came into force. REACH concerns the Registration, Evaluation, Authorisation and Restriction of Chemicals. Its aims include the provision of a high level of protection of human health and the environment from the use of chemicals and the imposition of responsibility on manufacturers and importers of chemicals, such that they understand and manage the risks associated with their use. A system of registration of relevant information applies. As well as the HSE website, further information on REACH can be found on the European Chemicals Agency (ECHA) website.

Control of Asbestos Regulations 2006 and 2012

The HSE estimates that about 4,000 people in Great Britain die each year from mesothelioma and asbestos-related lung cancer as a result of past exposure to asbestos. It is predicted that these figures could continue to rise in the future. Historically, exposure to asbestos was particularly high in industries like shipbuilding and railway engineering, and asbestos was a popular building material during the second half of the last century. Those presently at high risk include, perhaps obviously, asbestos removal workers, but also plumbers, carpenters and electricians. According to the HSE other less obvious occupations, such as computer installers and cabling engineers, are at risk of exposure.

The 2006 Regulations combine three earlier sets of regulations which covered prohibition of the use of asbestos, asbestos licensing and the control of asbestos at work. The importation, supply and use of any form of asbestos are banned, and a duty to manage asbestos in non-domestic premises is imposed. Guidance on the latter duty can be found in an approved Code of Practice, the Managing and Working with Asbestos, Control of Asbestos Regulations 2012 (2013).

The duty under the 2006 Regulations requires those in control of premises to take reasonable steps to determine the location and condition of materials containing asbestos and to keep up-to-date records of that information. They should presume materials contain asbestos unless strong evidence indicates the contrary and should assess the likelihood of anyone being exposed to asbestos fibres. A plan to manage these risks should be prepared. These duties fall on all those with responsibility for the maintenance and/or repair of non-domestic premises. This could apparently include landlords, tenants or a managing agent. Where employers or the self-employed are working with asbestos or doing work that may disturb asbestos, they should take steps to prevent exposure to asbestos fibres or ensure that exposure is kept as low as reasonably practicable. These steps should not rely on the use of respiratory protective equipment, but should prevent the spread of fibres.

The introduction of earlier Asbestos Regulations in 2002 provided a certain symmetry with the decision of the House of Lords in *Fairchild v Glenhaven Funeral Services Ltd* (2002) in the same year. Here their Lordships overruled a decision of the Court of Appeal that victims of mesothelioma were not entitled to recover compensation from former employers because they had had more than one employer, and it could not be scientifically established which specific employer was responsible for causing their exposure to the disease.

The 2012 Regulations brought limited changes, e.g. that some types of non-licensed work with asbestos now have additional requirements, i.e. notification of work, medical surveillance and record keeping.

Work at Height Regulations 2005

Several years ago, the HSE made the prevention of falls from height in construction one of its priorities. Falls from height are the most common cause of fatal and major injury. Poor management control, rather than equipment failure, is said to be the most likely cause. Thus, the HSE campaign resulted in its inspectors visiting workplaces to ensure that, where working at a height cannot be eliminated, fall prevention measures (e.g. guardrails, scaffolding and safe working platforms) are in place or that fall arrest systems such as safety harnesses are being used. Inspectors also check that equipment is maintained and inspected and that workers have been properly trained and are supervised. The Regulations require employers to ensure that all work at height is properly planned and organised, that those working are trained and competent and that risks from fragile surfaces and falling objects are properly controlled. Account should be taken of weather conditions when they might endanger health and safety.

As with most health and safety topics, useful guidance notes are available on the HSE website.

Construction regulations

The vast range of workplaces and working situations covered by the Factories Act 1961 meant that the rules, though detailed, could not be very specific. Each working situation will tend to have hazards peculiar to it. In the construction industry, statistics show that many accidents are caused by falls – from ladders, scaffolds, platforms or roofs – or by materials falling. Another significant cause of accidents, sometimes fatal, is the use of lifting equipment and machinery. Significant numbers of less serious accidents occur when employees step on or strike against objects. There is extra danger for workers in the industry when excavation and tunnelling take place. To take account of these special risks, there were several sets of Regulations in the construction industry:

- Construction (General Provisions) Regulations 1961
- Construction (Lifting Operations) Regulations 1961

- Construction (Working Places) Regulations 1966
- Construction (Health and Welfare) Regulations 1966

These Regulations were made by the Secretary of State for Employment under powers granted by the Factories Act 1961. In place of these old legislative measures there were more forward-looking regulations such as the Construction (Design and Management) Regulations 1994, the Construction (Health, Safety and Welfare) Regulations 1996 and the Control of Asbestos at Work Regulations 2002. Like the 'six pack', these owed much to a European influence. These provisions have largely been replaced or updated by either generic regulations (such as the Work at Height Regulations 2005) or construction industry-specific regulations as follows.

Construction (Design and Management) Regulations 2015

These Regulations, first introduced in 2007, combine two earlier sets of regulations, the Construction (Design and Management) Regulations 1994 and the Construction (Health, Safety and Welfare) Regulations 1996. The merged regulations seek to ensure that all duty holders work together in order to improve the planning, preparation and conduct of projects with the aim of ensuring the health, safety and welfare of all construction, maintenance and repair workers. Duty holders include the client, the designer, the principal, any other contractors and the workers. The principal contractor has a central role as the person who provides advice and assistance to the client on how to comply with the regulations, and who coordinates health and safety measures through the planning, preparation and construction stages of a project.

Where a contract is likely to involve more than 30 working days and more than 20 workers, or 500 person days, of construction work, the client must notify the HSE. There is an online process (Form F10).

The client is responsible for ensuring the appointment of appropriate duty holders to manage the project. The management will be carried out by the appointed principal contractor and key areas to be considered are:

- Falls from higher, of both people and objects
- Collapse of excavations
- Exposure to building dust
- Exposure to asbestos
- Electricity and gas
- Protection of members of the public

A duty imposed on the principal designer is the preparation and updating of a health and safety file containing information that those involved in carrying out construction or cleaning work on a structure in the future are likely to need, but could not be expected to know. This file is passed to the client at the completion phase.

Note that the approved code of practice, which supported the 2007 Regulations, has been withdrawn.

The regulations emphasise the need to involve workers and their representatives in planning, training and discussions about health and safety on a project. The 2015 Regulations, in a change from the 2007 version, have largely transferred the health and safety responsibility to the client.

Corporate Manslaughter and Corporate Homicide Act 2007

This long-called-for piece of legislation finally came into force in April 2008 after much debate, delays and procrastinations. As noted previously, accidents at work (or amongst the travelling public) are generally brought to public attention only when a major disaster occurs, such as the aforementioned Piper Alpha explosion in 1988 (167 deaths) or the capsizing of *The Herald of Free Enterprise* off Zeebrugge in 1987 (193 deaths). As a result of the latter incident, it was finally accepted that a corporation could be charged with the offence of gross negligence manslaughter, although on the facts of the particular case, the court was unable to identify the necessary 'controlling mind and will' to impose liability on the non-human owners of the vessel. This was so despite a finding that the shipping company's organisation was described as being infected with the disease of sloppiness from top to bottom. Although charges of gross negligence manslaughter have subsequently been successfully brought against corporations (the first being *R v OLL Limited and Kite* (1994) Winchester CrC), these cases have involved small private companies where it has been easy to identify the non-human corporation with its human managers and owners. Where a corporation has a complex hierarchical structure, there has been no comparable outcome.

With the enactment of the Corporate Manslaughter and Corporate Homicide Act 2007, more than ten years since a proposed offence of corporate killing was recommended by the Law Commission, Parliament eventually filled a significant gap in the sanctions available to deal with the deaths that result from not only high-profile disasters but from the all too frequent accidents at work. The Act provides that an organisation (including a corporation, public body, police force, partnership or trade union) is guilty of an offence if the way in which its activities are managed causes a person's death and amounts to a gross breach of a relevant duty of care owed by the organisation to the deceased. Relevant duties of care would include those owed by employers to their workers or to those employed by their subcontractors, as well as duties owed as occupiers of premises and the carrying on by the organisation of any construction or maintenance operations. However, organisations are only liable for gross breaches, namely those where the conduct alleged to amount to a breach of duty falls far below what can reasonably be expected of the organisation in the circumstances. It is the jury in any trial that must determine this question and, in doing so, the jury must consider whether the organisation failed to comply with any relevant health and safety legislation, and if so, how serious that failure was and how much of a risk of death it posed. The jury may also consider the organisation's attitude and approach to health and safety. Did it encourage or tolerate a lax regard for health and safety? A further qualification on an organisation's liability is that it is guilty of an offence only if the way in which its activities are managed or organised by its senior management is a substantial element in the breach of duty. Under the Act, senior management means the persons who play significant roles in the making of decisions about how the whole or a substantial part of an organisation's activities are to be managed or organised or the actual managing or organising of the whole or a substantial part of those activities.

The Act abolishes gross negligence manslaughter as far as organisations are concerned and imposes liability only on the organisation itself and not on its directors or managers. There is no offence of aiding, abetting, counselling or procuring the commission of an offence of corporate manslaughter. However, an individual might still be charged and convicted of gross negligence manslaughter at common law. The sanction for breach of the Act is a fine, which the Sentencing Council Guidelines have suggested should seldom be less than £500,000 and might be measured in millions of pounds and guidance is given based, among other things, on company turnover (2016). The court also has the power to order remedial action and to make a publicity order. The latter may involve publication of the fact of conviction, the particulars of the offence, the amount of any fine and the terms of any remedial order.

Surprisingly, the first cases prosecuted under the Act were of relatively small companies and did not result in high fines due to limited company resources or companies being wound up (e.g. the first conviction, *R v Cotswold Geotechnical Holdings Ltd* (2011) CA on the collapse of a soil testing pit, resulted in a fine of £385,000 payable over ten years) but a wide range of fines, into the millions, have since been applied with a record £2 million fine for metal recycling firm, Alutrade Limited, in March 2022 (Wolverhampton CrC).

Compensation for injuries at work

Where an employee is injured or is killed in an accident at work, they or their dependants may wish to sue for damages. Damages are only available where an employee can show that the employer was in some way at fault. Generally, this means that the employee must prove the tort of negligence. Although negligence is still developing as a tort, the example of the duty owed by an employer to take reasonable care for the safety of his employees was recognised very early. Note that s.69 Enterprise and Regulatory Reform Act 2013 expressly prohibits a civil claim purely on the grounds of a breach of statutory duty. Negligence and breach of statutory duty are considered in detail in the chapter dealing with the law of tort (Chapter 9). Apart from claims for damages, the injured employee may also be entitled to disablement benefits under the Industrial Injuries Disablement Benefit scheme. There may also be an entitlement to benefits under the employee's own private insurance arrangements or from a trade union.

When the court is considering a claim for damages from an employee, it will expect the employer to have taken reasonable care, given the state of knowledge and recognised practice in particular industries and trades. An employer will be expected to keep abreast of developing knowledge. The duties owed are not absolute, so the employer must weigh up the risks involved and set these against the effectiveness of precautions that could be taken, bearing in mind the cost and inconvenience of taking those precautions.

An employer owes the duty to take reasonable care of all employees as individuals. Where some employees are inexperienced or do not easily understand English or are newly recruited, the employer may have to take more precautions in their case than for other workers. However, the courts have said that they do not see the relationship of employer and employee as that of nurse and 'imbecile' [*sic*] child (*Qualcast (Wolverhampton) Ltd v Haynes* (1959) HL)!

The employer's duty to take reasonable care has been analysed regularly by the courts and they have been inclined to break down the duty into a number of broad heads:

- A duty to provide reasonably safe plant, equipment and premises.
- A duty to provide reasonably safe fellow workers.
- A duty to provide a reasonably safe system of work.

In all of these aspects the duty is ongoing. A safe system of work, once introduced, must be regularly checked and monitored. It may need to be changed to meet new standards created by new knowledge.

Employers are in breach of duty if they fail to take reasonable care. The duty is not absolute; they do not guarantee an employee's safety. It will be for the injured employee to prove negligence. In some cases, this will be admitted by the employer and the subsequent court action will concern itself only with the measure of damages. Employers may seek to defend themselves by pleading that the employee was partly to blame – contributory negligence. If that plea is successful, it will operate to reduce the damages payable by that proportion which the employee is found to be at fault.

Recent years have seen a growth in actions by employees on the grounds that they have suffered bullying or stress at work. HSE research indicates that nearly half a million people suffer work-related stress at a level which they believe makes them ill. In one of the earliest stress-related cases, *Walker v Northumberland County Council* (1995) QBD, a social worker successfully claimed damages in respect of mental stress which resulted in psychiatric injury as a consequence of ever-increasing workloads and the employer's failure to provide help despite the employee suffering an earlier breakdown. Subsequent appeal court decisions seem to take a contradictory approach to the extent of an employer's duty, but best practice would seem to suggest that employers should put into place arrangements to make work less stressful and to ensure that those arrangements are implemented, at least where an employee has suffered an earlier psychiatric illness. An employer may be expected to provide counselling or to have in place an occupational health department to deal with stress-related injuries at work. In at least one case, it was suggested that breach of the Working Time Regulations 1998 may be a factor in a stress claim. Failure to deal with claims of bullying may similarly result in civil claims either for breach of contract or for harassment under discrimination legislation. Indeed, it may even be possible to pursue a claim under the Protection from Harassment Act 1997, according to the House of Lords in *Majrowski v Guy's and St Thomas's NHS Trust* (2006) HL, although it has subsequently been suggested that the relevant conduct must be of an order which would sustain criminal liability.

Employers invariably must have Employers' Liability Insurance in accordance with the Employer's Liability (Compulsory Insurance) Act 1969 to ensure that funds are available to meet any claim for compensation (with amendments to the requirements of certificate display made by the Employers' Liability (Compulsory Insurance) (Amendment) Regulations 2008). Individuals also have rights under the Third Parties (Rights Against Insurers) Act 2010 to obtain satisfaction of a judgment directly against an insurer, which is particularly important if the company employing the individual has gone out of business. Insurance companies currently take in hundreds of millions in premiums to meet potential liabilities for accidents or diseases suffered as a result of employment. Insurers will also often undertake regular inspections of buildings and work processes to assess a company's compliance and knowledge of current health and safety legislation. This, together with their claims experience, i.e. how much they have to pay out in claims, will determine future premiums; bearing in mind that certain larger companies undertaking hazardous work will have premiums running into six figures, it is clear that businesses must give serious consideration to health and safety from an economic point of view as well as a legal and moral one.

Reporting of accidents and near misses

The current provisions for the reporting of accidents to the HSE are governed by the Reporting of Injuries, Diseases and Dangerous Occurrences Regulations 2013 (RIDDOR). There is a requirement to report (online through the HSE website) and keep a record of:

- work-related accidents which cause deaths;
- work-related accidents which cause certain serious injuries (reportable injuries) include a list of specified injuries, anything requiring seven consecutive days off work and injuries to the public/non-workers;
- diagnosed cases of certain industrial diseases and
- certain 'dangerous occurrences' (incidents with the *potential* to cause harm).

Full details of specified injuries, industrial diseases and dangerous occurrences are available on the HSE website. An accident book should be kept by all employers as good practice and is mandatory for those with ten or more employees, on construction sites with ten or more workers and for all mines and quarries (s.25(3) Social Security (Claims and Payments) Regulations 1979). As good practice recommended by the HSE, *all* accidents should be recorded, but it is mandatory to record those within RIDDOR requirements.

First Aid

There is a regulatory requirement, under the Health and Safety (First Aid) Regulations 1981, as amended, for first aid provisions in all places of work. These provisions include:

- managing the provision of first aid (first-aid kit, equipment, rooms etc.),
- requirements and training for first-aiders,
- requirements for appointed persons,
- making employees aware of first-aid arrangements and
- first aid and the self-employed.

Requirements as to equipment, training and number of first-aiders vary with the size and nature of operations, but reference should be made to the HSE guidance to ensure compliance and a safe working environment. Clearly, certain land and property-based operations, such as construction and agriculture, are among those which are considered hazardous when planning and managing first-aid provision.

The importance of health and safety at work

To conclude with a quotation from a BBC lecture by Sir Brian Appleton, technical adviser on the Cullen Report, commissioned after the Piper Alpha disaster:

> Safety is not an intellectual exercise to keep us in work. It is a matter of life and death. It is the sum of our contributions to safety management that determines whether the people we work with live or die.

Discussion Questions

1 Does the possibility of imprisonment act as an effective incentive to ensure health and safety compliance? Or does it simply serve as a punishment, where due?
2 What were the particular features of incidents which prompted the introduction of the Corporate Manslaughter and Corporate Homicide Act 2007?
3 Discuss the concept of 'reasonable practicability' in the context of s.2 of the Health and Safety at Work etc. Act 1974.
4 What features of the construction industry give rise to particular health and safety concerns? Consider this question with reference to the agricultural sector or any other industry of relevance to your studies.

11 Land law

Michael Fay

Classification of property

Land law is part of the wider discipline of property law and one of a number of types of property that can be owned. It is therefore useful to establish the meaning of the word 'property' and then to classify the different types. When considering the sale of goods, we saw that the word 'property' was used in the special sense of 'ownership'. The more general meaning of the word is, simply, anything capable of being owned.

The categories of property are diverse and cover both the tangible (such as land, jewellery and cars) and the intangible (including stocks and shares, patents and money). The law has developed different rules for each type to cover transfers of ownership. For example, if you buy a necklace in a high street jeweller, it can be handed over to you, but the same cannot happen when you buy land. A transaction involving land may be far more complex, especially when several people may have different, concurrent rights over the land. For example, suppose you take out a mortgage to help finance the purchase of a new home. To support your ongoing overheads, you rent out part of the property to a tenant. As the house is set in large, rural grounds, you also grant shooting and fishing rights and rights of way over the land to local clubs and societies. In this scenario, the landowners, mortgage provider, tenant and clubs and societies all have rights over the land at the same time. But in the example of the necklace, only you have a property interest.

English law categorises property as either 'real' or 'personal'. Historically, if a person had their freehold land wrongfully taken away, the law enabled them to bring a 'real' action to recover that land. If a person had any other type of property wrongfully taken away, the law merely enabled them to bring a 'personal' action for damages. Therefore, property came to be classified according to the type of court action that was needed to protect it.

What is land? *Que ad coeleum et ad inferos,* meaning 'the one who owns the surface owns up to the heavens and down to hell', but this is not a true reflection of the legal position.

Airspace

Owners of land may own the surface of the earth, the airspace above it and the ground below it, but this is not free of limitations. In terms of airspace, the court's decision in the case of *Bernstein v Skyviews and General Ltd* (1978) QBD demonstrates that a landowner only owns as much airspace as is necessary for the ordinary use or enjoyment of their land. In that case, a firm specialising in aerial photography flew an aircraft over Lord

DOI: 10.1201/9781003657101-11

Bernstein's house so that they could take a photograph and offer to sell it to him. The court held that the firm had not committed trespass, given that the pilot had flown the aeroplane over the house at a reasonable height. Most normal aircraft activity is now excused under s.76 of the Civil Aviation Act 1982, which precludes actions for nuisance or trespass if an aircraft flies over at a reasonable height (between 500 and 1,000 feet, with the latter being the minimum over congested areas such as cities and towns) and has complied with the relevant regulations.

In other situations, the threshold for determining the extent to which land encompasses the airspace above it will depend on the type of land and its potential use. For example, the court may consider that an owner of a grouse moor needs a considerably greater height of airspace for shooting purposes than the owner of a suburban bungalow, who may only wish to use the airspace above their property for erecting a TV aerial.

When the property is a leasehold, the extent of the airspace included will depend upon the nature and terms of the lease. In *Davies v Yadegar* (1990), the Court of Appeal said it was logical that a lease included the airspace above land. The High Court expressed a similar view in *Kelsen v Imperial Tobacco Co. Ltd* (1957) QBD about an advertising hoarding erected above a shop and in *H Waites Ltd v Hambledon Court Ltd* (2014) Ch in respect of a garage roof. But in *Rosebery Ltd v Rocklee Ltd* (2011) Ch, there was found to be no general rule that the lessor granted the full airspace when the lease included part of a roof, the use of which would affect tenants on other floors of the building.

It can be important to recognise that the airspace above land may have significant commercial worth, for example, a landowner may be able to rent out the airspace above their land to enable a marketing company to erect an advertising hoarding (for example, see *Stadium Capital Holdings v St Marylebone Properties Co. plc* (2010) CA). In the case of *Wollerton and Wilson Ltd v Richard Costain Ltd* (1970) Ch, a builder swung the arm of a tower crane over adjoining land without permission. When it became clear that they were trespassing, they offered to pay a weekly sum to the landowner. The landowner took the matter to court and was granted an injunction to prevent the builder from using the crane. The court, however, ordered the injunction to be postponed for several months, by which time the builder had finished the work that they were doing and no longer needed the crane. In such situations involving trespass in the airspace above land, the landowner may be entitled to a remedy even if their use of land is not affected.

Below the surface

The case of *Bocardo v Star Energy* (2010) SC provides some clarification as to the extent of a landowner's rights over the subterranean space below their land. It concerned a number of oil wells which were situated at least 800 feet below the claimant's land. The UK Supreme Court held that a landowner owns everything below the surface and the defendants, who had drilled into the wells from their adjoining land, had trespassed onto the claimant's land and ordered them to pay the claimants some compensation (damages).

A freeholder therefore owns the subterranean space below their land, but the position in a leasehold will, again, vary depending on the terms of the lease. In both *Lejonvarn and another v Cromwell Mansions Management Co. Ltd* (2011) Ch and *Gorst v Knight* (2018) Ch, the terms of the lease meant the tenants could not excavate the subsoil as it was not part of the demised property.

Section 205 of the Law of Property Act 1925 specifically includes any mines and minerals within its broad definition of land, but there are exceptions. For example, anyone finding potential treasure (as defined in s.1 of the Treasure Act 1996) must report it to the Coroner within a period of 14 days; otherwise, they will be committing a criminal offence. The Coroner may hold an inquest to determine whether the find is treasure and belongs to the Crown. Similarly, s.2 of the Petroleum Act 1998 entitles the Crown, rather than the landowner, to ownership of any mineral oil and natural gas found in the land, whilst any deposits of coal belong to the Coal Authority under s.7 of the Coal Industry Act 1994.

Structures

On the surface of land, there are likely to be structures such as houses, shops or offices. Buildings are included in the definition of 'land' in s.205 of the Law of Property Act 1925 and any fixtures within them are also part of the land. A fixture is something that is attached to, and regarded as forming part of, the land (for example, a fitted sink, a garden gate or a rose bush). A chattel is an item of personal property which can be removed easily and which is not regarded as forming part of land (for example, a television, fitted carpets or curtains). The same items may be classified as either fixtures or chattels in different contexts. For example, in *Holland v Hodgson* (1872) CEx, the court said a pile of stones in a builder's yard would constitute chattels, but the same stones in the form of a dry stone wall would be categorised as a fixture.

Fixtures and chattels

It can be important to determine whether an object is a fixture or a chattel for a number of reasons. For instance, when land is sold, a purchaser becomes the new owner of all of the fixtures, unless the parties have agreed otherwise. Similarly, when a tenant installs fixtures in a landlord's property, the landlord will become the new owner of them (unless they are classed as 'tenant's fixtures': see later on). Additionally, where land is mortgaged, any fixtures form part of the mortgage security, whether or not they are installed before or after the mortgage is created. Further, where a testator in a will leaves all of their real property to one person and all of their personal property to another, ownership of the fixtures will pass with the real property and ownership of the chattels with the personal property.

The question of whether an item is a fixture can also be relevant where a landowner becomes insolvent. In the case of *Lyon & Co. v London City & Midland Bank* (1903) KBD, the owners of some chairs had hired them out to the owner of a hall. To comply with the local byelaws on safety, the owner of the hall had fixed the rows of chairs to the floor. Following the owner's insolvency, the receiver was keen to show that the chairs formed part of the land, as fixtures, so that they could be counted as the owner's assets. The owner of the chairs was equally eager to demonstrate that the chairs still belonged to him, as chattels, because they would otherwise be limited to making a hopeless claim for their value in the insolvency proceedings. The court held that the chairs were chattels and so the owner of them was entitled to take them back.

The courts apply a two-stage test when determining whether an item is a fixture or a chattel. The first question is to what degree the item is annexed (attached) to the land. The second question is for what purpose it was annexed. The importance of the degree to which something is attached to the land depends on the item in question. An item does not need to be attached literally to class as a fixture, but if the item is easily detached, it will be a chattel. For example, in *Elitestone Ltd v Morris* (1997) HL, chalets held in place by their own weight were fixtures, but in *Chelsea Yacht & Boat Co Ltd v Pope* (2000) CA, a houseboat that was easily detached from its moorings was not. When determining the purpose of annexation, a court will consider what was intended when the item was installed, whether it is ornamental and attached simply to enable display, if it can be removed without damaging it or the land and whether it was installed by the original builder. The chalets in *Elitestone* were part of the land because they could not be removed without destroying them, but in *London Borough of Tower Hamlets v London Borough of Bromley* (2015) Ch, a statue was said to be a chattel because it was held in place by its own weight and could be removed without damage.

To avoid any disputes about whether items are fixtures or chattels, the parties to a sale of land will often reach an express agreement about which items are to be included in the purchase price and sold to the new owner and which items can be removed by the vendor and taken away with them. If vendors and purchasers always reached such agreements, there would be fewer disputes over items such as greenhouses, sheds, television aerials, dishwashers, carpets and garden ornaments!

Tenant's fixtures

In relation to a tenancy (see Chapter 12), the general rule is that any items installed by a tenant will belong to the landlord at the end of the tenancy. However, subject to any contrary term in the tenancy agreement, the tenant may be entitled to remove any items that are classified as 'tenant's fixtures' either before or within a reasonable time after the tenancy comes to an end. Tenant's fixtures are fixtures that are installed by the tenant, which are either: ornamental (such as paintings), trade fixtures (such as petrol pumps) or agricultural fixtures (such as machinery). In *Wessex Reserve Forces & Cadets Association v White* (2006) QBD, the High Court was asked to decide on the status of six structures at the end of a commercial tenancy. Applying the two-stage test outlined previously (see Fixtures and Chattels), the court concluded that two were the tenants' fixtures, three were chattels belonging to the tenant and one formed part of the land. In removing any tenant's fixtures, the tenant must repair any damage caused.

Wildlife and Plants

A landowner does not own any wild animals or birds that may be living on his land. However, subject to statutory restrictions (such as the Game Act 1831, the Wildlife and Countryside Act 1981 and the Hunting Act 2004), the landowner does have the exclusive right to kill such wildlife and, in doing so, claim ownership of it. For this reason, poaching is a trespass to land in tort (see Chapter 9), but it is also a criminal offence under s.30 of the Game Act 1831, the Night Poaching Act 1828 and if carried out with dogs, the Hunting Act 2004.

A landowner owns the trees and other plants growing on their land, so they are able to landscape their garden relatively freely. There are some statutory restrictions, and these will be of particular relevance in the management of large estates or woodlands. Trees may be affected by tree preservation orders (see Chapter 13). Woodlands can be designated as assets of community value under s.87 of the Localism Act 2011 or Sites of Special Scientific Interest under s.28 of the Wildlife and Countryside Act 1981. Of relevance to those of us who enjoy country walks or public parks, the picking, uprooting or destruction of specific wild plants, such as wild gladiola, and the intentional uprooting (but not picking) of wild plants by unauthorised persons is unlawful under s.13 of the Wildlife and Countryside Act 1981. Picking flowers from public parks is an offence under s.11 of the Theft Act 1968.

Water

A landowner does not own any water on their land but may have rights in respect of it, for example, to abstract quantities of it for use in connection with their land, again subject to any statutory restrictions. If the landowner is taking water from a river, then they must be careful to observe the rule that landowners further downstream are also entitled to the flow of the river unaltered in quantity and quality. If a landowner pollutes the water, then the owners of the neighbouring land or a watercourse can bring a claim in nuisance or trespass. This issue was recently before the courts in *Manchester Ship Canal Co. Ltd v United Utilities Water Ltd* (2024) SC, where the Supreme Court held that a statutory sewage undertaker that polluted a watercourse was not protected from nuisance or trespass claims by the Water Industry Act 1991.

Freehold and leasehold estates

People in England and Wales do not actually own land. Instead, all land belongs to the Crown and landowners possess an entitlement to land for a certain length of time. This entitlement is called an 'estate' and, since 1925, there have been only two types: the freehold estate and the leasehold estate. These are set out in s.1 of the Law of Property Act 1925.

This may seem to be largely a theoretical matter, but it provides the basis of the law of compulsory purchase, namely, that the state can acquire land for the public good for due recompense (a recent controversial example being the HS2 High Speed Rail Network) and also explains why land reverts to the Crown where someone dies and there is no-one to inherit under the intestacy rules (a process called 'escheat').

The major difference between the two estates in land is the length of time for which they will be held. A freehold estate, or 'fee simple absolute in possession', is the nearest equivalent to absolute ownership of land. Essentially, a freehold estate continues to exist as long as there are persons entitled to take it on the owner's death either under the terms of a will or, if the deceased has not made a will, on the application of the rules of intestacy. By contrast, an owner may grant a leasehold estate, or 'term of years absolute', to a tenant. Once a leasehold is created, a tenant is entitled to exclusive possession of the land for a particular period of time or 'term'. If the term of the agreement is quite short, it is often referred to as a tenancy (for example, a monthly tenancy). If it is longer, then it is usually referred to as a lease (for example, a 99-year lease), but there is no precise or rigid rule. Both freehold and leasehold estates can be sold, inherited either under a will or on intestacy or otherwise transferred, for example, by way of a gift.

The Leasehold and Freehold Reform Act 2024 contains several key reforms to this area of land law. Leaseholders are now able to extend their leases or purchase their freehold without first owning the property for two years, allowing leaseholders to act immediately upon acquiring their property. The eligibility criteria for Right to Manage schemes have also been expanded, allowing more leaseholders in blocks of flats to take control of their property's management (such as collecting service charges, communal area maintenance and insurance) without owning the freehold. More mixed-use buildings now qualify for

Right to Manage schemes as the limit on non-residential usage has increased from 25% to 50% and leaseholders are no longer required to cover the freeholder's legal fees in Right to Manage claims.

Reforms not yet in force at the time of writing include all leasehold extensions being extended for a term of 990 years, regardless of whether the property is a flat or a house (the current term extensions are 90 years for a flat and 50 years for a house); an outright ban on residential leaseholds, except for permitted leases set out in schedule 1 of the Act, meaning all property purchases would be freehold; and abolition of 'marriage value', so it is cheaper and easier for a leaseholder to extend their lease.

The message, therefore, is to watch this space.

Interests in land

Apart from the two legal estates (freehold and leasehold), other rights in land are called interests in land. Essentially, these fall into two types: legal interests and equitable interests.

Section 1 of the Law of Property Act 1925 provides that a limited number of rights in land may be legal interests if they are created by way of a deed. The most common of these are easements and profits à prendre (provided that their duration is equivalent to either a freehold or a leasehold) and charges by way of legal mortgage. The nature of each of these interests is considered later on. Any interest in land which is not a legal interest must be an equitable interest. Examples of equitable interests in land are: interests under restrictive freehold covenants; interests created without the use of a deed; interests under contracts to create legal estates or interests; and beneficial interests established under trusts of land.

Traditionally, the essential difference between a legal interest and an equitable interest is that a purchaser of land is bound by all of the legal interests in the land but may not necessarily be bound by the equitable interests. The position is regulated by the Land Charges Act 1972 (in respect of unregistered land) and the Land Registration Act 2002 (in respect of registered land). We will consider the extent to which a purchaser of land can discover and be bound by the major interests in land, and conversely, how an owner can ensure that their interests will bind a purchaser.

Easements

An easement is an interest in land which gives the owner of one piece of land (the 'dominant land') the right to use or restrict the use of another's land (the 'servient land'). Common examples of easements include: rights of way, storage rights, car parking rights, rights of light to land or buildings and rights of support to buildings.

Characteristics of an easement

For a right to be classified as an easement, it must satisfy the four criteria from the case of *Re. Ellenborough* (1955) CA.

1 First, there must be a dominant land which benefits from the easement and a servient land which is burdened by it. Therefore, a public right of way cannot be an easement because members of the public have the right to use it whether or not they own any dominant land.

2 Second, the right must benefit or 'accommodate' the dominant land. For a right to satisfy this criterion, the dominant and the servient land must be close to one another. In the case of *Bailey v Stephens* (1962) CCP, the court provided the example that an owner of land in Northumberland could not grant an easement for the benefit of land in Kent as the two pieces of land were not in sufficiently close proximity. The right must also benefit the dominant land itself, rather than its owner or their business. In the case of *Hill v Tupper* (1863) CEx, the claimant was a tenant of some land which adjoined a canal. They ran a business hiring out pleasure boats and argued that their right to put boats on the canal constituted an easement. Given that the right provided them with a purely commercial advantage, the court held that it could only constitute a personal licence rather than an easement.
3 Third, a right can only be an easement if the dominant and the servient owners or occupiers are different people. This means that a landowner cannot have an easement over their own land. However, if they grant a lease of their land to a tenant, they may grant related easements to the tenant, for example, to give the tenant access over the landlord's staircase to a first-floor flat.
4 Finally, the right must be capable of forming the subject matter of a grant. This requirement comprises a number of sub-criteria. The dominant and the servient owners must be legally competent, for example, of adult age. The right must also be sufficiently definite. In *William Aldred's Case* (1610) KB, the court held that a right to a view would be too vague to constitute an easement. The courts are also more likely to hold that a right is an easement if previous courts have found in favour of easements of this type. For a right to be classed as an easement, it should not generally require the servient owner to spend money. An exception to this general rule is the easement of fencing, which only arises in respect of agricultural land (see *Crow v Wood* (1970) or, more recently, *Churston Golf Club Ltd v Haddock* (2019) CA). Finally, the exercise of the right should not unduly deprive the servient owner of the benefit of their land. In the case of *Batchelor v Marlow* (2001) CA, the court held that the right to park six cars was too excessive to be an easement as it would unduly restrict the servient owner's use of their land.

Creation of an easement

Easements may be created in various ways. They can be express, implied or created by long use (known as 'prescription').

Express creation

Legal easements may be created by express agreement between the parties, by operation of statute or through a person's will. A servient owner may expressly 'grant' a legal easement over their land for the benefit of the dominant owner for a period equivalent to either a freehold or a leasehold estate. This type of easement must be documented by way of a deed. If a landowner sells off part of their land, they may also 'reserve' easements for the benefit of the land that they retain. This will often be documented in the same deed that is used to sell off the land. Easements may also be created by statute. For example, utility companies have rights to enter privately owned land to install supply pipes and cables (see the Gas Act 1986, the Electricity Act 1989 and the Water Industry Act 1991).

Finally, people may create easements through their will if they leave dominant land and servient land to be inherited by different people.

Implied creation

Easements may also be impliedly created in a number of ways.

1 *Necessity*
An easement may arise through necessity if the land cannot be used without it, for example, if there is no other means of access between the land and a publicly adopted highway. This was the issue in *Nickerson v Barraclough* (1981) CA, where the claimant bought a landlocked field.
2 *Common intention*
An easement may also be evidenced by the common intention of the parties. In the case of *Jones v Pritchard* (1908) Ch, the court held that there was an implied easement for the owners of semi-detached properties to make joint use of the properties' shared chimneys.
3 *The rule in* Wheeldon v Burrows
Following the case of *Wheeldon v Burrows* (1879) CA, an easement may additionally be implied if the right was being used when the land was sold and is necessary for its reasonable enjoyment and/or it has been used for a substantial period of time and can be detected by a reasonably careful examination of the land.
4 *Section 62 Law of Property Act*
Finally, if a right is being used when land is sold or leased, the deed may automatically convert such a right into an easement under s 62 of the Law of Property Act 1925. In the case of *Wright v Macadam* (1949) CA, a tenant had permission under a licence to store coal in her landlord's shed. When she renewed the lease, s 62 automatically converted her right into an easement.

Long user/prescription

As well as being expressly or impliedly granted or reserved, easements may also be acquired by long user or 'prescription', for example, through the Prescription Act 1832. For example, if a landowner uses a path to cross their neighbour's land for a prescribed period of time, the landowner may become entitled to an easement. An easement can be acquired via this method by either a freeholder or a leaseholder of dominant land against either a freeholder or a leaseholder of servient land. However, a tenant is unable to acquire an easement by prescription against his own landlord.

In order to acquire such an easement by prescription, the dominant owner must have used the servient owner's land as if they had a right to do so. If the dominant owner crossed their neighbour's land in secret, so that their neighbour would not discover them or if the dominant owner forced their way over their neighbour's land, for example, by damaging fences, they would not acquire an easement by prescription. In addition, the dominant owner must make regular use of the servient land.

There are three different methods of acquiring easements by prescription, and they each have their own requirements as to the period of time over which a right needs to be exercised before it can be classed as an easement. Whilst the technical details of these requirements fall outside the scope of this book, essentially, the right needs to be exercised for a minimum of 20 years before a claim for prescription can be considered.

Protection of an easement

The question as to whether the benefit of an easement needs to be protected on the Land Charges register (for unregistered land) or the Land Register (for registered land) depends on how the easement has been created. Some easements, such as those acquired by prescription, do not need to be protected by registration. Instead, they will automatically bind a purchaser of the land.

If there is an infringement of an easement, then, depending on the circumstances, the dominant owner may be able to use the self-help remedy of abatement. For example, if their right of way is blocked, they may be able to take steps to remove the blockage provided that they act promptly and reasonably, give notice to the servient owner (except in an emergency or where access to the servient owner's land is not required) and do not cause any unnecessary damage. It is, however, generally less risky for the dominant owner to consider bringing a claim in nuisance instead. They may seek damages from the servient owner and/or an injunction requiring them to remedy the problem and/or a declaration that the easement exists and defining its extent (to help avoid any issues arising in the future).

Extinction of an easement

Servient owners who seek to bring an easement to an end may do so in a number of different ways.

They may approach the dominant owner to see if they will agree to expressly release the benefit of the easement in return for payment of compensation. Such an express release would normally be documented by way of a deed.

The servient owner may be able to claim that the easement has been impliedly released because the dominant owner has not been exercising it, although long non-usage is often, of itself, not sufficient to evidence abandonment (*Benn v Hardinge* (1992) CA).

It might be established that the easement has come to an end due to a substantial change in the nature of the use of the land.

Finally, if the dominant and servient lands become owned and occupied by the same person, which is called 'unity of seisin', the easement will automatically come to an end. It is worth noting that, in such circumstances, the easement could be revived in the future if the two pieces of land are subsequently owned by different people.

Profits à prendre

A profit à prendre is an interest in land which enables someone to take something from another's land. Common examples include rights to fish and to graze cattle. Essentially, the legal rules on profits à prendre are identical to those which relate to easements (see earlier in the chapter), with two major differences. First, in relation to a profit à prendre, there is no requirement for the existence of any dominant land. Second, the only form of implied grant or reservation of a profit à prendre is through evidencing the parties' common intention.

Note: The Law Commission published a report on Easements, Covenants and Profits à prendre (Law Com No 327) in June 2011, and the government noted an intention to simplify the law in this area in February 2017. These laws may well change in the forthcoming years.

Restrictive freehold covenants

A restrictive freehold covenant is a promise that is given by one landowner (the 'covenantor') to another landowner (the 'covenantee') that they will not use their land in certain ways, for example, for business purposes. The covenantor bears the burden of the covenant and the covenantee is entitled to its benefit. Such covenants are contractual obligations and are usually documented by way of a deed. The requirements for a deed are set out in s.1 of the Law of Property (Miscellaneous Provisions) 1989 Act, namely:

- the document must be signed by both parties,
- each of the parties' signatures must be witnessed and
- the witness must also sign the document to attest to the parties' signatures.

If a party needs someone to sign the deed on their behalf, for example, because of illness, then two witnesses will be required.

As a consequence of the principle of privity of contract (see Chapter 5), the original covenantor who made the promise will always be liable for any breach of the covenant even after they have sold their burdened land to someone else. New owners of the burdened land may also be obliged to comply with the restrictive covenant in equity if the following four criteria arising from the case of *Tulk v Moxhay* (1842) Ch are satisfied.

1 First, the covenant must be negative. This means that it restricts the covenantor from taking action, for example, a covenant not to cause any nuisance from the property requires the covenantor to refrain from doing something.
2 Second, as in the case of easements (considered previously), the covenant must have been made for the benefit of the covenantee's land and not merely for the covenantee's personal benefit. The covenant may, for example, benefit the covenantee's land if it increases its value by prohibiting undesirable development or activities on the covenantor's neighbouring land. (Despite this general rule, certain bodies, such as the National Trust, are not required to own benefited land in order to enforce conservation covenants). This criterion also requires the covenantee to have owned the benefitting land on the date that the covenant was made.
3 Third, the original parties to the covenant must have intended that the burden of the covenant would be transferred to subsequent owners of the covenantor's land. There may be express evidence of their intention in the deed that created the covenant. If not, s.79 of the Law of Property Act 1925 implies such an intention (unless there is anything in the deed to suggest otherwise).
4 Finally, to be enforceable against all subsequent owners of the covenantor's land, the covenant must be protected by way of a land charge at the Land Charges Department (if the land is unregistered) or entered on the register at the Land Registry (if the land is registered).

If both the original covenantor and the original covenantee have sold their lands on to other people, it will be necessary to check whether the *benefit* of the covenant has similarly passed in equity to the new owners of the covenantee's land. If it has, they will be able to bring a claim against the current covenantors for an injunction or damages. A covenantee may transfer the benefit of a covenant to a subsequent owner of their land by expressly 'assigning' the benefit of that covenant in writing. For the benefit of a covenant to be transferred in this way, it would need to be expressly assigned each time the land is sold on to a new owner.

If there is no evidence that such an assignment has taken place, then the new owner may still be able to benefit from the covenant if they can demonstrate that it has been 'annexed' to their land. When the covenant was created, the terms of the deed may indicate that the original parties intended the benefit of the covenant to pass to subsequent owners of the covenantee's land through 'express annexation'. Even without such wording, s.78 of the Law of Property Act 1925 provides that the benefit has passed through 'statutory annexation' if the extent of the benefitting land is clear.

Particular rules apply to 'building schemes' or 'schemes of development' which are often put in place on housing estates. A building scheme exists when a developer makes each plot of land that they sell subject to the same set of covenants, which will benefit all of the other plots on the development. Each time a plot is sold, the new owners will automatically take on the burden of the covenants against their plot and simultaneously benefit from the covenants against the other plots.

It may be difficult for the current covenantee to bring a claim in equity against the current covenantor, for example, because the current covenantor is bankrupt. In such situations, the current covenantee may seek to bring a claim in common law against the *original* covenantor instead, provided that the original covenantor can be traced. To bring such a claim, the current covenantee must satisfy the four requirements from the case of *P & A Swift Investments v Combined English Store Groups* (1989) HL. First, the covenant must 'touch and concern' or enhance the value of the covenantee's land. Second, the original parties must have intended the benefit of the covenant to pass to future owners of the covenantee's land. (Even if such an intention is not expressly contained within the deed, it will be implied by s.78 of the Law of Property Act 1925). Third, at the time when the original parties entered into the covenant, the covenantee must have held a legal estate in the benefitting land. Finally, the current covenantee must demonstrate that their ownership of a legal estate in the land can be traced back to the original covenantee.

A covenant may come to an end in a number of ways:

> Similar to an easement, a covenant will automatically come to an end through unity of seisin (which we looked at previously) if both the burdened and benefitting lands are owned by the same person. (If the covenant exists under a building scheme, the covenant may be resurrected at some point in the future if both the burdened and benefitting lands fall into separate ownership once more.)

The covenantor may also ask the covenantee for an express release of the covenant by deed, usually in return for payment of some compensation.

The covenantor may claim that the covenant has come to an end through 'implied discharge' if they can demonstrate that the covenantee has failed to take action for a breach for a particular period of time. Unfortunately, it is unclear how long a period of time is required, and each case will turn on its own facts.

Finally, the covenantor may make an application to the Upper Tribunal (Lands Chamber) under s.84 Law of Property Act 1925 to discharge or modify a restrictive covenant.

In order to do so, the covenantor will need to prove one of a number of grounds. These include demonstrating that the covenant has become obsolete, for example, in the case of a covenant requiring the use of a private dwelling house in an area which is now dominated by business premises. An application under s.84 was the issue in *Alexander Devine Children's Cancer Trust v Housing Solutions Ltd* (2020) SC. A developer had built houses on land burdened by restrictive covenant not to build on it, then made an application under s.84. The UK Supreme Court overturned the Upper Tribunal decision, because the developer's cynical behaviour had presented a *fait accompli* and, had the application been made before any construction occurred, the developer would not have satisfied the necessary public interest grounds.

Where a covenant is discharged or modified, any entries relating to the covenant should be removed from, or amended on, the land charges register (where the land is unregistered) or the land register (where the land is registered).

Mortgages of land

The enduring quality of land makes it a valuable asset and a superior form of security for raising money. The owner of the land who borrows against it (the 'mortgagor') will naturally want to continue to use and enjoy their land, but the person lending the money (the 'mortgagee') must be given certain rights over the land in the event that the mortgagor defaults on making their repayments. These requirements are provided for in the rules governing legal mortgages.

In registered land, s.87 of the Law of Property Act 1925 provides for the creation of a mortgage by a charge by deed, expressed to be by way of legal mortgage. The rules ensure that a mortgagee has the necessary interest in the land for it to constitute effective security for the loan. This interest is reinforced by the rights granted to a mortgagee to realise this security if the mortgagor defaults on making their repayments. These include:

1 The mortgagee's 'right to take possession'. If a mortgagee has to sell the property in order to reclaim what they are owed, they will need to obtain vacant possession of it first. To do this, they will usually make an application to the court under the Civil Procedure Rules Part 55.
2 The mortgagee's 'right to appoint a receiver' under ss.101 and 109 of the Law of Property Act 1925. A mortgagee may appoint a receiver to manage or sell the mortgaged property, discharge the mortgage and pay any surplus over to the mortgagor. This is common in relation to mortgages of commercial property.
3 The mortgagee's 'right to sell' the property under s.101 of the Law of Property Act 1925. The case of *Cuckmere Brick Co v Mutual Finance Ltd* (1971) CA makes it clear that, when selling the property, the mortgagee is under a duty to obtain the best price reasonably obtainable (see also *Silven Properties Ltd v Royal Bank of Scotland plc* (2003) CA and *Burns v Bridge* (2024) Ch). Under s.105 of the Law of Property Act 1925, the mortgagee then pays off any costs in connection with the sale and the mortgage monies owed. They will use the remainder of the sale proceeds to discharge any subsequent mortgages and pay any surplus to the mortgagor. Case law, including *Rudge v Richens* (1873) CCP, *Bristol and West v Bartlett* (2003) CA and *Bank of Scotland plc v Burnley Borough Council* (2024) UTLC, demonstrates how, if there is a shortfall between the sale proceeds and the monies needed to discharge the mortgage, the mortgagee may

bring a claim against the mortgagor. Under s.20 of the Limitation Act 1980, the mortgagee must bring their claim within a period of 12 years in respect of any capital monies due and six years in respect of any arrears of interest.

A prospective purchaser must be careful to ensure that the land they are interested in will not be subject to an unredeemed mortgage. If the land is registered, any outstanding mortgages should be apparent from the land register (ss.27 and 32 of the Land Registration Act 2002). If the land is unregistered, then, under s.1 of the Law of Property Act 1925, a legal mortgage will be protected by the deposit of the title deeds with the mortgagee. As we discuss later in the section on conveyancing, any prospective purchaser of unregistered land will always need to see the title deeds. If they are not available, this will suggest that they are being held by a mortgagee as security to protect a loan. Any other mortgages will be classed as *puisne* mortgages and must be protected by registration at the Land Charges Department under s.2 of the Land Charges Act 1972 and will be revealed by the prospective purchaser's land charges search. The vendor can also be asked to disclose the existence of any outstanding mortgages in their replies to pre-contract enquiries.

Licences

A licence is a permission to use land which is granted by a licensor to a licensee. It does not confer any legal estate or interest in land but is merely a personal right that prevents the licensee from being a trespasser. Licences are used in a variety of situations, for example: in property guardianship arrangements where companies permit guardians to live in their unused premises in order to deter squatters and vandals; to permit fishing or grazing; and to enable given brands to temporarily trade in shopping centres. There are various different types of licences.

A *contractual licence* is granted by a licensor in return for payment of consideration, for example, a licence fee. Contractual licences are used in many situations, for example, when a hotel owner permits guests to stay or a sports centre authorises people to use its facilities. Conversely, if a landowner grants a licence without requiring the payment of any consideration from the licensee, it will be a *bare* or *gratuitous licence*. An example of such a licence would be an invitation to a friend's house for coffee. Whilst no money changes hands, the landowner gives permission to their friend to be on the premises.

Whilst legal interests in land, such as rights of way, cannot unilaterally be brought to an end by the person who bears the burden of them, the general rule is that licences are only personal rights which may be revoked by the licensor at any time. Similarly, whilst legal interests may benefit/burden future owners of the land, licences will automatically come to an end on any sale of the land. There is, however, a third type of licence (already discussed) which involves profits à prendre. As a profit à prendre comprises both a licence to enter another person's land and a property right to take something from that land, the licensor may not unilaterally revoke the right and future owners of the land will be bound by it.

Finally, if a landowner has unfairly led someone to believe that they have rights over the land and that person has acted in reliance on those rights, the court may hold that they have granted an equitable interest in the land through an *estoppel licence* and may award remedies accordingly.

When a landowner permits someone to access their premises, it is important to clarify the basis for that permission. Whether they decide to grant a lease or a licence will affect aspects such as how much control the landowner retains over the property, whether and how they can unilaterally bring the arrangement to an end and if the occupier has a right to renew the arrangement.

Trusts

Essentially, a trust exists when a person ('the trustee'), holds property for the benefit of another person ('the beneficiary'). The trustee holds the legal estate of the property and the beneficiary has an equitable interest which entitles them to the benefits of the property. The Trusts of Land and Appointment of Trustees Act 1996 (TLATA) sets out a number of powers and duties of the trustee. Section 6 provides the trustee with the same powers as if they were the landowner which includes the rights to either sell or mortgage the property. Before doing so, s.11 requires the trustee to consult with any beneficiaries, so far as it is practicable to do so. Section 12 of the Act enables any beneficiaries to occupy the property. If there is any dispute about the trust, any person with an interest in the land (including any trustee, beneficiary or mortgagee) can make an application to the court under s.14 for it to be resolved. Section 15 sets out a non-exhaustive list of factors which the court must take into account, including the welfare of any children who live at the property.

Trusts may be created in a number of ways and for a number of reasons. Section 53(1)(b) of the Law of Property Act 1925 enables a landowner to expressly create a trust of their legal estate in writing, either during their lifetime or through a will. For example, a widow owns a house. She has a son who is 8 years old. The terms of her will provide that, if she dies, the house is to be held on trust for her son by two trustees until her son reaches the age of 21.

If a landowner does not expressly create a trust of their legal estate, the court may imply that a trust has been created under s.53(2) of the Law of Property Act 1925. For example, suppose that the purchaser of a house agrees that their friend can move in with them and that she will become entitled to an interest in the house if she helps them with the mortgage repayments. Whilst the parties have not entered into any written agreement to create an express trust, the court may be prepared to imply a trust under which the purchaser holds the legal estate on trust for themself and their friend as beneficiaries.

There are two types of implied trust: a resulting trust and a constructive trust. The main difference between them is how the beneficiaries' respective shares are calculated when the property is sold. Under a resulting trust, the beneficiaries' shares are apportioned between them according to their respective contributions to the purchase price (see, for example, *Dyer v Dyer* (1788) ExCh and *Marr v Collie* (2018) PC). If there is a constructive trust, then, following the key case of *Stack v Dowden* (2007) HL, the court will take what is known as the holistic approach to quantifying each of the beneficiary's shares. They will take into account a variety of factors, including:

- any advice or discussions at the time of the conveyance or transfer that would indicate their intentions at that time as to their respective interests in the property;
- the reasons why they purchased the property jointly;
- the reasons why the survivor of the couple was authorised to give a good receipt for capital monies;
- the purpose for which they acquired the property;
- the nature of the parties' relationship;

- whether the couple had children for whom they both had responsibility to provide a home;
- how the purchase was financed, both initially and subsequently;
- how the parties arranged their finances, for example, whether their accounts were held in separate and/or joint names and
- how the couple paid their outgoings on the property and other household expenses.

Following *Stack v Dowden*, the court is more likely to imply a constructive trust in a domestic context which involves the family home. In circumstances which involve both domestic and commercial elements, the court will consider the relationship between the parties and the purpose for which they acquired the land. By way of an example, the case of *Laskar v Laskar* (2008) CA involved a mother and her daughter purchasing a property as a buy-to-let investment. Despite their familial relationship, the court held that the parties owned the property under a resulting trust.

Co-ownership

It is quite common for two or more people to share ownership of land at the same time. Examples include a couple who buy a house together or parties who jointly purchase business premises.

The law recognises two types of co-ownership:

- joint tenancies and
- tenancies in common.

Sections 1(6), 34(1) and 36(2) of the Law of Property Act 1925 together provide that co-owners will always own the legal estate as joint tenants. Co-owners may hold their equitable interests as either joint tenants or tenants in common. The case of *AG Securities v Vaughan and Others* (1990) HL provides that for a joint tenancy to exist, the following 'four unities' must be present:

1 Unity of possession which occurs when each co-owner is equally entitled to occupy and possess the whole of the land.
2 Unity of interest which requires each co-owner's interest in the land to be the same. For example, both parties may own a freehold interest.
3 Unity of title which is satisfied when each co-owner has acquired their title in the same way, usually through the same deed.
4 Unity of time which requires each co-owner to have acquired their interest at the same time.

In a tenancy in common, the only unity that needs to be present is the first unity, namely, the unity of possession. Another difference between these two forms of co-ownership is that, where there is a joint tenancy, the co-owners are regarded as owning the whole of the property together, whereas in a tenancy in common, the co-owners will each own a separate, quantifiable share in the property.

Determining whether the co-owners hold their equitable interest in the property as joint tenants or tenants in common can be important if one of the co-owners dies. Upon the death of a joint tenant, the right of survivorship applies. In *Harris v Goddard* (1983)

CA, the court explained that this means that the deceased's interest in the property is automatically transferred to the surviving joint tenant(s). For example, if a property is co-owned by a husband and wife as joint tenants and the husband dies, his interest automatically passes to his wife, who then becomes the absolute owner of the property. Conversely, if a tenant in common dies, their share will be transferred according to the terms of their will or, if they die without leaving a will, according to the rules of intestacy. Their interest may therefore be inherited by someone other than the surviving co-owners.

One problem associated with co-ownership is that there could potentially be many co-owners of one property, which could complicate transfers or other related transactions in respect of it. Therefore, whether co-owners own the property as joint tenants or tenants in common, they will hold their interests under a trust of land. A maximum of four trustee(s) will hold the legal estate (under s.34(2) of the Law of Property Act 1925) and the beneficiary/beneficiaries will hold an equitable interest in the property. Essentially, s.2(1) of the Law of Property Act 1925 (as amended by TLATA, sch. 3) provides that a purchaser who pays two trustees for the property will 'overreach' the beneficiaries' interests, which means that they will not be bound by them. In such a situation, the trustees will hold the proceeds of the sale on trust for the beneficiaries.

It should be noted that, in practice, the co-owners are often both trustees and beneficiaries (for example, where two people have jointly purchased a house). Where only one of the parties wishes to sell the property, they may need to apply to the court under s.14 of TLATA (as discussed earlier) to do so.

Particular rules apply in the context of matrimonial proceedings. For example, upon divorce, the court may rely on the Matrimonial Causes Act 1973 to adjust the relative interests of the spouses in the matrimonial home. The court has similar powers under the Civil Partnership Act 2004 where a civil partnership is being dissolved. Additionally, a spouse, civil partner or cohabitee may possess rights under s.30 of the Family Law Act 1996, whether or not they also have a beneficial interest in the property.

To determine whether co-owners own their equitable interests in the property as either joint tenants or tenants in common, the starting point is to check the terms of the conveyance or transfer to them. The parties may have made an express declaration in that deed as to whether they will hold their equitable interest as joint tenants or tenants in common. If the co-owners have not made such an express declaration, there is a general presumption that 'equity follows the law' (following *Stack v Dowden*). As mentioned, the law requires co-owners to hold the legal estate as joint tenants. The application of this presumption means that they will also hold their equitable interests as joint tenants. However, this presumption can be rebutted by evidence to the contrary.

For example, there may be other wording in the deed which indicates how the parties intend to own the property. The case of *Payne v Webb* (1874) CA demonstrates how a reference to the property being transferred to the parties 'in equal shares' would suggest a tenancy in common. The court may also find that the co-owners hold their equitable interest as tenants in common in the interests of fairness, for example, where the co-owners have made unequal contributions to the purchase price (*Jones v Kernott* (2012) SC) or have bought the property for commercial purposes. Recently, in *Nilsson v Cynberg* (2024) Ch, it was clarified that an express declaration can only be overridden by a subsequent, clear agreement (or conduct amounting to such) demonstrating a shared change in intention. In *Nilsson*, this 'subsequent agreement' was an informal one made on the parties' separation.

Even where the application of these tests indicates that the parties' equitable interests in the land are held under a joint tenancy, the case of *Harris v Goddard* (1983) CA explains how a party may 'sever' the joint tenancy in relation to their own equitable interest at any time and, in doing so, convert it into a tenancy in common. This would enable them to deal with their share in the property as they wish (for example, by way of mortgage, sale or gift) and also ensure that, on their death, it is transferred through their will to someone other than their co-owners. It is important to note that severance only converts the outgoing joint tenant's interest into a tenancy in common. The remaining joint tenant(s) interest will be unaffected. Section 36 of the Law of Property Act 1925 provides that an equitable joint tenant can sever their interest by either serving notice in writing on the other joint tenant(s) or by doing 'such other acts or things'. We will consider each of these methods in turn. The notice in writing need not be signed (*Re. Draper's Conveyance* (1967) Ch) but must indicate the outgoing joint tenant's immediate intention to sever (*Harris v Goddard* (1983) CA). Section 196 of the Law of Property Act 1925 requires the notice to be either left at or posted through the normal post to the last-known place of abode or business of the other joint tenant(s) or sent to them by registered post. Whichever method is used, the cases of *Kinch v Bullard* (1999) Ch and *Re. 88 Berkeley Road* (1971) Ch, respectively, demonstrate that the notice will still operate to sever the equitable joint tenancy even if the remaining joint tenant(s) never receive or read it.

The conveyancing process

Conveyancing is the process of transferring ownership of land, for example, by sale, gift, through the operation of a will or on intestacy. For the purposes of this section, we will focus on the example of a sale and purchase of land. There are two systems of conveyancing in England and Wales: the unregistered conveyancing system and the registered conveyancing system. The traditional system is unregistered conveyancing, but as all land in England and Wales is now subject to compulsory registration under s.4 of the Land Registration Act 2002, registered conveyancing has been gradually replacing it to the extent that 87% of all land in England and Wales is now registered (HM Land Registry, 2020). To determine whether either the unregistered or the registered system of conveyancing applies, a purchaser carries out an official search of the index map by sending a form to the Land Registry. In addition to confirming whether the land is unregistered or registered, the index map search result will also confirm any unique title number(s) for land which is registered.

Before the vendor puts the property on the market, they must ensure that they have a valid energy performance certificate for the property from an accredited assessor. The detailed requirements are set out under the Energy Performance of Buildings (England and Wales) Regulations 2012. Essentially, the vendor must provide a copy of the certificate, free of charge, to the prospective purchaser. The certificate will set out information about the energy efficiency of the property. Occupiers of non-dwellings that have a total useful floor area of over 500 square metres and a valid energy performance certificate and are frequently visited by members of the public and not exempt, are required to display their energy certificate in a prominent place. There are also obligations on those who have control of air-conditioning systems (with a maximum calorific output of more than 12 kW) to ensure that the system is inspected at least every five years by an energy assessor. The purpose behind these requirements is to enable the public to compare the energy performance of public buildings and to promote improved energy use.

The conveyancing of both unregistered and registered land involves a number of other stages which are summarised and distinguished in what follows.

Once the parties have agreed on the terms of the sale of either unregistered or registered land, the vendor drafts the contract and sends two copies to the purchaser for review. The purchaser may either agree to the draft or require amendments which are then negotiated between the parties.

Proof of title

The vendor must also prove that they own the title to the land by a process known as 'deducing title'. Traditionally, the vendor would deduce title to *unregistered* land by sending the purchaser an 'abstract of title' which summarised the title deeds. It is more common now for the vendor to provide the purchaser with an 'epitome' or chronological list of the title deeds, together with photocopies of them. The purchaser will need to review these documents to identify a 'good root of title' which is at least 15 years old (s.44 of the Law of Property Act 1925, as amended by s.23 of the Law of Property Act 1969) and which demonstrates ownership of all of the land being purchased. This document will usually take the form of a conveyance. For *registered* land, the vendor will deduce title by providing the purchaser with an official copy of the register that is kept by the Land Registry. The register is divided into three sections.

The Property Register contains a description of the property and will confirm whether it is either a freehold estate or a leasehold estate that has been granted for a term exceeding seven years.

The Proprietorship Register states the names of the registered proprietor(s) and, under s.9 of the Land Registration Act 2002, their class of title. The best and most common class of title is an absolute title. Where the Registrar is not satisfied that the registered proprietor owns the freehold estate or possesses a validly granted lease, they may register one of the following inferior classes of title: good leasehold title (for leasehold estates only) and possessory or qualified title (for either freehold or leasehold estates). The Proprietorship Register will also indicate whether the vendor's ability to convey ownership of the legal estate in the property is limited by way of a restriction (under ss.40–44 of the Land Registration Act 2002), for example, because they hold the property in trust for a beneficiary or because there is an outstanding mortgage.

The Charges Register, the final section, contains details of certain third-party interests in the property which will be protected by the entry of a notice under ss.32–39 of the Land Registration Act 2002. These include, for example, mortgages, contracts to convey or create legal estates, restrictive freehold covenants and equitable easements and profits à prendre. Following s.29 of the Land Registration Act 2002, a purchaser of registered land is only bound by interests that are entered on the register and 'overriding' interests. Overriding interests are listed in sch. 1 and 3 of the Land Registration Act 2002. They include leases that have been granted for terms not exceeding seven years and interests of beneficiaries under a trust of land where the beneficiary is in actual occupation of the property.

Searches and enquiries

Following the vendor's deduction of title, if the land is unregistered, the purchaser will conduct a search of the Land Charges Department. In practice, these searches are normally carried out by officials of the Department on the purchaser's behalf. This search may reveal a variety of registered land charges against the property, including: mortgages where the lender or 'mortgagee' does not hold the title deeds; options granted to a third party which entitle them to purchase the property at some point in the future; and restrictive freehold covenants and equitable easements that have been created on or after 1 January 1926. Whether or not the purchaser carries out a land charges search, s.198 of the Law of Property Act 1925 means that they will be bound by any interests that have been protected by entry on the register.

In terms of other types of interest that affect unregistered land, a purchaser will automatically be bound by any legal interests, for example, legal easements and profits à prendre. They will only be bound by equitable interests if they fail to prove that they are 'equity's darling', namely a bona fide purchaser for value of a legal estate without notice (*Pilcher v Rawlins* (1872) CA). The reference to 'without notice' means that a purchaser will not be able to demonstrate that they have equity's darling status if they have either actual, constructive or imputed notice of the equitable interest. Section 199 of the Law of Property Act 1925 provides that a purchaser will have 'actual' notice if they know about the interest; 'constructive' notice if they would have known about it had they properly examined the title deeds and physically inspected the property; and 'imputed' interest if their agent (for example, a solicitor) knows about the equitable interest.

Whether the property is unregistered or registered, the purchaser will also carry out a number of other pre-contract searches and enquiries. These include a search of the Local Land Charges Register, which is kept by the local authority. It is usual for purchasers to send in a search form, requesting the Registrar's staff to perform the search on their behalf. There are two types of local land charges: restrictions which limit the use of land, for example, planning permissions and financial charges in respect of the cost of making up roads that are to be adopted as public highways. If a local land charge is correctly registered, the purchaser is bound by it even if they do not carry out a search of the register. If either the local authority has failed to register a local land charge or the search fails to reveal a local land charge that has been correctly registered, the purchaser will still be bound by it but will be entitled to claim compensation from the local authority.

Using a standard search form which can include additional enquiries, the purchaser will also make a variety of enquiries of the local authority in relation to other matters. These may include issues relating to schemes for new roads and plans for railways. If the local authority's response to any enquiries is incorrect, the purchaser may be able to bring a claim against them in either negligence or for breach of contract.

The purchaser will also carry out a separate search with the local water company to make sure that the property enjoys mains water and drainage.

The purchaser may also carry out a number of additional searches depending upon the nature of the transaction and the property's location. By way of illustration, if the vendor is a company, the purchaser should carry out a search at Companies House to make sure that the company exists, has not gone into liquidation and has not been dissolved. Similarly, if the property is located in an area of past, present or future coal mining, the purchaser should carry out a search with the Coal Authority, for example, to check for issues relating to subsidence.

Pre-contract enquiries

In addition to these searches and enquiries, the purchaser also raises a variety of pre-contract enquiries of the vendor in writing, usually by sending them a standard form to which they can add any specific queries. Such enquiries normally relate to matters including easements and covenants, services enjoyed, fixtures and boundaries. The vendor is not obliged to reply to these enquiries and should be mindful that any incorrect information provided may make them liable to a claim for misrepresentation or prosecution for the criminal offence of fraud under s.1 of the Fraud Act 2006.

The purchaser will also often undertake a physical inspection of the property and have a survey of the property carried out. If the purchaser needs to fund the purchase price by way of a mortgage, the lender will usually conduct its own survey. Where this is the case, the purchaser should consider having their own survey carried out, given that the lender is only concerned to know whether the property is good security for the mortgage, not whether it is worth the purchase price.

Contract for sale

If the survey and the replies to all of the searches and enquiries are satisfactory and the purchaser is confident that they have the necessary finance to fund the purchase, the purchaser then proceeds to enter into a legally binding contract with the vendor to purchase the property. Section 2 of the Law of Property (Miscellaneous Provisions) Act 1989 provides that such a contract for the sale (or other disposition of any interest in land) is only valid if it is made in writing. The contract must incorporate all of the agreed terms and be signed by or on behalf of both parties. Once the terms of the contract have been agreed upon, the vendor produces two final copies of the document and sends them to the purchaser. The purchaser checks them and returns one to the vendor. The vendor and purchaser then each sign their own copy and exchange them. At this point of 'exchange of contracts', the contract becomes binding so that both the vendor and the purchaser become legally obliged to proceed with the sale and purchase of the property or risk being sued by the other for breach of contract. At the same time as exchanging contracts, the purchaser will usually pay a deposit to the vendor. Following the exchange of contracts, the vendor holds the property on trust for the purchaser. It is also important to note that, depending on the terms of the contract, the purchaser may bear the risk of insuring the property from this point.

Following the exchange of contracts and immediately before the vendor transfers ownership of the property to the purchaser on 'completion', the purchaser will conduct a number of further searches and enquiries, primarily to ensure that nothing has changed. These will include: another search of the registers kept by the Land Charges Department (where the land is unregistered); an official search with priority at the Land Registry (where the land is registered); a further physical inspection of the property; pre-completion requisitions of the vendor; and, if the vendor is a company, a further search at Companies House.

If the purchaser is satisfied with the results of these pre-completion searches and enquiries, the parties will proceed to completion. Section 52 of the Law of Property Act 1925 requires the transfer of ownership of the property from the vendor to the purchaser to be documented by way of a deed, which is referred to as either a conveyance or transfer.

The purchaser prepares two copies of the draft deed and sends them to the vendor. The vendor returns one copy to the purchaser, either approved as drafted or with amendments. When the terms of the deed are agreed, the purchaser creates a final copy, signs it and returns it to the vendor. On the date specified by the contract for completion, the purchaser transfers the purchase money to the vendor, the vendor arranges for any outstanding mortgages to be paid off or 'discharged', the vendor hands over the deed and other relevant documentation to the purchaser, and the purchaser becomes the new owner of the property.

Following completion, the purchaser may be obliged to pay stamp duty land tax on the purchase. Also, due to the system of compulsory registration mentioned previously, the purchaser must register their new ownership of the property at the Land Registry.

Rights and duties of owners of land

The rights of a landowner include the right to occupy the land without interference from others and this right is protected by the laws of both trespass and nuisance. Another important right of every landowner is the right of support for their land from adjacent land. Any withdrawal of such support, which causes their land to slip, for example, through excavation during building work, may mean that they can claim damages from the owner of that adjacent land. A landowner may also acquire a right of support for any buildings on the land through an easement, for example, a mutual right of support between terraced houses.

A landowner owes reciprocal duties not to use their land in a way which causes a nuisance and not to withdraw support for adjacent land. Their other duties include compliance with statute, including the planning laws that are principally contained in the Town and Country Planning Act 1990.

The statutory controls on a landowner range from restrictions on killing certain kinds of birds and animals (under the Wildlife and Countryside Act 1981) to positive obligations to repair dilapidated premises (under the Defective Premises Act 1972). Rights of access may be reserved over land by statute, for example, under the Countryside and Rights of Way Act 2000. Additionally, if a local authority wishes to acquire the land for redevelopment, they may be able to do so without the landowner's agreement through a compulsory purchase (in accordance with the Acquisition of Land Act 1981, as amended by the Planning and Compulsory Purchase Act 2004).

Access to Neighbouring Land Act 1992

A landowner may seek to carry out work that is reasonably necessary for the preservation of their own land. Examples of such work include repairing buildings, clearing drains and cutting hedges. If the work cannot otherwise be done or would be substantially more difficult to do without accessing neighbouring land, the landowner could seek their neighbour's permission and document the agreement by way of a licence (discussed further in the following). If their neighbour refuses to give permission, the landowner may rely on s.1 of the Access to Neighbouring Land Act 1992 to make an application for a court order to provide them with access. The court will not make such an order if the works would interfere with or disturb the use or enjoyment of the neighbouring land or cause hardship to an occupier of it.

Section 2 of the Act enables the court to specify a variety of matters, including: the nature of the work that can be done, the areas of the neighbouring land that can be accessed, the date(s) of entry, the days and hours during which the work may be carried out, the persons who may enter the neighbouring land to carry out the work, the manner of the work (including controls to avoid or limit damage, loss or injury and loss of privacy) and requirements for payment to the neighbour. If the landowner breaches the terms of a court order, the neighbour may be able to bring a claim against them for damages under s.6 of the Act.

An order binds those affected by it and may, if it is protected by registration, bind subsequent owners or tenants of the neighbouring land. Either the landowner who applied for the order or those affected by it may apply to the court for a variation.

A landowner cannot unreasonably refuse to grant access to their land if the work is necessary for the preservation of a neighbouring property. In *Prime London Holdings 11 Ltd v Thurloe Lodge Ltd* (2022) Ch, the claimant asked the defendant for access to a passageway to erect scaffolding so their property could be re-rendered and repainted, but the defendant refused. The High Court found this to be an unreasonable refusal and made an order under the Act to allow the claimant access.

Party Walls etc. Act 1996

The Party Walls etc. Act 1996 applies to 'party walls' and 'party structures', including walls that separate terraced or semi-detached houses and structures that divide upper and lower flats in the same building. Essentially, s.8 of the Act gives the owner of one property a conditional right to enter the other property to carry out certain works, including repairs. The owner seeking to carry out the work will need to notify their neighbour beforehand (s.8) and potentially carry out works to protect their property and/or pay them compensation (s.7). In *Taylor v Jones and another* (2024), the Court of Appeal held that compensation is only payable for damage caused by the work and not for preexisting issues brought to light as a result of it. In *Taylor*, the defendant did not have to compensate the claimants for longstanding structural issues with their mews, likely present since the 1970s, including the underpinning of the property's rear wall, which was unsupported by the foundations.

Acquiring land other than through purchase, inheritance or gift

Land is normally acquired through purchase, inheritance or gift and, as seen in Chapter 5, contracts and transfers of land require written form (s.2 Law of Property (Miscellaneous Provisions) Act 1989). There are, however, provisions whereby land can be acquired through circumstances other than compliance with the formalities. This is not to be taken as a simple overriding of statutory requirements, as very particular criteria are required to be fulfilled. The two principal mechanisms where this is seen are acquisition through adverse possession and the operation of the equitable doctrine of proprietary estoppel.

Adverse possession

In the leading case of *J A Pye (Oxford) Ltd v Graham* (2002) HL, a farmer was able to claim 'squatters' rights', or adverse possession, over some land by grazing cattle and managing the land for a significant period of time after a licence, which allowed lawful occupation of the land, had expired. This led to a successful claim, in adverse possession, of around 23 hectares of development land in Berkshire, which was worth around £10 million! (Having failed to retain the land in the House of Lords, the original landowner then queried whether the law of adverse possession was contrary to Protocol 1, Article 1 of the European Convention on Human Rights (see Chapter 1; *J A Pye (Oxford) Ltd v UK* (2007) ECHR GC).

To establish a claim for adverse possession, a squatter needs to prove four elements:

1 First, they must demonstrate that they had **factual possession** of the land; namely, that they had sufficient control over the land and treated it as if they owned it. This is a context-dependent test. For example, in the case of *Red House Farms (Thorndon) Ltd v Catchpole* (1977) CA, the court held that a squatter's use of marshland for shooting wildfowl was enough to establish factual possession of the land because the land could only be used for that purpose.
2 Second, the squatter must demonstrate that they had an **intention** to **possess** the land to the exclusion of everyone else, for example, by enclosing the land and padlocking the only gate to it (see, for example, the case of *Buckinghamshire County Council v Moran* (1990) CA).
3 Third, the squatter's possession of the land must be without the landowner's consent or **adverse**. As happened in *J A Pye (Oxford) Ltd v Graham* (2002) HL, a person who has previously lawfully occupied land under a licence may be able to claim adverse possession of it if they continue to occupy the land after the licence has come to an end.
4 Finally, the squatter must prove that they have adversely possessed the property for a particular period of time. If the land is *unregistered* (or it is *registered* and the squatter has been in adverse possession of the land for a period of at least 12 years before 13 October 2003), the relevant rules are set out in both the Limitation Act 1980 and the Land Registration Act 1925. Essentially, the squatter must demonstrate that there has been a continuous period of adverse possession for at least 12 years (s.15 Limitation Act 1980). In such situations, s.17 of the Act automatically extinguishes the landowner's right to reclaim ownership of their land.

   If the land is *registered*, sch. 6 of the Land Registration Act 2002 provides that the squatter only needs to evidence a ten-year period of adverse possession.

Despite this shorter period, it is more difficult for squatters to bring successful claims for adverse possession of registered land for a number of reasons. Schedule 6 of the Land Registration Act 2002 requires a squatter of registered land to proactively make an application to the Land Registry, upon which the Land Registry will give the landowner, any mortgagee and any landlord the opportunity to challenge the squatter's claim. If the landowner does make a challenge, the squatter will only be able to claim ownership of the land if they can prove one of the grounds under sch. 6 para 5 of the Land Registration Act 2002. These are:

- estoppel,
- some other entitlement to the land or
- a reasonable mistake as a boundary.

We will consider each of these in turn.

First, the squatter may still be able to claim ownership of the land if they can prove estoppel; namely, that they have relied to their detriment on the landowner's assurance that the squatter owns the land and it would be unfair to deny the squatter's claim (namely, the landowner is 'estopped' from going back on their assurance).

Second, the squatter may be able to demonstrate that they have some other right in respect of the land, for example, they have inherited the land under the terms of a will.

Finally, the squatter could argue that their claim for adverse possession is based on their reasonable mistake as to a boundary, for example, the positioning of a fence between neighbouring properties. In *Brown v Ridley* (2025), the UK Supreme Court clarified that the squatter must have had a reasonable belief of ownership for a period of ten years *before* the claim, but that this period did not need to be immediately *prior* to the claim, allowing for parties to seek alternative dispute resolution in boundary cases before resorting to costly legal action.

If the squatter fails to prove one of these three grounds, they are still entitled to bring a claim two years later, the landowner will be unable to challenge it and the squatter will be able to apply to be registered as the new owner. To avoid this happening, the landowner could grant the squatter permission to be on the land, for example, through a licence, which would mean that their occupation of the land would no longer be adverse.

Proprietary estoppel

In Chapter 5, the concept of promissory estoppel was introduced. Proprietary estoppel is a close relation, but a key distinction is that proprietary estoppel can be used as a cause of action, i.e. a claimant can sue on the basis of proprietary estoppel. Promissory estoppel can only be used as a defence or, to use the equitable maxim, can be used only as a shield and not a sword.

Broadly, where a person makes representations, promises, to another (the promisee) that the promisee will have rights to property, they should not later be allowed to deny those rights where the promisee has reasonably acted upon the promise/understanding to their detriment, i.e. it would be unconscionable to go back on the promise.

This principle was seen operating in the nineteenth century in *Ramsden v Dyson* (1866) HL and a somewhat stringent set of criteria (or probanda, as they were termed) was established in *Willmott v Barber* (1880) Ch. But the modern law is generally seen to have been clarified in *Taylor Fashions v Liverpool Victoria Trustees* (1982) Ch.

A key, illustrative case is *Gillett v Holt* (2001) CA, where a partner in a farming business, Geoffrey Gillett, had worked for the managing partner and landowner, Kenneth Holt, for nearly 40 years, since he was at school. Their relationship was far closer, for most of this period, than business partners, with Holt paying school fees for one of Gillett's sons, and each attending the other's family events such as christenings, Christmas celebrations and the like. Holt had made many clear, often public, promises that Gillett would be given a land interest and Gillett had acted on these promises, over the decades, by not taking work elsewhere, not going away to college (as he had wished), working at a lower than market salary and not acquiring his own home. The relationship broke down and Holt cut Gillett out of his will and tried to evict him and his wife from their house. It was relatively easy to establish that Holt had made the promises and that Gillett had acted upon them to his detriment.

Importantly, once the criteria have been satisfied, the court then has discretion as to the remedy to be awarded, which will be the 'minimum equity to do justice' (*Crabb v Arun DC* (1976) CA). In *Guest v Guest* (2020), the UK Supreme Court affirmed that the starting point for a remedy is the promisee's expectation, not the detriment they have suffered. This may or may not be the extent of the award requested by the claimant (see *Davies v Davies* (2016) CA, where the final award was £500,000 rather than the farmland and business which had been claimed). In the case of *Gillett*, the claimant was awarded the farmhouse, just over 100 acres of farmland and related buildings and £100,000, by way of compensation for his exclusion from all other parts of the extensive farming business.

Two cases heard within six months in the House of Lords illustrate the point that whilst in particular circumstances, somewhat oblique, verbal promises may suffice as representations or promises (*Thorner v Major* (2009), the doctrine cannot simply be used as a means to oust the usual legislative requirements that contracts for land must be in writing (*Cobbe v Yeoman's Row* (2008). In *Cobbe*, oral promises had been made regarding a commercial transaction in property. The House of Lords held that there were no circumstances in which the usual requirements for contracts of land could be dispensed with.

The doctrine of proprietary estoppel is particularly encountered in family and farming settings where long-standing agreements and living and working arrangements are not always formalised. In *Spencer v Spencer* (2023) Ch), the court recognised that the non-financial detriment of someone who has committed their working life to a farm is an important factor in such claims (see also *Guest v Guest* (discussed previously), *Cleave v Cleave* (2024) Ch and *Armstrong v Armstrong* (2024) Ch). However, the doctrine is not to be seen as a reliable way of enforcing promises and formal documentation should always be sought.

Discussion questions

1 **Ownership and Estates:** What is the difference between a freehold and a leasehold estate, and how does this distinction affect the rights and responsibilities of the owner?
2 **Possession and Ownership:** How does the law distinguish between mere possession of land and legal ownership, and what remedies are available for trespassers or wrongful occupiers?
3 **Co-ownership:** What are the key differences between joint tenancy and tenancy in common, and how can disputes between co-owners be resolved under English law?
4 **Registered and Unregistered Land:** How does the Land Registration Act 2002 impact the protection of interests in land, and what are the advantages and limitations of land registration?
5 **Easements and Covenants:** What is the legal significance of easements and restrictive covenants, and how can they affect the use and development of land?
6 **Adverse Possession:** Under what conditions can someone acquire ownership of land through adverse possession, and how does this doctrine balance fairness with the protection of legal title?

12 Landlord and tenant

Mark Simcock and Michael Fay

Introduction

The relationship between any landlord and a tenant is usually regulated through a contract commonly referred to as a lease or tenancy agreement. The lease contains all the provisions agreed between the parties prior to its commencement and governs the tenants' occupation of the property to which the lease refers. The lease will typically set out how some of the following matters are dealt with, amongst other provisions:

- the rent payable,
- the mechanism for reviewing that rent,
- responsibilities for repairs,
- responsibilities for insurance,
- agreement as to what the property can and cannot be used for and
- the mechanism for resolving disputes between the parties.

A lease is one of the two legal estates in land recognised by the Law of Property Act 1925, but unlike a freehold interest in property, which exists in perpetuity, a leasehold interest in property is for a limited duration. A lease will only exist for a period of time agreed by the parties or permitted by the law.

Typically, there are two types of leases:

- fixed term leases and
- periodic tenancies.

A fixed term lease, such as a seven-year lease or a 99-year lease, expires when the term ends (i.e. after, respectively, seven or 99 years). This is sometimes described as expiry by 'effluxion of time'. A periodic tenancy, such as a weekly, monthly or yearly tenancy, may be terminated at the end of a period. If not terminated, the tenancy continues for another period. Thus, a weekly tenancy might end after one week or might continue for several years.

A lease must usually have some essential elements that distinguish it from a licence agreement, which is a mere permission to do something in relation to land that would otherwise be a trespass. A licence does not create any interest in the land itself, whereas a lease does. There are three types of licenses: a bare license, a contractual license and a license coupled with an interest. If you go to a party at a friend's house, you have a bare license to enter the property; if you visit the cinema, buying a ticket gives you a

DOI: 10.1201/9781003657101-12

contractual licence to be on the premises; and if you have a right to, for example, harvest timber, you would also have a license to enter the land and cut the timber. Establishing if an agreement is a lease or licence is very important as property let under the terms of a lease can acquire protection under the law, which we will see later in this chapter, where a licence does not. It has been known for parties to enter into a licence only to see that convert to a lease through the action of the parties, irrespective of what the written agreement sets out.

Lease or licence?

A well-known case considered to have established the differences between a lease and a licence was *Street v Mountford* (1985) HL, which concerned an agreement, described as a licence, under which two rooms of a house were let for a weekly rent. The parties had even agreed that the contract did not give Mrs Mountford a tenancy protected by the Rent Act 1977. However, the House of Lords decided the agreement was, in fact, a weekly tenancy. Lord Templeman explained that an occupier of residential property had to be either a tenant or a licensee. A tenant, he said, was an occupier who had been 'granted exclusive possession for a fixed or periodic term certain in consideration of a premium or periodical payments'.

Exclusive possession

Exclusive possession is the right of a tenant of a property to 'keep out strangers and keep out the landlord', subject to any access rights the landlord might retain in the lease agreement. It is the key distinction between a lease and a license – the latter does not grant the licensee the ability to exclude the landowner (the licensor). In *Street v Mountford*, the House of Lords held that the agreement gave the occupier exclusive possession of the property and, consequently, it was a lease to which the protections of the Rent Act 1977 applied. This means that if a license agreement grants exclusive possession to the tenant(s), it could potentially be a lease.

Definite term

The length, or 'term', of a lease can range from a matter of hours to thousands of years. The term does not have to start on the date the lease agreement is made and s.149(3) of the Law of Property Act 1925 allows the parties to agree a start date up to 21 years in the future. However, when the term begins, the maximum duration must be defined at the outset. In *Prudential Assurance Co. Ltd v London Residuary Body* (1992) HL, a London council tried to grant a lease for a parcel of land until it was required for widening Walworth Road. The lack of certainty about the maximum term meant the lease was not valid and the House of Lords decided it was instead a yearly periodic tenancy. The case confirmed that a valid lease has either a defined maximum duration (i.e. 99 years) or is a periodic arrangement (i.e. monthly) capable of being ended by serving notice. If a periodic agreement places too many restrictions on one party's ability to end it, the term will be too uncertain to be a lease (see *Mexfield Housing Co-operative v Berrisford* (2012) SC.

Payment of rent

In *Street v Mountford* (1985), Lord Templeman stated that a lease was the grant of exclusive possession in consideration for a premium or rent. Consideration is an important aspect of a contract – it is something of value exchanged by the parties to create a binding agreement. In short-term residential tenancies, rent is given in consideration for exclusive possession for a defined term. A premium is an upfront sum the tenant pays to the lessor to 'buy' a lease for the term, although they may still pay a relatively low rent thereafter. But rent is not an essential ingredient of a lease. Section 205(1)(xvii) of the Law of Property Act 1925 defines a lease as existing with or without rent and there is no requirement for consideration if a lease is created by deed (see *Ashburn Anstalt v WJ Arnold & Co.* (1989) CA. In very long leases, the rent might be a nominal sum or even a 'peppercorn rent' – a token consideration that is not usually demanded in practice. Some examples include a single rose, one crab per annum, a daffodil and, of course, a peppercorn.

Applying Street v Mountford

Essentially, the decision in *Street v Mountford* (1985) means that if a court believes the effect of an agreement is to give the occupier exclusive possession, then, in the absence of special circumstances, the agreement will be treated as a lease. This will be the case even if the agreement is described as a licence and its wording does not give the occupier exclusive possession. This does not mean the content of an agreement is irrelevant, but that the courts will consider the whole of the case. In *AG Securities v Vaughan* (1990) HL, a group of four people sharing a flat were held to be licensees because each had an independent agreement, made on different dates, with different terms and rents. But in *Antoniades v Villiers* (1990) HL, the license agreement, which gave the landlord a right to sleep in a tiny flat with a cohabiting couple, was said to be a tenancy. On paper, the couple did not have exclusive possession, but the House of Lords decided the clause was a pretence inserted to avoid the protections of the Rent Act 1977. Because the status of an agreement is defined by the rights it grants, it means unscrupulous landlords cannot avoid statutory protections, like the Rent Act 1977, by mislabelling or inserting sham clauses into agreements. However, in *Clear Channel UK Ltd v Manchester City Council* (2005) CA, the court indicated the labels used by the parties may be of greater significance in commercial agreements made between two parties of equal bargaining power, with the benefit of full legal advice.

Exceptions

In *Street v Mountford*, the House of Lords did accept that there may be special circumstances when an agreement which gives, or appears to give, exclusive possession is only a licence. Some examples which occur commonly include:

- **Lodgers and hotel guests** – Such occupants normally only have a licence. They do not have exclusive possession of the premises, as they receive services which require licensors or their agents to have unrestricted access to the room.
- **Occupation as necessary for employment** – An employee who is required to occupy premises as a necessary part of the proper performance of contractual duties only has a licence, even though he or she has exclusive possession of the premises.
- **Social, domestic or charitable** – An agreement made between relatives or friends or with charitable intent, which gives exclusive possession, may only amount to a licence if creation of a lease was not intended.

A significant number of cases involving the distinction between a lease and a licence have come before the courts in recent years, and each tends to turn on its own individual set of facts. In every case, the court is seeking to determine the intention of the parties, particularly whether the tenant truly has exclusive possession; however, the landlord may have tried to disguise it.

Assignment and subletting of obligations under the lease

Under a lease, both the landlord and the tenant possess interests in land. The tenant holds a leasehold interest whilst the landlord holds a reversionary freehold interest to resume occupation of the premises when the tenancy ends. A landlord who owns the freehold of the let premises may sell their interest and a purchaser would become the owner of the reversion and also the tenant's new landlord. A tenant may also sell their leasehold interest in a similar way. This is technically called an assignment of the lease. The purchaser of that leasehold interest becomes the new tenant. It may be, however, that the lease prohibits assignment by the tenant, either absolutely or without the landlord's consent. Where the lease prohibits assignment without the landlord's consent, s.19 of the Landlord and Tenant Act 1927 provides that consent must not be unreasonably refused. What is unreasonable is a question of fact, but in *International Drilling Fluids v Louisville Investments (Uxbridge) Ltd* (1986) CA, the court indicated the question was whether there was one that would be reached by a reasonable person in the circumstances. Undue delay in giving consent or notifying the tenant of a refusal of consent may render the landlord liable in damages to the tenant (Landlord and Tenant Act 1988). If the tenant assigns without consent, they have a defence to an action brought against them by the landlord if the landlord's refusal was unreasonable.

In relation to tenancies granted on or after 1 January 1996, other than residential tenancies, the lease may specify circumstances in which the immediate landlord may refuse to consent to an assignment of the lease and may also specify conditions to which the giving of consent will be subject (see Landlord and Tenant (Covenants) Act 1995). Refusing consent in the specified circumstances or giving consent subject to such conditions will not be unreasonable.

The effect of an assignment is that the tenant hands over their interest in the property entirely. The original tenant (the tenant assigning the lease, known as the assignor) retains no interest in the property and effectively the new tenant steps into the shoes of the new tenant (known as the assignee). In practice, it may be mutually beneficial for any landlord to consent to an assignment. The original tenant may be struggling financially or may want to move to a different part of the country, so securing a new tenant to take on the obligations of the old tenant would be in the interests of the landlord.

As an alternative to assigning the lease, the tenant may choose to sublet (and so become the head tenant) and receive rent from their own tenant, referred to as a sub-tenant. The lease may similarly prohibit subletting, either absolutely or without the landlord's consent. Again, where subletting is prohibited without the landlord's consent, consent must not be unreasonably refused (s.19 Landlord and Tenant Act 1927). Likewise, the landlord may be liable in damages to the tenant where there is undue delay in giving consent or notice of refusal. The effect of a sublet is different. Here the original tenant creates a smaller leasehold interest in the property and so lets the property to a sub-tenant for a term shorter than the term the original tenant has with the landlord. The original tenant, or head tenant, effectively becomes a landlord themselves, a landlord to the sub-tenant. Importantly, however, unlike an assignment, the head tenant retains the reversionary interest in the property following the end of the term comprising the sublet.

Covenants in leases

The covenants, or terms, in a lease set out the obligations of both the landlord and tenant during the period the tenant occupies the property. There are two types of covenants: express covenants and implied covenants. Express covenants are expressly agreed upon by the parties to the lease and are set out in the lease document. Implied covenants do not result from express agreement by the parties to the lease but are implied into it by common law or statute, unless the parties agree otherwise. However, certain covenants implied by statute into a lease cannot be excluded by agreement of the parties.

Implied covenants (other than repairing covenants)

Tenant's implied covenants

Other than repairing covenants (which are examined later on), three major examples of implied covenants that impose an obligation on the tenant are:

- to pay rent,
- to pay taxes relating to the property and
- not to use the property for illegal purposes.

In relation to rent, in the absence of an express covenant or contrary agreement, a covenant may be implied if the wording of the lease implies that the payment of rent is intended. In relation to taxes, in the absence of an express covenant or contrary agreement, the tenant is under an implied obligation to pay rates and taxes relating to the property for which the landlord is not personally liable. If the landlord has undertaken within the lease to carry out repairs to the property, then there will be an implied obligation on the tenant to grant the landlord access to the property to carry out those repairs.

Landlord's implied covenants

Other than repairing covenants, two major examples of a landlord's implied covenants are:

- the implied covenant not to derogate from their grant and
- the implied covenant for quiet enjoyment.

Non-derogation – In the absence of an express covenant or contrary agreement, the landlord is under an implied obligation not to do things which make the leased premises less suitable for the purpose for which they were let, assuming that the landlord knows the purpose. For example, in *Harmer v Jumbil (Nigeria) Tin Areas Ltd* (1921) CA, the land was leased to store explosives, but the landlord's decision to build on adjacent land would have made such a use illegal, and in *Platt v London Underground Ltd* (2001) Ch, a kiosk was leased near the south exit of Goodge Street station, but the exit was only open to the public during morning rush hour, discouraging trade.

Quiet enjoyment – Quiet enjoyment, despite its name, is not really about noise or lack of it. It covers any actions of the landlord that would stop the tenant from being able to enjoy exclusive possession of the premises. In the absence of an express covenant or contrary agreement, the landlord is under an implied obligation not to interfere with the tenant's quiet enjoyment of the premises. In *Southwark London Borough Council v Mills* (2001) HL, Lord Hoffmann states that, while the local authority flat was 'not quiet and the tenant is not enjoying it', the purpose of the covenant is that a tenant's lawful possession should not be substantially interfered with by the landlord (for example, by unlawful entry or eviction).

Regarding both of these implied covenants, the landlord may also be liable in respect of acts done in breach of the covenants by persons claiming under them, such as tenants of other neighbouring property which they owns.

Express covenants (other than repairing covenants)

Tenant's express covenants

Examples of express covenants which may be made by a tenant are a covenant:

- to insure,
- to pay taxes relating to the property,
- only to use the leased premises for certain purposes,
- not to assign or sublet and
- to pay rent.

Insurance – A tenant may agree to insure the premises, the extent and nature of the policy being determined by the covenant.

Taxes – An express covenant may impose upon the tenant an obligation more or less extensive than or identical to that which would otherwise be implied.

Usage – The tenant may agree to restrictions on their use of the premises. For example, they may agree only to use the premises as a single private dwelling, which would prevent both use for business purposes and subletting as flats.

Rent – The tenant will normally expressly covenant to pay a specified amount of rent on a specified day. Rent is normally payable in advance unless the parties have agreed otherwise. A tenant may be entitled to deduct sums from the rent before paying it to the landlord. The lease may specifically permit or prevent the making of such deductions. Where the lease is silent, the basic principle is that the landlord is entitled to the entire rent. The lease may contain a term which suspends the tenant's obligation to pay rent in circumstances such as damage or destruction of the premises by fire. The lease may contain a rent review clause, which permits regular rent reviews throughout the course of the tenancy (i.e. once every three years). The lease will normally contain a formula to be used in determining the revised rent, such as an open market rent to be fixed by agreement, arbitration or a form of index linking, possibly to the retail prices index.

Landlord's express covenants

Examples of express covenants which may be made by a landlord are a covenant:

- to insure,
- to pay taxes relating to the property and
- for quiet enjoyment.

Insurance – If the landlord covenants to take out an insurance policy, it may be that the tenant agrees to pay the insurance premiums.

Taxes – The landlord may covenant to undertake liability which would otherwise fall within the tenant's implied obligations.

Quiet enjoyment – The requirements of an express covenant may be wider in some respects, but more limited in others, than those of an implied covenant. For example, an express covenant for quiet enjoyment continues to bind the party who made it throughout the duration of the lease, whereas an implied covenant for quiet enjoyment only binds a landlord whilst he remains landlord.

Options and provisos

A lease may, in addition, contain several options and provisos, which are conditions or qualifications attached to any agreement. These will have been typically negotiated between the parties prior to the lease agreement being completed.

Examples of options are:

- options to purchase the property at some point during the lease and
- options to renew the lease.

Option to purchase – This may entitle the tenant to purchase the landlord's reversion, which may be the freehold or may be a lease, depending upon whether the landlord owns the freehold or are themselves a tenant. The tenant may be entitled to purchase at a fixed price or, alternatively, the lease may lay down a mechanism and formula to be applied in determining the purchase price. Section 27 of the Leasehold and Freehold Reform Act 2024 removed the requirement that a tenant live in a property for two years before they could purchase the landlord's freehold reversion.

Option to renew – This may entitle the tenant to renew the lease by serving notice on the landlord on a specified day. The tenant will be entitled to require a new lease for a term specified by the option. The lease may specify a fixed rent or, alternatively, may lay down a mechanism and formula to be applied in determining the new rent. A tenant of a house on a long lease has a statutory right to extend the duration by 50 years (s.14 Leasehold Reform Act 1967). Section 27 of the Leasehold and Freehold Reform Act 2024 has likewise removed the two-year qualifying period for lease extensions and, though it is not yet in force, s.33 will increase the extended duration of a lease to 990 years.

Options should be registered to bind purchasers of the landlord's estate.

Examples of provisos are:

- provisos for forfeiture and
- break clauses.

Forfeiture – The right of forfeiture is the right the landlord has to remove the tenant from the property and bring the lease to an end early due to the tenant not fulfilling their obligations under the lease. If the 'usual' covenants on rent, land taxes and repair are implied into a lease, then a landlord can exercise the right of forfeiture for non-payment of rent, but not in respect of any other breach of covenant, unless that right is reserved within the lease agreement.

Break clauses – A break clause is a proviso which empowers one or both parties to a lease to terminate the lease at certain specified dates by serving notice on the other.

Covenants to repair the property

When negotiations for a lease are taking place, it is important to establish the responsibilities for the repair of the property during the term of the lease. The parties may expressly agree on responsibility for repairs, such agreement forming the 'express repairing covenants' of the lease. Where no such express agreement has been reached, the parties may find that there are repairing obligations implied by common law or by statute.

Express repairing covenants

A lease may expressly impose repairing obligations upon the landlord or tenant in the form of express repairing covenants. The wording of these covenants is very important as it can impose expensive and significant obligations on any incoming tenant. For example, a lease might require the tenant:

- *to put premises in repair* at or within a reasonable time of the commencement of the tenancy;
- *to leave premises in repair* (i.e. to hand them over in repair) at the end of the tenancy or
- *to keep premises in repair* throughout the tenancy (i.e. to put the premises in repair at the commencement of the tenancy, keep the premises in repair throughout the tenancy and hand them over in repair at its end).

Where the landlord covenants to repair, they are not obliged to repair until they have notice of the disrepair, i.e. from the tenant or from their own inspection of the premises. Once repairs are undertaken, the standard required is 'such repair as, having regard to the age, character and locality of the premises, would make them reasonably fit for the occupation of a reasonably minded tenant of the class who would be likely to take them'. This was established by Lord Esher in *Proudfoot v Hart* (1890) CA and is likewise found in s.11(3) of the Landlord and Tenant Act 1985. The criteria are applied by looking at the age, locality and the like as they would have been at the start of the tenancy. This can be important where, for example, the locality becomes less desirable during a long lease. The standard of repair required is not then reduced because the locality would no longer attract the same sort of tenant.

A covenant to repair can include responsibility for painting and decorating to prevent the property from falling into disrepair. Landlords of short-term residential tenancies are usually responsible for redecoration, except where the damage is due to the tenant's wilful or negligent actions. When the property is a long-term leasehold, the tenant will usually be responsible for keeping the property in good repair, including redecorating. A lease may also contain an express covenant to paint, which may specify the materials to be used and the frequency of the work.

Express repairing covenants may specifically exclude liability to repair in respect of disrepair resulting from 'fair wear and tear', but the exclusion is also implied into every short-term residential tenancy. Fair wear and tear is disrepair caused by normal and reasonable use by the tenant for proper purposes or by the normal action of time and weather (*Regis Property Ltd v Dudley* (1959) HL). A tenant is expected to undertake repairs necessary to prevent the premises from falling into further disrepair, but s.11(4) of the Landlord and Tenant Act 1985 invalidates any express repair covenant by the tenant that relates to the matters contained in the landlord's implied covenant (see later on).

A covenant to repair does not normally impose an obligation to make improvements or to renew the premises in their entirety. However, such a covenant will normally impose an obligation to rebuild in the case of destruction of premises by fire, unless the lease expressly excludes such liability.

Implied repairing covenants

Where a lease fails to lay down express repairing obligations, repairing obligations may be implied at common law. Further, repairing obligations may be imposed by statute, regardless of the terms of a lease.

Implied repairing covenants – relating to the landlord

Fitness for Human Habitation

Section 8 of the Landlord and Tenant Act 1985 contained an implied covenant requiring a landlord to keep a house in a state 'fit for human habitation' at the beginning of and throughout the term of the lease. The covenant only applied to houses up to a maximum rental value and it excluded leases for a fixed term of three years or more where the tenant had agreed to put the premises into a state reasonably fit for human habitation.

The Homes (Fitness for Human Habitation) Act 2018 amended ss.8–10 of the Landlord and Tenant Act 1985, restricting the application of s.8 to Wales, and only to lets that are not occupation contracts under the Renting Homes (Wales) Act 2016. By virtue of s.9 of the Landlord and Tenant Act 1985, the s.8 covenant also continues to apply to agricultural workers in Wales where the house is part of their remuneration.

The implied covenant in s.9A of the Landlord and Tenant Act 1985 applies to dwellings (instead of merely houses) in England leased either for a fixed term of less than seven years or as a periodic agreement. A landlord is to keep such dwellings (including common areas) fit for human habitation at the time the lease is granted (or at the date it commences, if this is in the future) and throughout the whole of the term. Section 9C applies the covenant to agricultural workers in England where the dwelling is part of their remuneration.

Under s.10 of the Landlord and Tenant Act 1985, a dwelling is unfit for human habitation if it is defective in one or more of the following areas: repair, stability, freedom from damp, internal arrangement, natural lighting, ventilation, water supply, drainage and sanitary conveniences and facilities for the preparation and cooking of food and for the disposal of wastewater. In England, the list also includes the 29 hazards set out in the Housing Health and Safety (England) Regulations 2005.

While it appears that s.8 only requires the landlord to repair where it is possible to make the premises fit for habitation at reasonable expense, s.9A(2) and (3) define the circumstances where the landlord is not required to undertake repairs. These are:

- Where the tenant is responsible because of the duty to use the premises in a tenant-like manner or an express covenant of substantially the same effect.
- Rebuilding or reinstating the dwelling following destruction or damage by fire, storm, flood or other inevitable accident.

- Keeping in repair or maintain anything the tenant is entitled to remove from the dwelling.
- Carrying out works or repairs that would put the landlord in breach of any legal requirements.
- Carrying out works or repairs requiring the consent of a third party (such as a superior landlord) where consent has not been obtained despite reasonable attempts.
- The unfitness is wholly or mainly down to the tenant's breach of covenant.
- The landlord is not obliged to make the repair because of an exclusion or modification to the lease authorised by the county court under s.12 of the Landlord and Tenant Act 1985.

A covenant is also implied in s.9A(7) allowing the landlord (or a third party with the landlord's written authorisation) to enter the premises at reasonable times of the day to view their state and condition upon giving at least 24 hours' notice, in writing, to the tenant or occupier. Though s.9A is silent on the matter, it is likely that the landlord's liability is consistent with s.11 (see later on) and does not arise until they have received notice of the disrepair. A landlord's breach of the requirements in ss.8 or 9A may entitle the tenant to vacate the premises without paying rent and to recover damages.

These provisions apply to tenancies of less than seven years commenced on or after 20 March 2019, to new secure, assured and introductory tenancies commencing on or after 20 March 2019, to all tenancies renewed for a fixed term on or after 20 March 2019 and to all periodic tenancies commenced on or after 20 March 2020. Under s.9A(4), a landlord cannot exclude or limit their liability under the covenant, nor penalise the tenant for enforcing or relying on those obligations.

The Social Housing (Regulation) Act 2023 further amended the Landlord and Tenant Act 1985. Sections 10A and 10B create an implied term on the remedying of hazards in dwellings let on social housing leases. Social landlords are now required to investigate hazards within 14 calendar days and commence remediation within an additional seven days. Emergency repairs must be completed within 24 hours.

Fitness for habitation of furnished houses

Subject to express provision to the contrary in the lease, there is an implied condition at common law that a furnished house will be fit for habitation when the tenancy begins. If this is not the case, for example, because of defective drains, the tenant can either repudiate the lease or sue for damages. The landlord is not required to keep the premises fit for habitation throughout the tenancy if they are fit for habitation at its start. If the lease is for a short-term residential let, then the common law duty will likely operate alongside s.9A of the Landlord and Tenant Act 1985.

Repair of the structure of dwelling-houses (s.11 Landlord and Tenant Act 1985)

Section 11 of the Landlord and Tenant Act 1985 implies a covenant into short residential leases (i.e. a fixed term of less than seven years or a periodic agreement) requiring the landlord to keep in repair the structure and exterior of the dwelling (including guttering, drains and external pipes), installations for the supply of water, gas, electricity and sanitation (including baths, basins, etc.) and installations for heating and hot water. The landlord is not obliged to make repairs until they have notice of the disrepair (i.e. from the tenant or from their own inspection of the premises) and is not in breach of covenant unless they fail to carry out repairs within a reasonable time of receiving the notice.

The covenant implied by s.11 do not require the landlord to do repairs for which the tenant is liable either under the duty to use the premises in a tenant-like manner or an express covenant to the same effect; to rebuild or reinstate the premises in the case of destruction or damage by fire, storm, flood or other inevitable accident or to repair the tenant's fixtures.

The standard of repair required by the covenant implied by s.11 is assessed with regard to the age, character and prospective life of the dwelling-house and the locality in which it is situated. Section 11(6) entitles the landlord (or third parties with the landlord's written authorisation) to enter the premises at reasonable times of the day to view their state and condition on giving 24 hours' written notice to the tenant. Section 11 also invalidates any express covenant made by the tenant that relates to a matter in the landlord's implied covenant, including to put in repair or deliver up in repair; to paint, point or render, to pay money in lieu of repairs; or to pay money on account of repairs by the landlord (s.11(5)). So, while in *Regis Property v Dudley* (1959) HL it was said the tenant could have replaced a roof tile to prevent further damage, as the exterior of the property is now a s.11 matter, the tenant would promptly inform the landlord of the defect instead.

Section 14 provides that s.11 does not apply to:

- leases granted to existing tenants or to former tenants remaining in possession, where s.11 did not apply to the previous lease;
- tenancies of agricultural holdings;
- leases granted on or after 3 October 1980 to various specified bodies, such as local authorities and
- leases granted to a government department or to the Crown (unless the lease is managed by the Crown Estate Commissioners).

The landlord is not merely obliged to carry out repairs on structures, installations, etc., that form part of the property let to the tenant, but the structure of all parts of the building in which he has an estate or interest. Equally, they are required to maintain installations that serve the dwelling, provided that they form part of the building in which they have an estate or interest, are owned by them or are under their control. For example, where a tenant's flat is heated by radiators, but the boiler is in the basement and this does not form part of the property let to the tenant, the landlord is obliged to maintain both the radiators and the boiler. Section 12 provides that an agreement or term of a lease is void to the extent its effect is to modify or exclude the provisions of s.11, unless it was authorised by the county court. The county court may authorise such an agreement or term with the consent of the parties if it is reasonable to do so. Unless such an agreement or term is authorised, any covenant made by the tenant to carry out repairs is of no effect to the extent it requires the tenant to undertake repairs that the landlord is obliged to carry out under s.11.

Repair and maintenance of common parts of blocks of flats

Subject to express provision to the contrary in the lease, there is an implied condition at common law that landlords will take reasonable steps to repair and maintain common parts of the premises which remain in their control, such as lifts and stairs. This implied condition basically appears only to apply in relation to residential blocks of high-rise flats, though it may apply in other special situations in which maintenance of the common parts by the landlord is essential to make the tenancy agreement effective.

Implied repairing covenants – relating to the tenant

Obligation not to commit waste

The law of tort imposes upon tenants an obligation not to commit waste. Voluntary waste comprises damage to or destruction of premises resulting from deliberate or negligent acts (such as demolition or the making of alterations), but does not encompass damage or destruction which results from their reasonable use.

Ameliorating waste comprises acts which improve the value of land or premises (such as constructing an extension of existing premises). The courts will not normally be prepared to award any remedy in respect of ameliorating waste. Permissive waste comprises damage to or destruction of premises resulting from omission to undertake necessary repairs (such as failure to replace slates resulting in the loss of a roof). Periodic tenants (except, perhaps, yearly tenants) are not liable for permissive waste. Where a tenant commits waste, the landlord may bring an action for damages and (except in the case of permissive waste) may claim an injunction to prevent continuing or repeated acts of waste.

The tenant's duty to use the premises in a tenant-like manner

Where the lease does not expressly impose repairing obligations, it is implied that the tenant will use the premises in a tenant-like manner. A fixed term tenant, being liable for permissive waste, should hand the premises over in repair at the end of the tenancy, though the tenant is not required to carry out repairs in respect of 'fair wear and tear'. A yearly tenant should carry out minor repairs necessary to keep the premises windproof and watertight. A periodic tenant for a period of less than a year should perform small tasks that any reasonable tenant would perform (such as cleaning the windows). Whatever the length of the tenancy, the tenant should carry out repairs required in consequence of voluntary waste.

Remedies for breach of repairing covenants

1 ***Damages*** – Where a party to a lease fails to perform their repairing obligations under the lease, they may be liable in damages. Where the landlord is in breach of his repairing obligations, the tenant cannot recover damages in respect of the period prior to the date when the landlord was notified of the disrepair. If the landlord fails to repair within a reasonable time of receiving such notice, the tenant may sue for damages or carry out the repairs and recover their cost from the landlord. Where the tenant is in breach of a covenant to put, keep or leave the premises in repair, s.18(1) of the Landlord and Tenant Act 1927 provides that the damages cannot exceed the amount by which the value of the landlord's reversion was diminished in consequence of the breach. Where the tenant is obliged to leave the premises in repair (subject to the exclusions in s.11 of the Landlord and Tenant Act 1985), no damages are recoverable if, at or shortly after the end of the tenancy, the premises are to be demolished or structural alterations that

would render the repairs valueless are to be made. In the case of a lease for a fixed term of or exceeding seven years, of which at least three years remain unexpired, s.1 of the Leasehold Property (Repairs) Act 1938 provides that, before seeking damages or forfeiting the lease because the tenant has breached a covenant to keep or leave the premises in repair, the landlord must first serve notice on the tenant under s.146 of the Law of Property Act 1925. This notice must, among other matters, inform the tenant of his or her right to serve a counter-notice within 28 days of the service of the s.146 notice. If the tenant serves a counter-notice, the landlord cannot sue for damages or forfeit the lease without the leave of the court. The court may give leave if:

a the value of the landlord's reversion has been substantially diminished by the breach or will be so diminished unless the breach is immediately remedied;
b the breach must be immediately remedied to give effect to the provisions of a statute, a bye-law or of delegated legislation;
c in those cases where the tenant is not in occupation of the whole of the premises, the immediate remedying of the breach is required in the interests of the occupier;
d the expense required in order to remedy the breach is small compared to the much greater expense, which would probably be required if the work required was postponed or
e there are special circumstances which make the giving of leave just and equitable.

2 ***Forfeiture*** – A landlord can forfeit a lease if the tenant is in breach of a repairing covenant and the lease contains a proviso giving the landlord a right of re-entry if the tenant breaches those obligations. In residential lettings, s.2 of the Protection from Eviction Act 1977 requires the landlord to apply to court for a possession order before re-entering the property. The landlord must issue a notice under s.146 of the Law of Property Act 1925, setting out:

a the specific breach;
b the remedy, if the breach is one capable of being remedied and
c the compensation, if such is appropriate or required.

The courts have explained that the point of the s.146 process is to give the tenant a final chance (*Expert Clothing Service & Sales Ltd v Hillgate House Ltd* (1986) CA. If the breach of obligation can be remedied, then the landlord is entitled to compensation under s.146(1). But if the breach cannot be remedied in a reasonable time or is irremediable, then the landlord can seek a possession order and the tenant can apply for relief under s146(2). It is important to note that when a residential lease is 21 years or longer, ss.168–169 of the Commonhold and Leasehold Reform Act 2002 place certain restrictions on the use of s.146.

3 ***Landlord's right to repair at the tenant's expense*** – Where the lease empowers them to do so, the landlord may enter the premises, carry out repairs and recover their cost from the tenant.

4 ***Specific performance*** – A tenant may obtain an order of specific performance against the landlord compelling them to carry out their repairing obligations. A landlord can only obtain such an order against a tenant in very exceptional circumstances (see *Rainbow Estates v Tokenhold* (1999) Ch.

5 ***Tenant's set-off of repair costs against rent*** – If the landlord fails to repair within a reasonable time of receiving notice of disrepair and the tenant carries out the necessary repairs, the tenant may be entitled to deduct their cost from rent payments.

6 ***Receiver*** – Upon application by the tenant, the court may appoint a receiver to manage the property if it is just and convenient to do so. This may be the case where the landlord refuses to perform his repairing obligations. The general power of the High Court to appoint a receiver is contained in s.37 of the Supreme Court Act 1981. A specific power relating to buildings containing two or more residential flats is given to the county court by Part II of the Landlord and Tenant Act 1987.

Repairing liabilities imposed on the parties to any agreement by statute

Whilst a landlord may not be liable in tort or at common law, a duty of care can be imposed upon a landlord through statute in the form of the Occupiers' Liability Act 1957 and the Defective Premises Act 1972.

Occupiers' Liability Act 1957

Under this Act, the occupier of the premises owes a duty of care to all visitors and, for the purposes of this Act, the occupier is the person in overall control of the premises. This is typically the tenant who enjoys exclusive possession, but can, in some situations, be a landlord. Where a landlord has control over common parts of the building, they will owe a duty of care to any visitors. Equally, a landlord with repair obligations for a building would be responsible for the risks created by any disrepair they failed to address. In *Wheat v Lacon & Co. Ltd* (1966) HL, the licensee of a pub argued the brewery was responsible for their injury because the light illuminating a set of steps was in disrepair. The court accepted that both the owner (landlord) and occupier (tenant) could both owe a duty under the Occupiers' Liability Act 1957 at the same time, but since the licensee was responsible for the day-to-day maintenance of the pub, repairing the light was their responsibility.

Defective Premises Act 1972

Section 1(1) imposes a duty on persons who undertake work to a dwelling to do it properly, in a professional manner, with good materials and to leave the property fit for human habitation. This duty is owed both to the person for whom the building is provided and also to anyone acquiring the property subsequently. Section 4 of the Act provides that where a property is let and the landlord is under an obligation to the tenant to maintain or repair the property under the terms of the tenancy, then the duty of care that arises from that extends not just to the occupier but to all persons who may be reasonably be expected to be affected by any defect with the property.

Security of tenure: termination of a tenancy at common law

During the period of the lease the tenant has what is referred to as security of tenure, that is, protection from eviction for the duration of the term, provided they fulfil all the obligations under the lease agreement. At common law, a tenancy may terminate in a variety of ways. Until it terminates, both parties are bound by its terms. The purpose of the present section is to examine briefly these common law methods of termination. It should be noted at this point, however, that the termination of business tenancies is regulated by Part II of the Landlord and Tenant Act 1954. Similarly, the ability of a landlord to obtain possession of a dwelling-house let under a regulated tenancy is restricted by the Rent Act 1977, and a dwelling let as an assured tenancy or an assured shorthold tenancy by Part I of the Housing Act 1988. These statutory provisions are examined later, after a discussion of the common law principles.

Methods of termination at common law

Expiration of a fixed term

At common law a fixed term tenancy terminates when the term expires. Thus, for example, a seven-year lease expires after seven years have passed.

Service of notice under a break clause

A break clause is a proviso that empowers one or both parties to a lease to terminate the agreement at certain specified dates. The party wishing to trigger the break clause serves notice on the other. For example, a 14-year lease might contain a proviso empowering the tenant to terminate the lease at the end of the seventh year by prior service of notice.

Forfeiture

A breach of covenant by the tenant does not automatically entitle the landlord to forfeit a lease – that is, re-enter the premises and resume possession. However, the common law does contain a right of re-entry where the tenant is in breach of a condition of the lease. A condition is a clause that shows a clear intent on the part of the landlord that, not only will the tenant be personally liable, but the landlord will have the right to terminate the lease. For this reason, non-payment of rent is treated somewhat differently from other breaches of covenant. To trigger the common law right of forfeiture, a landlord must usually first issue a formal demand for payment, except where the lease contains a proviso to forfeit for nonpayment of rent or the rent is six or more months in arrears (s.210 Common Law Procedure Act 1852). Once a landlord seeks to forfeit the lease, the tenant might be able to claim relief, pay the arrears and the landlord's costs and be reinstated. The extent of the tenant's entitlement to relief, and the powers of the court in respect of an application, vary depending upon whether the action is brought in the High Court or the county court (s.212 Common Law Procedure Act 1852 and s.138 County Courts Act 1984). The rules here are complicated and, at times, illogical. Thus, in practice, most leases will contain an express proviso reserving the right to forfeit the lease for non-payment of rent 'whether lawfully demanded or not', avoiding the strictness of the common law procedure.

A landlord can only forfeit the lease for all other breaches of covenant if it contains a proviso that gives them a right of re-entry upon the breach in question. The option of forfeiture might, typically, be reserved by a landlord where a tenant has been in arrears of rent for 21 days or the tenant uses the property for unlawful purposes (such as operating a brothel: *Rugby School (Governors) v Tannahill* (1935) KBD). If a landlord is aware of a breach that entitles them to forfeit the lease and then expressly or by implication affirms the continued existence of the lease, they waive their right to forfeit it, unless it is a continuing breach (such as a state of disrepair). For example, acceptance of or demand for rent falling due *after* the breach will give rise to such waiver. But rent accepted during the currency of a s146 notice (discussed in what follows) does not, because the right to forfeit arises only once the notice has expired (*Greenwich London Borough v Discreet Selling Estates* (1990) CA).

Other than in the case of forfeiture for non-payment of rent, the landlord must issue a notice under s.146 of the Law of Property Act 1925, setting out:

a the specific breach;
b the remedy, if the breach is one capable of being remedied and
c the compensation, if such is appropriate or required.

The courts have explained that the point of the s.146 process is to give the tenant a final chance (*Expert Clothing Service & Sales Ltd v Hillgate House Ltd* (1986) CA). If the breach of obligation can be remedied, then the landlord is entitled to compensation under s.146(1) if this is sufficient to remedy the harm suffered. But if the breach cannot be remedied in a reasonable time or is irremediable, then the landlord can seek a possession order. In residential lettings, other than where the tenant willingly vacates the property, s.2 of the Protection from Eviction Act 1977 requires the landlord to apply to the court for a possession order before re-entering the premises. Where s.146 applies to a forfeiture, it entitles the tenant to apply to the court for relief against forfeiture. The court will consider both the conduct of the parties and all other relevant circumstances, including the fact that the tenant has remedied the breach or has promised to do so, and can then grant or refuse relief as it thinks fit. Special rules apply under s.146(4) to sub-tenants, including mortgagees.

Special rules also exist in relation to forfeiture where the tenant disputes the amount of a service charge (s.81 Housing Act 1996 and ss.168–169 Commonhold and Leasehold Reform Act 2002).

Merger

Merger occurs when a tenant acquires a landlord's reversion and intends that the two legal estates, the freehold and leasehold interests, merge into one, thereby terminating the tenancy. So effectively the tenant is acquiring the landlord's interest in the property. Such an intention will normally be expressed in the assignment of the landlord's reversion to the tenant.

Surrender

Surrender may occur in either of two situations:

1 ***Surrender by express agreement of the parties*** – Express surrender occurs when landlord and tenant agree to end a lease before it is due to expire. Express surrender will normally be by deed, though an agreement in writing is sufficient in the case of terms of three years or less. Naturally there may be conditions to be agreed by the parties in the event of an express surrender upon matters like dilapidations, etc.
2 ***Surrender by operation of law*** – Surrender by operation of law occurs when both parties to a lease act in such a way that the surrender of the lease before it is due to expire is implied. This is the case where the conduct of the parties is such as permits no explanation other than that the lease has been surrendered. For example, surrender may be implied where a tenant abandons possession and his landlord re-lets the

premises to a new tenant. Again, it may be implied when a landlord grants his tenant a new lease before the old one has expired. It will probably not be implied where a tenant abandons possession and the landlord leaves the premises vacant in anticipation of the tenant's return. Surrender, whether express or by operation of law, does not invalidate a lawful sublease.

Service of notice to quit

At common law, a periodic tenancy may be terminated by service of notice to quit. The terms of a tenancy may expressly indicate the length of notice required in order to terminate it. Where this is not the case, the basic rule is that a tenancy of six months or less may be terminated by service of one period's notice and, consequently:

- to terminate a weekly tenancy, one week's notice is required;
- to terminate a monthly tenancy, one month's notice is required;
- to terminate a quarterly tenancy, one quarter's notice is required.

In the case of a yearly tenancy or a periodic tenancy the period of which exceeds one year, six months' notice is required.

In general, notice must be served so as to terminate a tenancy on the anniversary of its creation. In the case of a yearly tenancy, this would be one or more years from the date of grant. In the case of a weekly or monthly tenancy, this would be, respectively, one or more weeks or months from the date of grant. A quarterly tenancy must normally be terminated on a quarter day (25 March, 24 June, 29 September or 25 December).

Frustration

The doctrine of frustration is applicable, though in limited circumstances, to leases. This would be where the tenant is frustrated or prevented from occupying and utilising the property as they would wish, through no fault of their own. The circumstances need to make the tenancy impossible, unlawful or radically change it, such as destruction of the dwelling by fire.

Offences related to the recovery of possession of premises

Section 6 of the Criminal Law Act 1977 provides that it is an offence, without lawful authority, to use or threaten violence for the purpose of securing entry into premises, knowing that someone opposed to the entry is present on the premises. Section 1(2) of the Protection from Eviction Act 1977 provides that it is an offence to unlawfully deprive or attempt to deprive the residential occupier of premises (namely, the person who occupies those premises as a residence under contract, statute or rule of law) of his or her occupation of all or part of the premises in the absence of belief, with reasonable cause, that the residential occupier has ceased to reside there.

The only lawful method of obtaining possession of premises occupied by a residential tenant is by obtaining a court order, unless either the tenant peaceably vacates the premises or the tenancy is an excluded tenancy under the Housing Act 1988. Section 1(3) of the Protection from Eviction Act 1977 makes it an offence for any person to commit acts likely to interfere with the peace or comfort of a residential occupier of any premises or members of their household or to persistently withdraw or withhold services reasonably required to occupy the premises as a residence, with the intention to cause the residential occupier to give up possession of the premises or refrain from exercising rights or pursuing remedies in respect of it. Section 1(3A) specifically applies the preceding offence to the landlord of a residential occupier or his agent, except where the person who did the acts or withdrew the services had reasonable grounds for so doing. A tenant who gives up possession of premises as a result of conduct falling within s.1(2) or (3) may be entitled to recover damages from the landlord.

Section 2 of the Protection from Eviction Act 1977 provides that where a person resides in premises which are let as a dwelling, a right of re-entry or forfeiture can only be lawfully enforced by proceedings in court. Section 3 provides that where a lease of a dwelling-house comes to an end and the occupier continues to reside in it, the landlord can only lawfully regain possession by court proceedings unless the tenancy was a protected tenancy or an excluded tenancy.

Business tenancies (Landlord and Tenant Act 1954, Part II)

Having examined some of the common law principles relating to the operation of leases, there are specific legal provisions in law for different classes of property. For leases pertaining to property occupied for the purposes of a business use, the relevant legal provisions are contained in Part II of the Landlord and Tenant Act 1954. These provisions only apply to property occupied under the terms of a lease and not a licence.

The tenant must be in occupation of the premises and that must, at least in part, be for the purposes of a business carried on by the tenant or for those and other purposes (s.23). Business includes a trade, profession or employment and includes any activity carried on by a body of persons, whether corporate or unincorporated, and so can cover a wide variety of enterprises, from shops and offices to hospitals and clubs. In the majority of cases, it will be straightforward in determining if the property is occupied for a business purpose. A conflict can sometimes arise when there are properties being occupied for a mixed use, but the statutory codes relating to business and residential property are mutually exclusive. In *Royal Life Saving Society v Page* (1978) CA, a doctor took a maisonette to live in and the landlord had granted permission for the doctor to work from the property in addition to living there in addition to working from their consulting rooms in Harley Street. The court heard that both the addresses from Harley Street and the maisonette featured in the doctor's records, but, in fact, he only rarely saw clients at the maisonette. In that case, the court found the work that the doctor carried out at the maisonette was only incidental to his residential use and found that the occupation of the property was regulated by the Rent Act 1977. In *Graysim Holdings Ltd v P & O Property Holdings Ltd* (1996) HL, the tenant had leased a market hall and sublet to individual stalls to traders. The court held that the tenant was not in occupation for its own purposes, as the subtenants were carrying on their businesses independently. However, *Royal Borough of Kensington & Chelsea v Mellcraft Ltd* (2024) Ch found that the sole company director occupied the property as both a residence and a place of business, managing various properties from the flat, including the collection and forwarding of sub-rents. The court ruled that this constituted an occupation for business purposes under the Act.

Under s.23, the tenant must occupy at least part of the premises. This does not require the tenant to occupy the premises personally. Rather, it may be sufficient if the premises are occupied by the tenant's agents or employees. The tenant need not be in sole occupation of the premises. The tenant must, however, retain a sufficient degree of control over the premises. For example, where a business tenant sublets the premises and acts as manager, providing services to the subtenants, this may be sufficient to amount to occupation of the premises by the tenant. The premises may include land without a building, such as an undeveloped area used as a car park.

The Act does not apply to a tenancy if the lease prohibits business use by the tenant unless the landlord has agreed to the business use. As seen in *Royal Life Saving Society v Page*, the Act also does not apply where premises are mainly used for residential purposes and any business use is subsidiary to this. This may be so, for example, where a residential tenant takes in a lodger.

Types of tenancy to which the Act does not apply

Part II of the Landlord and Tenant Act 1954 does not apply to the following types of lease:

- Lettings of agricultural holdings (s.43). These have their own provisions usually under the Agricultural Holdings Act 1986 or the Agricultural Tenancies Act 1995 and are beyond the scope of this book.
- Mining leases (s.43).
- Service tenancies (s.43), i.e. tenancies granted to the tenant for the purpose of the tenant's employment or an office or appointment which the tenant holds.
- Residential tenancies (s.23), which are discussed later in this chapter.
- Leases of premises required for public purposes (s.57) or regional development (s.60).
- Leases of premises required for national security purposes (s.58).
- Fixed term tenancies not exceeding six months (s.43). If the tenant has been in occupation for more than 12 months or the lease contains a term permitting extension or renewal of the tenancy beyond six months, then the Act does apply.
- Extended tenancies granted under the Leasehold Reform Act 1967 (s.16).
- Tenancies of premises where the parties have agreed to a new tenancy of the premises (s.28).

Contracting out of Part II of the Landlord and Tenant Act 1954

Section 38 of the Act allows both the landlord and tenant to agree that the provisions contained within ss.24–28 of the Act (which relate to termination of the tenancy and essentially automatically entitle the tenant to continue in occupation of the property at the end of the fixed term and offer the tenant the option of requesting a new lease) *will not* apply to the tenancy. The landlord must serve notice on the tenant before the tenancy is completed, informing the tenant of the rights he or she is giving up. This notice must be in the form set out in Schedule 1 of the Regulatory Reform (Business Tenancies) (England and Wales) Order 2003 and the requirements specified in Schedule 2 of that order must also have been met. This notice to the tenant must have been served not less than 14 days before the tenant enters into the lease agreement. If the tenant agrees to the exclusion of the provisions of ss.24–28 they must either make a declaration in the form set out in Part 7 of Schedule 2 of the order or if the 14-day notice period is not met, then the ingoing tenant can make a statutory declaration in the form set out in Part 8 of Schedule 2 of the order setting out that they have received the notice and understood it.

Protection under Part II of the Landlord and Tenant Act 1954 for the Business Tenant

Where the provisions of Part II of the Landlord and Tenant Act 1954 do apply to a tenancy, the Act may entitle the tenant to automatically continue the existing tenancy when the contractual term expires until terminated by either party under the condition sets out under the Act, to apply for a new lease and permit the landlord to serve an effective notice to quit on the tenant at the end of the fixed term. Section 24 provides that a business tenancy to which the Act applies can only be terminated by:

- forfeiture,
- surrender,
- merger,
- service of notice to quit by the tenant or
- by satisfying special requirements laid down in the Act.

Thus, where a fixed term tenancy expires or the landlord serves notice to quit (in relation to a periodic tenancy or under a break clause), then, unless the special requirements laid down by the Act have been satisfied, the tenancy does not terminate. Rather, the tenancy continues until terminated in one of the five ways referred to previously. A tenancy which continues in this way is described as a continuation tenancy. Basically, a business tenant has the right of automatic continuation, which amounts to a continuation of the contractual tenancy and provides security of tenure for a business tenant beyond the fixed term. The terms of the continuation tenancy are the same as the terms of the contractual tenancy, except for any terms permitting the landlord to terminate the tenancy or creating personal obligations which do not run with the land. Whilst the tenancy continues in this way, the tenant continues to have the right to remove the tenant's fixtures and to use the easements given to them by the lease.

Under a continuation tenancy, the landlord cannot increase the rent except by applying to the court (s.24A). They can only do this if either he has commenced the special procedure which the Act lays down for terminating the tenancy or the tenant has commenced the special procedure which the Act lays down for applying for a new tenancy. The court may be prepared to grant an interim rent, which will remain in force until the continuation tenancy ends. A tenant can also apply to the court for an interim rent. The court is not bound to grant such a rent, but if it chooses to do so, the rent must be an open market rent fixed in accordance with principles laid down by the Act. The principles are basically those which the Act lays down in relation to determining the rent payable under a new tenancy (s.34). Additionally, the court is required to consider the rent payable under the terms of the tenancy and to determine the rent upon the basis of the assumption that the tenancy is a yearly tenancy (s.24A). If a continuation tenancy ceases to be a business tenancy, the landlord may terminate it by giving the tenant between three and six months' notice or in accordance with the terms of the tenancy (s.24).

Termination by the tenant

The Act permits a business tenant to terminate their lease either by surrender or by merger. The Act also allows the tenant to terminate the tenancy by serving notice to quit (in relation to a periodic tenancy or under a break clause proviso) (s.24). In relation to the expiration of a fixed term tenancy, the tenant must give the landlord written notice that he or she does not wish the tenancy to continue, either at least three months before the date when the tenancy would expire if the Act did not apply or at least three months before any day following that date (s.27).

Termination by the landlord

Subject to the discretion of the court to grant relief, the Act allows the landlord to forfeit the tenancy where breach of a condition or covenant gives rise to a right of re-entry (s.24). Other than this, the landlord can only terminate the tenancy if they satisfy special requirements which the Act lays down (see later on). To rely upon the expiration of a fixed term or service of notice to quit (in relation to a periodic tenancy or under a break clause), the landlord must satisfy these special requirements. If they do not do so, then a *continuation tenancy* will result (s.24). Similarly, where a landlord wishes to terminate a continuation tenancy, their options are either to forfeit, if appropriate, or to satisfy the special requirements laid down by the Act.

The *special requirements* which the Act lays down for termination of a tenancy by the landlord are as follows:

1 Under s.25 of the Act the landlord must serve a written notice (a 'Section 25 notice') on the tenant in the form laid down by the Landlord and Tenant Act 1954, Part 2 (Notices) Regulations 2004. The practical effect of this notice is to commence the lease renewal procedure by establishing a set of time limits within which the parties can negotiate. The notice must specify the date of termination and must be served on the tenant between six and 12 months before the date which it specifies (s.25). If it has not already terminated, the contractual tenancy will terminate on this date (though the current tenancy may continue as a continuation tenancy beyond this date). In relation to the expiration of a fixed term tenancy, the date of termination cannot be earlier than the date when the tenancy would have expired had the Act not applied to it.

   Regarding the termination of a tenancy by notice to quit (in relation to a periodic tenancy or under a break clause), the date of termination cannot be earlier than the earliest date upon which the tenancy could have been ended, had the Act not applied to it, by notice to quit served by the landlord. Provided that the s.25 notice conforms with the requirements of the tenancy agreement, the landlord need not serve a separate notice to quit on the tenant. In relation to the termination of a continuation tenancy, the date of termination can be any date chosen by the landlord. In the s.25 notice, the landlord must declare whether they are opposed or agree to the grant of a new lease. If the landlord does not wish the tenant to be granted a new tenancy, the notice must state this and must also indicate the grounds upon which the landlord objects to the grant of a new tenancy. The landlord may only oppose the grant of a new tenancy on grounds laid down by s.30. If the landlord is not opposing renewal, they must set out in their s.25 notice their proposals for the new tenancy.
2 If the tenant does not want a new tenancy, they need to do nothing and the tenancy will terminate on the date specified by the notice.
3 If the tenant does want a new tenancy, they must apply to the court for a new tenancy if the parties cannot agree upon one themselves (s.29). If the tenant wishes to apply to the court, they must do so by the date specified in the s.25 notice as the termination date, otherwise the right to apply is lost. The parties may agree to extend this deadline. The relevant court will be the county court for the district in which the property is situated.

Once the tenant applies to the court for a new tenancy, s.64 provides that the current tenancy cannot terminate within three months of the disposal of the tenant's application by the court. This is so whether or not the court grants a new tenancy. Thus, if the termination date specified by the landlord's s.25 notice falls within this three-month period, the current tenancy will not terminate until the three months have passed.

The court will order the grant of a new tenancy unless the landlord establishes one or more of the grounds upon proof of which the court may or must refuse the tenant's application (s.29). If the tenant has a change of mind they may apply to the court for revocation of the order within 14 days of its making (s.36). If the tenant does this the current tenancy will continue for a period agreed by the parties or determined by the court, thus giving the landlord a reasonable opportunity to re-let or dispose of the premises (s.36).

If the court grants a new tenancy, it will determine the terms of the new lease if the parties cannot reach an agreement upon them. In the absence of agreement, the Act specifies in ss.32–35 rules about:

- length of the tenancy, up to a maximum of 15 years,
- property to be included in the tenancy,
- rent payable, which will be an open market rent fixed in accordance with statutory principles and
- any other terms of the new tenancy.

4 Under s.30 there are statutory grounds on which landlords can object to the grant of a new tenancy. Proof of the certainty of these grounds means that a court *must* refuse the tenant's application for a new tenancy. In other cases, the court *may* refuse to grant a new tenancy. These mandatory grounds are set out in the following.

Mandatory grounds for refusing the tenant's application for a new lease

There are three grounds which, if the landlord can prove, the court *must* refuse the tenant's application:

- **Ground (d)** – The landlord has offered and is willing to provide the tenant with alternative accommodation which is suitable for the tenant's requirements on terms which are reasonable. In assessing whether the alternative accommodation is suitable for the tenant's requirements (including the preservation of goodwill), the court should consider the time when the new accommodation will be available for the tenant, the nature and class of the tenant's business and the situation, extent and facilities of the accommodation provided by the current tenancy.
- **Ground (f)** – The landlord intends to demolish or reconstruct the premises or to undertake substantial construction work and could not reasonably do so without obtaining possession of the premises. The landlord must prove at the time of the court hearing that they have the necessary genuine intention. The landlord must also prove that there is a reasonable prospect of this event happening. They will do so if, on the evidence, a reasonable person would think that the landlord had a reasonable prospect of carrying out his intention. In *Cunliffe v Goodman* (1950) CA, the court set out that the project must have 'moved out of the zone of contemplation . . . and into the valley of decision'. This has proved to be a challenging ground for a landlord to prove. In *S Franses Ltd*

v The Cavendish Hotel (London) Ltd (2018) SC, the tenant had occupied the ground and basement floors of a building used as a hotel. The landlord had sought to oppose the renewal under ground F, proposing a redevelopment plan. The court found that the landlord's intention was not genuine, that the proposed works were impractical and designed solely to defeat the tenant's renewal rights. So, this case reiterates that landlords must have a firm and settled intention to carry out the works. Similarly, in *Sainsbury's Supermarkets Ltd v Medley Assets Ltd* (2024) Central London CC the court ruled that the landlord had not demonstrated a genuine and settled intention to carry out the works that they proposed. A landlord must also prove that they not only need physical occupation of the property, but that to carry out the works, they also need legal possession of the property. Physical possession could, for example, be taken through a right of re-entry to carry out works usually contained within the lease. A tenant could be successful in resisting an attempt under ground F to take possession by agreeing to take a tenancy of the property that permits the landlord to still carry out the works listed by agreeing to give up part of the property that could be unaffected by the landlord's works.

- **Ground (g)** – The landlord intends to occupy the premises either as their residence or for the purposes of a business to be carried on by them. This ground is not normally available to a landlord who acquired the freehold or a tenancy of the premises by purchase within five years of the date of termination specified by his s.25 notice, so effectively the landlord must have owned the property for at least five years. Again, the landlord must prove at the time of the court hearing that he has the necessary intention.

Discretionary grounds for refusing the tenant's application

There are four grounds upon proof of which, by the landlord, the court *may*, but is not obliged to, refuse the tenant's application:

- **Ground (a)** – The tenant ought not to be granted a new tenancy in consequence of the state of repair of the premises resulting from the tenant's failure to comply with his or her obligations under the current tenancy to repair and maintain them. The breach must be serious enough to persuade the court not to grant a new tenancy, an undertaking made by the tenant to remedy the breach being a relevant factor.
- **Ground (b)** – The tenant ought not to be granted a new tenancy in consequence of his or her persistent delay in paying rent due under the current tenancy. This requires either a failure to pay rent due on several occasions or a very long delay in paying rent due. The court will consider factors such as the reason for the delay and how, if a new tenancy is granted, the landlord may be protected against breaches of the tenant's covenant to pay rent.
- **Ground (c)** – The tenant ought not to be granted a new tenancy in consequence of other substantial breaches of his or her obligations under the current tenancy or for any other reason connected to the use or management of the premises. The court will consider factors such as the seriousness of the breach and whether the tenant has remedied or proposes to remedy it.
- **Ground (e)** – The tenant ought not to be granted a new tenancy where:
  - the tenant is a subtenant of part of the property let by the landlord under a superior tenancy,

- the landlord might reasonably expect to achieve a substantially better rent by letting the property as a whole and
- the landlord requires possession of the premises to let or dispose of the property let under the superior tenancy as a whole.

The court will take into account factors such as the fact that the landlord consented to the subletting. The ground is only of value to the landlord where the superior tenancy is due to terminate shortly after the sub-tenancy.

Tenant's request for a new tenancy

A tenant who has not been served a s.25 notice by the landlord may wish to request a new tenancy, which they can do under s.26 of the Act. The requirements which must be satisfied are as follows:

1 The current tenancy must have been granted for a term exceeding one year. This includes a tenancy granted for a fixed term and thereafter from year to year.
2 The tenant must serve an s.26 notice, requesting a new tenancy, on the 'competent landlord'. A tenant cannot serve an s.26 notice if the landlord has served an s.25 notice on him or her or he or she has given the landlord notice to quit or notify the landlord (under s.27) that he or she does not wish the fixed term tenancy to continue when its expiration date arrives. The notice must be in writing and in the form of Form 3 of the Landlord and Tenant Act 1954, Part 2 (Notices) Regulations 2004. It must state the date upon which the new tenancy is to commence. The current tenancy will automatically terminate on this date. The date must be between six and 12 months after the making of the s.26 request to the landlord, and cannot be earlier than the date on which, had the Act not applied to it, the tenancy would have expired or could have been terminated by notice to quit served by the tenant. The notice must contain the tenant's proposed terms of the new tenancy, including the property to be comprised in the tenancy and the rent to be paid.
3 If the landlord does not wish to oppose the tenant's application for a new tenancy, he need not respond to the s.26 notice served on him by the tenant. He thus loses his right to oppose the tenant's application. If the landlord does wish to oppose the tenant's application, he must serve a written counter-notice on the tenant indicating that he opposes the application. They must serve this counter-notice on the tenant within two months of the serving of the s.26 notice on them by the tenant, otherwise they lose their right to oppose the tenant's application. The notice must indicate the grounds upon which the landlord objects to the grant of a new tenancy. The grounds upon which the landlord may so object, laid down by s.30 which have already been outlined.
4 Once the tenant serves the s.26 notice, he or she becomes entitled to apply to the court for a new tenancy if the parties cannot agree upon one themselves (s.29). If the tenant wishes to apply to the court, he or she must do so before the date specified in the s.26 notice. The parties can agree to extend this deadline. The landlord also has the right to apply to the court for renewal or an order terminating the tenancy. Once the tenant applies to the court for a new tenancy, s.64 provides that the current tenancy cannot terminate within three months of the disposal of the tenant's application by the court. This is so whether or not the court grants a new tenancy. Thus, if the termination date

specified by the tenant's s.26 notice falls within this three-month period, the current tenancy will not terminate until the three months have passed.

The court will order the grant of a new tenancy, unless the landlord establishes one or more of the grounds upon proof of which the court may or must refuse the tenant's application (s.29). If the tenant has a change of mind, he or she may apply to the court for revocation of the order within 14 days of its making (s.36). If the tenant does this, the current tenancy will continue for a period agreed by the parties or determined by the court, thus giving the landlord a reasonable opportunity to re-let or dispose of the premises (s.36).

If the court grants a new tenancy, it will determine the terms of the new lease if the parties cannot reach an agreement upon them.

In the absence of agreement:

- the length of the tenancy will be that which the court considers to be reasonable in the circumstances, to a maximum of 15 years (s.33),
- the property included in the new tenancy will be that part of the property included in the current tenancy which the court designates in the circumstances (s.32),
- the rent payable will be an open market rent, fixed in accordance with principles laid down by the Act (s.34) and
- the other terms will be determined by the court having regard to the terms of the current tenancy and all other relevant circumstances (s.35).

5 If the landlord establishes one of the statutory grounds upon which he is entitled to object to the grant of a new tenancy, it may be that the tenant will not be granted a new tenancy. Upon proof of certain grounds by the landlord, the court must refuse the tenant's application for a new tenancy. Upon proof of other grounds, the court has discretion to grant or refuse to grant a new tenancy.

Tenant's compensation

Under s.37 Landlord and Tenant Act 1954, Part II

Compensation under s.37 of the Act is available in two cases:

1 Where the tenant's application to the court for a new tenancy is refused, i.e. the landlord is successful, because the landlord intends to demolish the premises (ground (f)), because the landlord intends to occupy the premises himself (ground (g)) or because the landlord wants to amalgamate a subtenant's property with other property in order to get a better rent by letting as a whole (ground (e)). In these cases, the tenant is being required to leave the property through no fault of their own.
2 Where either the landlord's s.25 notice or his counter-notice served in response to the tenant's s.26 notice states any or all these three grounds of opposition (but no others) and the tenant either fails to apply for a new tenancy or makes and then withdraws an application. (Note: the leave of the court is required to withdraw an application.)

The amount of compensation is determined by reference to the rateable value of the property multiplied by an appropriate multiplier. This multiplier is set by regulation. Currently that multiplier is one as prescribed by the Landlord and Tenant Act 1954 (Appropriate Multiplier) Order 1990.

Compensation for improvements under s.1 Landlord and Tenant Act 1927, Part I

Part I of the 1927 Act applies to premises held under a lease which are used for trade or business purposes, but does not apply to mining leases, leases of agricultural holdings or leases which express in writing that the premises are let to the tenant as holder of an office, appointment or employment which he or she continues to hold (s.17). An agreement depriving a tenant of the right to compensation under Part I is invalid unless made for valuable consideration before 10 December 1953 (s.9 as amended by s.49 Landlord and Tenant Act 1954).

Where Part I of the 1927 Act applies, it entitles the tenant, on vacating a holding at the end of the tenancy, to compensation from the landlord in respect of improvements made by the tenant or his/her predecessors in title (s.1). Where premises are used only partly for trade or business purposes, compensation is only payable in relation to improvements if and to the extent to which they relate to the trade or business (s.17). It is the tenant's immediate landlord who is required to pay compensation. To give rise to this right to compensation, an improvement must not be a fixture which the tenant is entitled to remove, and it must add to the letting value of the holding at the end of the tenancy (s.1). Moreover, the improvement must not have been made for valuable consideration (s.2).

Further, a tenant is not entitled to compensation under Part I unless the requirements laid down by s.3 have been satisfied. Prior to making the improvement, the tenant is required to notify the landlord of his or her intention, providing the landlord with a specification and plan of the improvement. Unless the making of the improvement is required by statute, the landlord is entitled to serve notice of objection to the improvement on the tenant within three months of receiving the tenant's notice. If the landlord does not serve a notice of objection within three months, he loses his right to do so. The tenant can lawfully go ahead with the improvement even if it is prohibited by the terms of the lease. If the landlord does serve a notice of objection within three months, the tenant can apply to the court for a certificate that the improvement is proper. When such an application is made, all superior landlords must be notified of the tenant's intention and have a right to be heard before the court. The court must give the tenant a certificate if it is satisfied that the improvement will add to the letting value of the holding at the end of the tenancy, is reasonable and suitable to the character of the property and will not diminish the value of any other property belonging to the landlord.

The landlord is entitled to offer to make the proposed improvement themselves in return for a reasonable increase in rent or an increase determined by the court. In this case the court cannot grant a certificate unless it is subsequently shown to the court's satisfaction that the landlord has failed to do so. Before giving the tenant a certificate, the court is entitled to make such modifications to the specification or plan as it thinks fit and to impose such conditions as it thinks reasonable. Once the certificate has been issued, the tenant can lawfully carry out the improvement, even if it is prohibited by the terms of the lease. He or she must comply with conditions imposed by the court and must complete the work within time limits agreed with the landlord or fixed by the court.

Claiming compensation

The tenant's claim for compensation for improvements, made in the prescribed form, must be served on the landlord within the time limits laid down by s.47 in Part 3 of the Landlord and Tenant Act 1954. Where the claim relates to the giving of notice to quit, whether by the landlord or the tenant, the claim must be made within three months of the giving of the notice.

If the claim for compensation is in relation to termination following the service of a s.26 notice by the tenant, it must be made within three months of the date upon which the landlord served the counter-notice. If the claim is in relation to termination by expiration of time, it must be made between three and six months before the end of the tenancy. Where the claim is in relation to forfeiture, it must be made within three months of the effective date of a court order for possession or, in the case of re-entry without a court order, within three months of the date of re-entry.

Amount of compensation (improvements)

In the absence of agreement between the parties to the lease, the amount of compensation is determined by the court. The amount of compensation cannot exceed either:

- the net addition to the value of the holding which directly results from the improvement or
- the reasonable cost of carrying out the improvement at the end of the tenancy, less the cost of putting the improvement into a reasonable state of repair (unless or to the extent to which the lease obliges the tenant to carry out such repairs).

In determining the net addition to the holding's value, account must be taken of:

- the intended use of the premises after the end of the lease,
- any intention to demolish or reconstruct and
- the likely time lapse between the date of termination and any change of use, demolition or reconstruction.

If the landlord does not carry out his intentions within the time specified, the tenant is entitled to make a further application for compensation. The amount of compensation will be reduced in accordance with the value of any benefits received by the tenant or predecessors from the landlord or predecessors in consideration of the making of the improvement (s.2(3)).

In March 2023 the Department for Levelling Up, Housing and Communities asked the Law Commission to review the provisions of Part 2 of the Landlord and Tenant Act 1954 in an attempt to possibly create a more adaptable and efficient leasing framework for landlords and tenants. At the time of writing, a second consultation paper is due to be published in the autumn of 2025, setting out detailed recommendations. Some of the suggested areas for the review to examine have been:

- To look at modernising the security of tenure model by examining the current process for lease renewals.
- To streamline the contracting out process and maybe make the default lease an opted-out one, so that parties would have to opt into the security of tenure provisions.
- To look at ways to incorporate sustainability and environmental objectives into the agreement.
- To look at enhancing dispute resolution techniques.

Residential tenancies

There are a number of statutory regimes that can apply to properties occupied on residential tenancies and the date the tenancy commenced will be key to deciding which set of legal rules applies. The differences between these provisions are significant and the implications for the landlord and tenant substantial. In general, residential property let prior to 15 January 1989 will be a regulated tenancy under the Rent Act 1977; tenancies created on or after 15 January 1989, but before 28 February 1997, could be either assured tenancies or assured shorthold tenancy under the Housing Act 1988; and those let after 28 February 1997 will, by default, be assured shorthold tenancies under the Housing Act 1988 (as amended by the Housing Act 1996). It is important to note that the Renters (Reform) Bill (discussed later on) promises significant change in this area of law.

Rent Act 1977 (regulated tenancies)

The Rent Act 1977 applies to private residential tenancies entered before 15 January 1989, though certain tenancies created after that date do still fall within the provisions of the Act. A tenancy to which the Rent Act applies is known as a *regulated tenancy* and consists of two stages: a contractual or protected tenancy (s.1) and a statutory tenancy (s.2). The tenancy must be of a dwelling-house, which may be a house or part of a house, including a flat, maisonette or even a single room (*Curl v Angelo* (1948) CA), but not a houseboat (*Chelsea Yacht and Boat Co. Ltd v Pope* (2000) CA). The dwelling-house must be let as a separate dwelling, and the tenant or joint tenants must have exclusive possession of rooms where they conduct the main activities of living, especially cooking (although cooking facilities are not necessary: *Uratemp Ventures Ltd v Collins* (2001) HL), eating and sleeping. Other rooms, such as bathrooms, may be shared (s.22). There is no set definition of a dwelling-house, but the central question appears to be whether the premises can reasonably be described as someone's home or, as Lord Greene put it in *Curl v Angelo*, whether it is a place where a person 'moves and has their being'. So, a home could be a bedsit or a hotel room, but the agreement must be in the form of a lease, not a licence, for the Act to apply.

Tenancies incapable of being protected

The tenancy *must not* be one of the types which are incapable of being protected tenancies. The following are examples of types of tenancy which are incapable of being protected tenancies:

- Tenancies with high rateable values (being above specified limits).
- Tenancies paying a low rent, effectively to exclude long leaseholders from protection.
- A shared ownership lease granted under Part V of the Housing Act 1985 (s.5A) or a lease granted by a housing association which satisfies the requirements of s.5A(2).
- A dwelling-house let with land other than the site of the dwelling-house (s.6), though where land is let with a dwelling-house and the primary purpose of the agreement is the letting of the dwelling-house, the Act may apply.
- A dwelling-house let at a rent which includes payments for genuine board and attendance (s.7).
- In relation to a student studying a course provided by a specified educational institution, a tenancy granted to the student by a specified body (s.8).

- A tenancy for which the purpose is to let the dwelling-house to the tenant for a holiday (s.9).
- A tenancy under which the dwelling-house is comprised in an agricultural holding (see Agricultural Holdings Act 1986) and is occupied by the person who controls the farming of the holding (s.10).
- A tenancy of premises with an 'on-licence' for the sale of intoxicating liquors (s.11).
- A tenancy of a dwelling-house which forms part of a building, where the landlord from the time the tenancy was granted has occupied another dwelling-house in the same building as his residence (s.12). The tenancy is not rendered incapable of being a protected tenancy by the presence of a resident landlord if the dwelling-house forms a flat in a purpose-built block of flats. (Note that the complex provisions of s.12, relating to resident landlords, fall outside the scope of this book.)
- A tenancy under which the interest of the landlord belongs to certain types of public body, such as a county council, development corporation or housing action trust (s.14).
- Business tenancies which fall within Part II of the Landlord and Tenant Act 1954 (including premises with mixed business/residential use) (s.24(3)).

Security of tenure under the Rent Act

Security of tenure under the Rent Act 1977 works in two ways: first, it automatically continues the tenancy after the expiry of the fixed term; second, it only permits a landlord to regain possession of the property in restricted circumstances. A protected or contractual tenancy comes to an end when the term expires (if the lease is for a fixed term) or when the tenancy is otherwise terminated at common law. Following termination of the protected tenancy, s.2 converts the former protected tenant into a statutory tenant of the dwelling-house provided and for as long as the dwelling-house remains their residence. A statutory tenancy can also arise out of a protected tenancy by succession. Under Schedule 1, Para 2, a surviving spouse or civil partner (including those who were living as if they were married or civil partners), becomes a statutory tenant when the original tenant dies. If the original tenant died before 15 January 1989, a family member could also have succeeded to the tenancy and become a statutory tenant, provided they met certain conditions relating to residency and relationships with the deceased. When the death occurs after 15 January 1989, then the successor will acquire an assured tenancy under the Housing Act 1988 and not a statutory tenancy (schedule 1, para 3 Rent Act 1977).

A statutory tenancy is not an interest in land; rather, it is a statutory right to remain in possession of a dwelling-house. The terms of the statutory tenancy are those of the protected tenancy except to the extent to which they are inconsistent with provisions of the Act (s.3). Thus, terms of the protected tenancy entitling the landlord to recover possession of the dwelling-house will not become terms of the statutory tenancy. A statutory tenancy may therefore only come to an end where either the property ceases to be the main home of the tenant or the landlord is able to prove one of the mandatory or discretionary grounds for possession as set out in the Act.

- *Termination of a statutory tenancy by act of the tenant*

  A statutory tenancy will not arise or will come to an end if the dwelling-house ceases to be the tenant's residence (s.2). The fact that the tenant is frequently absent from the dwelling-house for lengthy periods does not necessarily mean that the dwelling-house has ceased to be his or her residence.

Similarly, where a tenant has two houses, the fact that they occupy one house more regularly than the other will not necessarily make the former house a residence for this purpose. In each case, the primary consideration in determining whether a dwelling-house is the tenant's residence is the intent.

A statutory tenant may terminate their statutory tenancy by serving notice on the landlord (s.3). If the protected tenancy was a fixed term tenancy, the tenant must give the landlord at least three months' notice for this purpose. If the protected tenancy was a periodic tenancy, the length of notice required is either the same as under the protected tenancy or four weeks, whichever is the greater.

A statutory tenant may surrender statutory tenancy expressly or impliedly.

- *Termination of a statutory tenancy by the landlord*

  A landlord can only obtain possession of a dwelling-house which is subject to a statutory tenancy by obtaining a court order (s.98). To obtain a court order, the landlord must prove one or more of the discretionary or mandatory grounds for possession which are laid down by the Act. If the landlord can prove a discretionary ground, the court may then order possession if it considers it reasonable to do so. The court possesses discretion to stay or suspend the execution of the order or postpone the date of possession for such period as it thinks fit. On proof of one or more of the mandatory grounds, the court must order possession.

  The landlord also needs a court order to obtain possession of a dwelling-house which is subject to a protected tenancy. In addition to proving one of the discretionary or mandatory grounds, a landlord must also, under the Act, satisfy the court that suitable alternative accommodation is available for the tenant and that the court considers it reasonable to make an order for possession. The suitable alternative accommodation must be suitable in that no tenant is going to want to move to an inferior property or a property over which they have lesser security of tenure. Effectively this means that the alternative accommodation will need to be a full Rent Act protected tenancy. It will need to be in a similar proximity to work as the former dwelling and no more expensive to live in and maintain. The requirement for reasonableness is an overriding requirement except where a mandatory ground is proven, upon which the question of reasonableness does not arise.

Discretionary grounds for possession (Rent Act 1977)

Schedule 15, Part I of the Rent Act 1977 sets out the following discretionary grounds for a landlord obtaining possession.

Case 1 applies where the tenant has failed to pay rent due or has broken or failed to perform an obligation of the tenancy. For outstanding rent, the key date is the date upon which legal proceedings are issued.

Case 2 applies where the tenant has been guilty of conduct, which is a nuisance or annoyance to adjoining occupiers or has been convicted of using the dwelling-house or allowing its use for immoral or illegal purposes.

Case 3 applies where the condition of the dwelling-house or common parts has deteriorated due to acts of waste by or the neglect or default of the tenant, a lodger or a subtenant.

Case 4 applies where the condition of any furniture provided under the tenancy has deteriorated owing to ill-treatment by the tenant, a lodger or a subtenant.

Case 5 applies where the tenant has given his landlord notice to quit and, consequently, the landlord has contracted to sell or let the dwelling-house.

Case 6 applies where the tenant has, without the landlord's consent, either assigned or sublet the whole of the dwelling-house.

Case 7 was repealed by the Housing Act 1980.

Case 8 applies where the dwelling-house was let to the tenant in consequence of his or her employment by the landlord or a former landlord and the tenant is no longer in that employment, provided that the dwelling-house is reasonably required by the landlord for a person engaged in full-time employment by him or by one of his tenants.

Case 9 applies where the dwelling-house is reasonably required by the landlord as a residence for himself, for a child of his over 18 years old, for his father or mother or for his father-in-law or mother-in-law. This case is of limited relevance today as it does not apply if the landlord became landlord by purchasing the dwelling-house or an interest in it after 23 March 1965. Part III of Schedule 15 provides that the court shall not make an order under case 9 if greater hardship would be caused by granting the order than by refusing to grant it.

Case 10 applies where the tenant has sublet part of the dwelling-house under a protected or statutory tenancy at a rent more than the maximum rent recoverable under Part III of the Rent Act 1977.

Case 10A, inserted by the Immigration Act 2016, applies where a person is disqualified as a result of their immigration status from occupying the property.

Mandatory grounds for possession under Part II of Schedule 15, Rent Act 1977

In order to obtain possession under cases 11–15 and 20, the landlord must have notified the tenant not later than the beginning of the tenancy that possession might need to be recovered under the appropriate case (though, in relation to cases 11, 12 and 20, the court may dispense with these requirements where it is just and equitable to do so).

Case 11 applies where the owner, prior to letting the dwelling-house, occupied it as his residence.

Case 12 applies where the owner intends to occupy the dwelling-house as his residence when he retires from regular employment.

Case 13 applies to fixed term tenancies for a term not exceeding eight months if, at some time within the 12 months prior to commencement of the tenancy, the dwelling-house was let as a holiday home.

Case 14 applies to fixed term tenancies for a term not exceeding 12 months if, at some time within the 12 months prior to commencement of the tenancy, the dwelling-house was let to a person pursuing or intending to pursue a course of study provided by a specified educational institution and the tenancy was granted by a specified body.

Case 15 applies where a dwelling-house is held for the purpose of being available for occupation by a minister of religion as a residence from which to perform his or her duties and the court is satisfied that the dwelling-house is required for this purpose.

Cases 16, 17 and 18 relate to premises once occupied by agricultural workers, which are required for agricultural employees, provided that notice was served prior to commencement of the tenancy that possession might be required under the relevant case, and that the other requirements of the relevant case are satisfied.

Case 19 may be applicable where a dwelling-house was let under a protected shorthold tenancy (see consideration of the similar provisions of the Housing Act 1988).

Case 20 applies where a dwelling-house was let after 28 November 1980 by a person who was a member of the regular armed forces when he acquired it, if he can satisfy requirements like those in case 12.

Rent control under the Rent Act 1977

Rents under the Rent Act 1977 are effectively regulated. Either or both parties to a protected tenancy may apply to a public official known as the Rent Officer to have a fair rent registered for two years. Usually, no further application can be made within the two-year period. This registered rent is the maximum which the landlord can recover, though if it exceeds the current rent, no increase in rent is authorised unless permitted by the terms of the tenancy. If no such application is made, the rent payable is that which is authorised by the lease or agreed by the parties.

Similarly, either or both parties to a statutory tenancy may apply to the Rent Officer to have a fair rent registered for two years. This is the maximum which the landlord can recover, though if it exceeds the current rent, no increase in rent is authorised unless permitted by the terms of the tenancy. If no such application is made, the maximum rent payable is that which was payable under the last rental period of the protected tenancy (or, if a fair rent was registered under the protected tenancy, that which was so registered).

In determining a fair rent, s.70(1) states the rent officer will have regard to all relevant circumstances (but not personal circumstances), such as:

- the age, character and locality of the property,
- state of repair of property,
- quantity, quality and condition of furniture and
- any premium which has lawfully been required or received on the grant or renewal of the tenancy.

In determining a fair rent, s.70(3) states certain matters must be disregarded, namely:

- disrepair or defects attributable to the failure of the tenant or a predecessor in title to comply with the obligations under the tenancy,
- improvements carried out by the tenant or a predecessor in title other than under the terms of the tenancy and
- improvement of or deterioration of furniture due to the conduct of the tenant, a lodger or a subtenant.

In practice, the best method of assessment is normally comparison with registered fair rents of comparable dwelling-houses. Landlords have attempted to challenge the lack of market comparables and in *Spath Holmes v Greater Manchester Rent Assessment Committee* (1995) CA and *Curtis v London Rent Assessment Committee* (1999) QBD, the courts agreed that the Rent Officer had to use comparable market rents as a starting point. This success was short-lived, however, as the 1997 Labour government limited fair rent increases in the Rent Acts (Maximum Fair Rent) Order 1999. The Order was challenged before the European Court of Human Rights in *Spath Homes v United Kingdom* (2002) ECHR under Article 1 of Protocol 1 (peaceful enjoyment of possessions), Article 6 (right to fair trial), Article 13 (right to an effective remedy) and Article 14 (freedom from discrimination, but the case was ultimately dismissed).

Housing Act 1988 (Assured Tenancies)

The majority of tenancies created after 15 January 1989 will either be an *assured tenancy* or an *assured shorthold tenancy* agreement under the Housing Act 1988. Assured tenancies were originally the default under the Act and landlords had to serve a s.20 notice to create an assured shorthold tenancy. The Housing Act 1996 amended this position and the default tenancy granted after 28 February 1997 is an assured shorthold tenancy (s.20A Housing Act 1988). The key difference is in the landlord's ability to regain possession of the property. Under an assured tenancy, the landlord must prove a ground under s.8 and Schedule 2 of the 1988 Act to regain possession, while for an assured shorthold tenancy, the landlord needs only to prove that the tenancy has come to an end and the requisite notice has been served on the tenant. The Renters' Rights Act (discussed in what follows) promises significant reform to this area of law, namely the repeal of Chapter II of the Housing Act 1988, which will mean an end to assured shorthold tenancies.

The Housing Act 1988 only applies to leases and not to licences. The tenancy must be of a dwelling-house, the purpose of the letting being residential. A dwelling-house may be a house or part of a house (s.45) and can include a flat, maisonette or even a single room (*Curl v Angelo* (1948) CA), but still not a houseboat (*Chelsea Yacht and Boat Co. Ltd v Pope* (2000) CA). The dwelling-house must be let as a separate dwelling, and the tenant or joint tenants must have exclusive possession of rooms where they conduct the main activities of living, especially cooking (although cooking facilities are not necessary: *Uratemp Ventures Ltd v Collins* (2001) HL), eating and sleeping.

In order to be within the protection of the Act, the tenant or one of the joint tenants must be an individual (i.e. not a company) and must occupy the dwelling-house as their only or principal home.

Tenancies incapable of being an assured tenancy

Under the 1988 Act, the following kinds of tenancies are incapable of being an assured tenancy:

- A tenancy entered into before 15 January 1989.
- A tenancy with a high rateable value above certain defined limits.
- A tenancy under which either no rent is payable or the rent payable falls below specified limits.
- A business tenancy within Part II of the Landlord and Tenant Act 1954.
- A tenancy under which agricultural land exceeding two acres is let with the dwelling-house.
- A tenancy granted to a student at a specified educational institution by a specified body.
- A tenancy the purpose of which is to let the dwelling-house to the tenant for a holiday.
- A tenancy of a dwelling-house which forms part of a building, where the landlord is an individual who occupies another dwelling-house in the same building as his only or principal home. The tenancy is not necessarily rendered incapable of being an assured tenancy by the presence of a resident landlord if the dwelling house forms a flat in a purpose-built block of flats. (Note that the complex provisions of Sched. I, para. 10, relating to resident landlords, fall outside the scope of this book.)
- A tenancy under which the interest of the landlord belongs to a local authority or other specified body.

- A protected tenancy (see previous discussion of the Rent Act 1977), a housing association tenancy (see Rent Act 1977) or a secure tenancy (see Housing Act 1985).

Security of tenure under the 1988 Housing Act

Like the Rent Act 1977, the Housing Act 1988 works in a similar way to provide security of tenure to a tenant. It restricts the landlords' ability to regain possession of the property without a court order (s.5) and the court will only grant such an order for assured tenancies if one of the grounds in Schedule 2 is proved. Until that happens, the tenant has a continuing right to possession. However, it is possible for a tenant to leave voluntarily by agreement with the landlord, so going to court is not always necessary.

Before taking proceedings in court, the landlord must normally serve on the tenant a notice that satisfies the requirements of s.8 of the Act. The notice must be in the prescribed form, which currently is:

- England – Assured Tenancies and Agricultural Occupancies (Forms) (England) (Amendment) Regulations 2015
- Wales – Form 3 of the Assured Tenancies and Agricultural Occupancies (Forms) Regulations 1997

The notice must set out that the landlord intends to bring possession proceedings, and it must state the landlord's grounds for possession from Schedule 2 of the Act. The notice must also indicate the earliest date on which the proceedings can begin (which must be at least two weeks from the date of service of the notice) and indicate that the proceedings will not begin later than 12 months from the date of service of the notice. Where the notice includes any of the statutory grounds 1, 2, 5, 6, 7, 9 or 16, the proceedings cannot begin earlier than the date at which the landlord could have terminated the tenancy at common law by serving notice to quit and, in any event, not earlier than two months from the date of service of the notice. If proceedings are not issued within 12 months, then a fresh s.8 notice will need to be served.

Mandatory grounds (Part I of Schedule 2, Housing Act 1988)

To obtain possession on grounds 1–5, the landlord must have notified the tenant not later than the beginning of the tenancy that possession might be recovered on the ground in question. In relation to grounds 1 and 2, the court can dispense with the notice requirement where it is just and equitable to do so.

Ground 1 – Applies either where the landlord at some time prior to the beginning of the tenancy occupied the dwelling-house as his only or principal home or where the landlord requires the dwelling-house as his or his spouse's only or principal home and the landlord now wishes to return. The landlord must have given the tenant written notice of this possibility prior to commencement of the tenancy agreement.

Ground 2 – Applies where the dwelling-house is subject to a mortgage granted before the tenancy began and the mortgagee requires vacant possession to exercise a power of sale.

Ground 3 – Applies to fixed term tenancies for a term not exceeding eight months if, at some time within the 12 months prior to commencement of the tenancy, the dwelling-house was let as a holiday home.

Ground 4 – Applies to fixed term tenancies for a term not exceeding 12 months if at some time within the 12 months prior to commencement of the tenancy the dwelling-house was let to a person pursuing or intending to pursue a course of study provided by an educational institution specified by the Secretary of State and the tenancy was granted by that institution or by another specified institution or body of persons.

Ground 5 – Applies where a dwelling-house is held for the purpose of being available for occupation by a minister of religion as a residence from which to perform his duties and the court is satisfied that the dwelling-house is required for this purpose.

Ground 6 – Applies where the landlord intends to demolish or reconstruct the whole or a substantial part of the dwelling-house or to carry out substantial works.

Section 11 provides that where the court makes an order for possession on ground 6 or ground 9 (but not any other ground), the landlord must pay to the tenant the reasonable expenses likely to be incurred by the tenant in moving from the dwelling-house. The amount will either be determined by agreement between landlord and tenant or by the court.

Ground 7 – Applies to a periodic tenancy which has devolved under the will or intestacy of the former tenant (but not by statutory succession) if proceedings for recovery of possession are commenced either not later than 12 months after the death of the former tenant or, if the court so directs, after the date on which the landlord became aware of the former tenant's death. For this ground, acceptance by the landlord of rent from a new tenant after the former tenant's death does not create a new periodic tenancy unless the landlord agrees in writing to change a term or terms of the tenancy.

Ground 7A – Inserted by the Anti-Social Behaviour, Crime and Policing Act 2014. This ground will be satisfied where one or more of the five conditions are met. These include:

- the tenant has been convicted of a serious offence,
- where the tenant has breached the provisions of an injunction under the Act,
- that the dwelling-house has been the subject of a closure order under the Act or
- an offence has been committed under the Environmental Protection Act 1990.

Ground 7B – Inserted by the Immigration Act 2016, is only applicable to property in England. It relates to tenants who become disqualified from occupying the property due to their immigration status.

Ground 8 – Covers serious rent arrears and applies when both at the date of service of the s.8 notice and at the date of the court hearing:

- At least eight weeks' rent is unpaid in cases where the rent is paid weekly or fortnightly.
- At least two months' rent is unpaid in cases where the rent is paid monthly.
- At least one quarter's rent is more than three months in arrears in cases where the rent is paid quarterly.
- At least three months' rent is more than three months in arrears in cases where the rent is paid yearly.

The rent must be lawfully due from the tenant. A tenant in receipt of a s.8 notice specifying that the landlord is relying on this ground could avoid eviction by paying off some or all of the arrears before the hearing date.

Discretionary grounds (Part II of Schedule 2, Housing Act 1988)

Ground 9 – Applies where suitable alternative accommodation is available for the tenant or will be available when the order for possession takes effect.

Part III of Schedule 2 provides that a certificate of the local housing authority for the district in which the dwelling-house is located, certifying that the authority will provide suitable alternative accommodation for the tenant by a specified date, is conclusive evidence that suitable alternative accommodation will be available by that date. Otherwise, Part III of Schedule 2 provides that alternative accommodation is suitable if:

a it will either be let on an assured tenancy (other than an assured shorthold tenancy or an assured tenancy in relation to which notice is given that possession may be required on any of grounds 1–5) or the terms of the letting will give the tenant security of tenure reasonably equivalent to that provided by such an assured tenancy;
b the accommodation is reasonably suited to the needs of the tenant and family in relation to proximity to the place of work;
c the accommodation is similar in relation to rent and extent to that provided in the neighbourhood by any local housing authority for persons whose needs in relation to extent are similar to those of the tenant and family or is reasonably suited to the means of the tenant and to the needs of the tenant and family in relation to extent and character and
d if any furniture was provided under the assured tenancy, furniture is provided which is either similar to that formerly provided or is reasonably suited to the needs of the tenant and family.

Accommodation is not suited to the needs of the tenant and family if their occupation of it would render it an overcrowded dwelling within the meaning of Part X of the Housing Act 1985.

Section 11 provides that where the court makes an order for possession on ground 6 or ground 9 (but not any other ground), the landlord must pay to the tenant the reasonable expenses likely to be incurred by the tenant in moving from the dwelling-house. The amount will either be determined by agreement between landlord and tenant or by the court.

Ground 10 – Applies where rent is unpaid on the date on which the proceedings for possession are begun and (except where the court considers it to be just and equitable to dispense with the requirement that notice be served under s.8) was in arrears at the date of the service of the s.8 notice. This is different from the more serious ground 8 as the landlord only has to show that there are some rent arrears, no minimum amounts are specified and the landlord does not have to show rent was outstanding at the hearing, only on the date of service of proceedings.

Ground 11 – Applies where (whether or not rent is in arrears on the date on which proceedings for possession are begun) the tenant has persistently delayed in paying rent.

Ground 12 – Applies where any obligation of the tenancy, other than one related to the payment of rent, has been broken or has not been performed.

Ground 13 – Applies where the condition of the dwelling-house or common parts has deteriorated due to acts of waste by or the neglect or default of the tenant or any other person residing in the dwelling-house. Where the waste, neglect or default is that of a lodger or subtenant, it must be shown that the tenant has not taken such steps as ought reasonably to have been taken in order to remove the lodger or subtenant.

Ground 14 – Applies where the tenant or any other person residing in the dwelling-house has been guilty of conduct which is or is likely to cause a nuisance or annoyance to persons residing in, visiting or engaging in a lawful activity in the locality or has been convicted of using the dwelling-house or allowing its use for immoral or illegal purposes or an indictable offence has been committed in it or in its locality. The Anti-Social Behaviour Act 2003 amended the Housing Act 1988 to instruct the court, when considering ground 14, to consider the effect that any nuisance or annoyance has had, and could continue to have, on any affected persons.

Ground 14ZA – Inserted by the Anti-Social Behaviour, Crime and Policing Act 2014. The court may give possession under this ground, where it considers it reasonable, where the tenant has been convicted of an indictable offence which took place at the scene of a riot in the UK.

Ground 14A – Applies where the landlord is a registered social landlord or a charitable housing trust and the tenant, spouse or partner has left the dwelling-house in consequence of violence or threats of violence by his/her spouse or partner towards him/her or a member of his/her family, and is unlikely to return.

Ground 15 – Applies where the condition of any furniture provided under the tenancy has deteriorated owing to ill-treatment by the tenant or by any other person residing in the dwelling-house. In the case of ill-treatment by a lodger or subtenant, it must be shown that the tenant has not taken such steps as ought reasonably to have been taken to remove the lodger or subtenant.

Ground 16 – Applies where the dwelling-house was let to the tenant in consequence of his or her employment by the landlord (or a previous landlord under the tenancy) and the tenant is no longer in that employment.

Ground 17 – Applies where the tenant or a person acting at the tenant's instigation induced the landlord to grant the tenancy by a false statement made knowingly or recklessly.

Housing Act 1988 (Assured Shorthold Tenancies)

The assured shorthold tenancy (AST) has been a cornerstone of the private rental sector since its introduction in 1988. ASTs were popular with landlords because they had significantly less security of tenure than regulated tenancies under the Rent Act 1977 or assured tenancies (discussed previously). The key provision separating ASTs from assured tenancies is the s.21 notice – the so-called 'no fault' eviction. Concerns about security and stability for renters and a rise in homelessness have led successive governments over a ten-year period to promise reforms in this area. The Renters' Rights Act has repealed Chapter II of the Housing Act 1988, abolishing ASTs. This section, therefore, sets out the provisions in place at the time of writing and is followed by a closer look at the reforms contained in the Act.

An assured shorthold tenancy (AST) could have been created either before 28 February 1997 or after that date. ASTs created before 28 February 1997 had to satisfy the requirements of an assured tenancy and the notice requirements in s.20(1) of the 1988 Act. The tenancy had to be for a fixed term of not less than six months and a landlord could not hold the power to terminate it before the expiry of those six months. A notice to that effect had to be served on the tenant prior to the beginning of the tenancy, and that notice had to be in the prescribed form.

After 28 February 1997, the Housing Act 1996 came into force, amending the 1988 Act. These amendments made ASTs the default tenancy created under the 1988 Act, so a landlord no longer needed to serve a s.20 notice on the tenant prior to the beginning of the lease, unless it fell into one of the exceptions in Schedule 2A of the 1988 Act.

The main advantage from a landlord's viewpoint is that possession of a property occupied under an AST can be obtained without the landlord proving one of the grounds set out in Schedule 2 (discussed previously). Section 21 of the Housing Act 1988 provides a 'no-fault' ground for possession where the landlord is entitled to possession of the property simply by serving a notice. But s.21 requires careful examination, as there are certain requirements, both before and during the tenancy, that a landlord must comply with for a 'no-fault' notice to be valid. It is also important to note that there has recently been some divergence on housing policy between the English and Welsh administrations and, again, readers are advised to check the position at the time of their reading.

After the Housing Act 1996, an order for possession of a property let on an AST cannot be made under s.21 of the 1988 Act any earlier than six months from the beginning of the lease agreement. In effect, a landlord could serve notice on the tenant at the beginning of an AST but could not regain possession of the property until six months later. Section 36 of the Deregulation Act 2015 amended s.21 in relation to ASTs in England. A tenant now cannot be served with a 'no-fault' notice in the first four months of an AST and proceedings for possession cannot begin if a period of six months has elapsed from the date the s.21 notice was served.

For fixed term AST agreements, the court will make an order for possession under s.21(1) where it is satisfied that the fixed term of the AST has come to an end and the landlord has given the tenant at least two months' notice in writing, stating they require possession of the property. For s.21(1) notices served in England, these must be in the prescribed form.

In the case of ASTs that are periodic agreements from the start (as opposed to ASTs that are initially fixed term and have become periodic), the landlord must rely on a notice served under s.21(4), which requires a minimum of two months' notice in writing, stating they require possession and ending on a specified date coinciding with the last day of a period of the tenancy – in effect, the end of the week, month and so on. In England, s.21(4ZA) removes the need for the date specified in a s.21 notice to coincide with the end of a period of the tenancy.

Where the court is satisfied that the correct notice has been served, either under s.21(1) or s.21(4), then it has no discretion and must make an award for possession in favour of the landlord, although it does have some discretion around the actual date of possession, especially where a tenant can demonstrate exceptional hardship.

Though it may appear s.21 grants the landlord an unrestricted power to repossess a property let on an AST, provided they comply with the relevant notice provisions, there are several important restrictions on its use.

Section 33 of the Deregulation Act 2015 prevents 'retaliatory evictions', where a tenant makes a complaint about the condition of the dwelling and the landlord issues a s.21 in response. Once the tenant has made a written complaint about the property's condition, the landlord cannot issue a s.21 notice if they do not respond within 14 days of the date the complaint was given or if they give an inadequate response (an adequate response is a description of the landlord's proposed action and a reasonable timescale: s.33(3)). A s.21 notice is also invalid if served within six months of a local authority improvement notice under the Housing Health and Safety Rating System.

Section 21A to the 1988 Act (added by s.38 of the Deregulation Act) invalidates a s.21 notice where the landlord has failed to comply with 'prescribed requirements'. In England, these requirements are set out in ss.2–3 of the Assured Tenancy Notices and Prescribed Requirements (England) Regulations 2015.

First, a landlord must ensure gas appliances are checked annually, must provide the tenant with a copy of a valid gas safety certificate (GSC) before they occupy the property, then within 28 days of each annual gas safety inspection (s.2). If the landlord has a valid GSC prior to the start of the AST but does not give a copy to the tenant, this is remediable by providing a copy of the GSC at a later date (*Trecarrell House Ltd v Rouncefield* (2020) CA). But if the landlord does not have a valid GSC *before* the tenant occupies the property, this prevents the landlord from ever serving a s.21 notice on the tenant (*Byrne v Harwood-Delgado* (2022) Luton CC).

Second, a landlord must provide a prospective tenant with a copy of a valid Energy Performance Certificate (EPC), which must be renewed every ten years (s.2) and a copy of the 'How to rent: the checklist for renting in England' either in hardcopy or via email, so long as the tenant is content to accept electronic documents (s.3).

Third, a landlord receiving a damage deposit must protect that deposit through one of the recognised tenancy deposit schemes within 30 days of receipt (s.213(3) Housing Act 2004). The penalty for non-compliance is a fine up to three times the value of the deposit. Failing to protect the deposit also forfeits the landlord's right to rely on s.21 for possession, even if they later protect the deposit (s.213 and s.215(1) Housing Act 2004). A s.21 notice is also invalid if the landlord has protected the tenant's deposit but not given the tenant details of the scheme (s.213(6) and s.215(2) Housing Act 2004). To again be able to serve a valid s.21 notice, a non-compliant landlord can repay the tenant's deposit in full or with deductions agreed between the parties (s.215(2A) Housing Act 2004) or issue the prescribed information about the tenancy deposit scheme (where the deposit was protected within 30 days of receipt) (s.215(2) Housing Act 2004). Otherwise, a landlord would need to prove one or more of the grounds for possession set out in Schedule 2 of the 1988 Act.

Renters' Rights Act 2025 – Changes to the nature of residential lettings in England

Reform is on the horizon for residential tenancies. The Act is lengthy and a complete analysis of the legislation is beyond the scope of this book; thus, this section sets out the major changes, not all of which are yet in force.

The major headline is that s.2 of the Act repeals Chapter II of the Housing Act 1988. This enacts two significant reforms to the law on residential tenancies. The first is the end of ASTs. All tenancies under the 1988 Act would become assured tenancies, with a rent period not exceeding one month, and would automatically renew on a rolling basis (clause 1). Any term of a lease stating it is for a fixed term or a term exceeding the one-month rent period would have no effect. Tenants would have the right to terminate an assured tenancy by providing a minimum of two months' notice.

The second effect of the repeal of Chapter II of the 1988 Act is the end of landlords' ability to rely on s.21 'no-fault' notices (s.2 Renters' Rights Act). As with existing assured tenancies under the 1988 Act (discussed previously), a landlord would need to serve a s.8 notice and prove one of the grounds in Schedule 2 of the 1988 Act in order to bring the tenancy to an end. Section 3 and Schedule 1 of the Act clarify and expand the grounds for possession in Schedule 2 of the 1988 Act. Proposed changes include:

- **Ground 1** – Occupation by landlord or family is amended and seeks to redefine those who would be eligible under this ground.
- **Ground 1A** – Additional ground where the landlord intends to sell the property.

- **Ground 1B** – Additional ground for possession after a rent-to-buy agreement.
- **Ground 2** – Sale by mortgagee is amended so the mortgage does not have to be granted before the beginning of the tenancy.
- **Ground 2ZA** – Additional ground for possession where a superior lease ends.
- **Ground 2ZB** – Additional ground for possession in cases where there is a superior lease of 21 years or more.
- **Ground 2ZC** – Additional ground for possession within six months of the end of a superior lease where the superior lease was held by a provider of social housing, assisted accommodation or an agricultural tenant.
- **Ground ZD** – Additional ground for possession within six months of the end of a superior lease for a fixed term of 21 years or more.
- **Ground 3** – Holiday accommodation is repealed.
- **Ground 4** – Student accommodation is amended.
- **Ground 4A** – Additional ground for possession of student accommodation for the occupation of students.
- **Ground 5** – Amended so the landlord no longer has to notify the tenant before the beginning of the tenancy that possession might be recovered for a minister of religion.
- **Ground 5A** – Additional ground for possession for occupation by an agricultural worker.
- **Ground 5B** – Additional ground for possession for occupation by a person who meets employment requirements.
- **Ground 5C** – Was formerly Ground 16 (employee of the landlord) and would become a mandatory ground.
- **Ground 5D** – Additional ground for possession for end-of-employment requirements.
- **Ground 5E** – Additional ground for possession for occupation as supported accommodation.
- **Ground 5F** – Additional ground for possession of a dwelling house occupied as supported accommodation.
- **Ground 5G** – Additional ground for possession for tenancy granted for homelessness duty.
- **Ground 5H** – Additional ground for possession of stepping stone accommodation.
- **Ground 6** – The landlord intends to redevelop the property is amended.
- **Ground 6A** – Additional ground for repossession of alternative accommodation provided during redevelopment.
- **Ground 6B** – Additional ground for possession to allow compliance with enforcement action.
- **Ground 7** – Relating to the death of a tenant is amended.
- **Ground 8** – Rent arrears is amended so that a tenant has up to 13 weeks or three months in arrears and any amount of unpaid rent arising from the non-payment of universal credit is ignored.
- **Ground 18** – An additional ground for possession where a tenant of supported accommodation has unreasonably refused to cooperate with support services.

Section 6 of the Act seeks to permit landlords to increase the rent once each year, giving the tenant at least two months' notice of the increase. Rent increases will be linked to market rates and tenants will be able to challenge excessive rent increases through the First-Tier Tribunal (Property Chamber). Sections 9 and 10 also prohibit payment of rent in advance, with the exception of the first rental period up to a maximum of one month or 28 days.

Section 12 of the Act introduces three new sections to the 1988 Act: 16A, 16B and 16C, introducing a provision (in the form of an implied term) to allow tenants to request permission, in writing, to retain a pet. A landlord would not be allowed to unreasonably withhold permission, but would be able to require that insurance covering pet damage be purchased. Insurance for pet damage can be purchased by the tenant or the landlord, who can pass on the reasonable costs, including the premium and any excess, to the tenant.

Sections 65–75 of the Act introduce a Landlord Redress Scheme, providing a mechanism for complaints made by a tenant against a member of the scheme to be independently investigated and determined, with the exact nature of such a scheme to be determined by the Secretary of State in subsequent regulations. The Act combines this with the creation of a Private Rented Sector Database of residential landlords (ss.76–97), and the Bill seeks to prevent persons from marketing or letting a residential property unless the landlord has an active entry on the database.

The Bill also addresses discrimination in the rental market, prohibiting restrictions on enquiries, viewing and tenants relating to children (s.34) and benefits status (s.35). It will also prohibit any such terms in mortgages and insurance policies and reserves a power for the Secretary of State to protect others (s.40). The Bill contains similar provisions for Wales (ss.44–46) and Scotland (ss.51–53).

Section 56 of the Act addresses the matter of 'rent bidding'. When advertising a dwelling for let, it will be required for the landlord to state a 'proposed rent' and inviting, encouraging or accepting an offer to pay rent in excess of that amount will be prohibited. Enforcement of this provision would be for local authorities, who would have the power to impose financial penalties up to £7,000.

The Bill also seeks to amend the Housing Act 2004 in respect of the decent homes standard and will also establish additional powers for the Secretary of State to set additional standards for residential property regarding the state of repair, things to be provided for use by the occupant and the means of keeping the premises to a suitable temperature.

Housing and Planning Act 2016 – recovery of abandoned premises in England

Part 3 of this Act provides for the recovery of abandoned residential properties in England let under the terms of an AST, where a number of provisions are met.

- first, the unpaid rental condition must be met,
- second, the appropriate notices must be served and
- third, there must have been no responses from the tenant to the landlords' requests for contact.

Unpaid rent – This condition is where a certain amount of rent has not been paid:

- weekly tenancies – eight weeks unpaid rent,
- monthly tenancies – two months unpaid rent,
- quarterly tenancies – one quarter's rent being in arrears for at least three months and
- yearly tenancies – three months' rent being in arrears for at least three months.

Notices – The process requires the service of three notices. The first two notices must be served on the tenant, any named occupiers and any deposit payers. They must set out:

- that the landlord believes that the property has been abandoned,
- that the landlord asks for a response from the tenant by a certain date and
- that the landlord is seeking to bring the tenancy to an end.

The second notice can only be served once the unpaid rental condition has been met. The third notice must then be attached to the property. There is the possibility for any tenant to have the tenancy reinstated if they can convince the court that they did not respond to the landlord's notice for valid reasons.

Tenancy deposits

From 6 April 2007, it has been a requirement of landlords to what is termed 'safeguard' the deposit any tenancy pays in relation to an AST (s.212 Housing Act 2004). Within 30 days of receiving the deposit, a landlord must register or safeguard that deposit with a government-approved deposit protection scheme and provide the tenant with what is described as prescribed information about the particular scheme (the Housing (Tenancy Deposits) (Prescribed Information) Order 2007). In the event of an end-of-tenancy dispute over the return of the deposit to the tenant or it being retained by the landlord, the deposit protection scheme will adjudicate and resolve the dispute based on the evidence. Prior to 6 April 2012, landlords had to protect a tenant's deposit within 14 days of receipt.

A deposit taken in respect of a fixed term tenancy before 6 April 2007 or any periodic tenancy created following the expiry of a fixed term before that date can be protected, but does not need to be. Where the fixed term was created before 6 April 2007 but expired and became a periodic term after that date, which still existed on 26 March 2016, that tenancy has to be protected (s.215A Housing Act 2004). For tenants granted an additional fixed term tenancy after 6 April 2007, then, naturally, those deposits, even if initially received before 6 April 2007, will need to be protected.

The prescribed information that the tenancy must have in their notice from the landlord includes:

- the name and contact details for the deposit scheme provider,
- the procedure for repayment of the deposit,
- the procedure to be followed if the landlord or the tenant cannot be traced and
- details of the procedure to be followed in the event of a dispute.

Also included should be:

- the amount of the deposit,
- the property address,
- the contact details of the parties and
- details of the circumstances where a landlord can retain all or some of the deposit being protected.

There are essentially two types of deposit schemes. One is a *custodial scheme* where the landlord transfers the actual deposit to the scheme provider, who then holds it on behalf of the parties until such time as the tenant leaves the property. The second is an *insurance backed scheme* whereby the landlord retains the actual deposit but pays a fee to the scheme provider.

The Tenant Fees Act 2019 bans most letting fees and caps tenancy deposits in the private rented sector in England, for new or renewed agreements signed on or after 1 June 2019.

Rent control under the Housing Act 1988

Unlike the Rent Act 1977, tenancies under the Housing Act tend to reflect the market; they are not regulated in the same way.

FIXED TERM ASSURED TENANCIES

The Act does not regulate the amount of rent which may be charged or the extent of rent increases under a fixed term tenancy. The parties are free to negotiate the rent to be paid and the tenant will be obliged to pay that rent for the fixed term, provided the lease does not allow for a review during the fixed term. When the fixed term expires, if a statutory periodic tenancy commences, the provisions of s.13 of the Housing Act 1988 (considered in what immediately follows) will apply. The s.13 procedures only apply to periodic tenancies.

PERIODIC ASSURED TENANCIES

Section 13 provides for rent increases under assured periodic tenancies (including statutory periodic tenancies). The minimum period between such increases is one year. The s.13 procedure is not applicable where the lease contains a rent review clause that is binding upon the tenant and it does not prevent the parties from reaching an agreement of their own. Section 13 also applies to periodic ASTs granted on or after 28 February 1997. If the landlord wishes to review the rent during a periodic term, then they must serve a notice upon the tenant in the prescribed form, proposing the new rent and the date from which it will take effect. This notice period is:

- for a yearly tenancy – six months,
- where the tenancy period is less than one month – one month and
- in all other cases – one period of the tenancy.

A tenant has the right to refer the matter of the rent to the Rent Assessment Committee (Wales) or the First-tier Tribunal (England) via a prescribed form and before the date upon which the new rent is due to take effect. If a tenant takes no action, then the new rent will take effect as set out in the notice. The RAC or Tribunal will consider that rent that might reasonably be expected to be paid for such a dwelling-house on similar terms to those currently enjoyed, but disregarding the effect of any tenants' improvements or any neglect permitted by the tenant. Naturally the parties remain free to negotiate a new rent for the property without reference to the RAC or the Tribunal.

Succession by tenant's spouse to an assured periodic tenancy

The succession provisions of the Housing Act 1988 do not apply to ASTs. Where the tenant under an assured periodic tenancy dies, the spouse or partner of the tenant can take over the tenancy under s.17, provided that the spouse or partner occupied the dwelling-house as his or her only or principal home immediately before the death of the tenant. This rule can only operate where the deceased tenant was not already a successor, as the Act only provides one right of succession, e.g. by virtue of being the survivor of joint tenants or by inheriting under the will or intestacy of the deceased. The landlord could then seek possession of the property under Ground 7, one of the mandatory grounds for possession under Schedule 2 of the Act.

Long leases of residential premises: Leasehold Reform Act 1967

This Act and the Housing and Urban Development Act 1993 are Acts which grant long leaseholders the right to enfranchisement by acquiring the landlord's freehold or the right to extend their lease. For a tenant to acquire the right of enfranchisement:

- they need to be the tenant of a house as defined by the Act and
- it must be a long tenancy.

The previous requirement that the tenant had lived in the property for at least two years was removed from 31 January 2025 by s.27 of the Leasehold and Freehold Reform Act 2024, so a tenant has a right to enfranchisement or extension immediately upon moving into the property.

The right of enfranchisement can also fall to both the personal representatives (s.6A Leasehold Reform Act 1967) or the resident family members of a deceased tenant (s.7), as well as to any assignees.

The Act applies to tenancies which satisfy the following conditions:

1 The tenancy must be a long tenancy. This basically means that the tenancy must be for a fixed term exceeding 21 years.
2 The tenancy must be a tenancy of a house (but does not include a flat created by horizontal division of a building).
3 The tenancy must be for a low rent (s.1). The Housing and Regeneration Act 2008 removed this test but only for leases granted after 7 September 2009, so for many existing leases the low rent condition will still apply. There are statutory formulae in place to calculate what a low rent is, with a different formula for properties let before and after 1 April 1990.
4 The rateable value of the house must fall within specified limits (if the tenancy was granted after 1 April 1990, a statutory formula is applicable).

Where the requirements of the Act are satisfied, the tenant may claim the freehold of the house (s.8) or a 50-year extension to the term of the lease (s.14) by serving notice on the landlord at any time during the lease. That notice must be in the prescribed form (Form 1, Leasehold Reform (Notices) Regulations 1997 in England and the Welsh equivalent). The effect of the notice is as if the parties had freely entered a contract of sale for the property. A tenant can also withdraw their intention, should they feel that the price eventually determined through the Act is not agreeable to them, but may incur costs of compensation towards the landlord.

Following receipt of notice from the tenant, the landlord has two months to reply, again in the prescribed format (being Form 3, Leasehold Reform (Notices) Regulations 1997 or the Welsh equivalent) and should set out if they agree or not to the tenant's claim, along with the grounds of any objection. These grounds are limited and, essentially, formulate around the tenant's eligibility, as well as objecting to the inclusion of parts of the property if that is relevant.

There are detailed provisions in s.9 of the Act setting out how the purchase price should be calculated. The first approach applies where the rateable value of the property is less than £1,000 (Greater London) and £500 (elsewhere) and the second approach applies where the rateable value is above these limits.

It is important to note that this area of law is in a state of flux as the provisions of the Leasehold and Freehold Reform Act 2024 are enacted. Though not yet in force, s.33 will uplift the extension to the term of a lease from 50 to 990 years, and s.37 will link the price payable upon enfranchisement or extension to a property's market value (defined in Schedule 4) and other compensation (restricted by Schedule 5 to the reduction of value, loss or damage to the freeholder's interests in other property).

RATEABLE VALUE LESS THAN £500 (£1,000 IN GREATER LONDON)

The valuation is based on an assessment of the amount that the premises might be expected to make if sold on the open market by a willing seller, although a number of assumptions are stipulated in s.9 of the Leasehold Reform Act 1967, one of which is that the property is subject to the existing tenancy and, furthermore, that the current tenancy has been extended, that is for another 50 years. When it comes into force, s.37 of the Leasehold and Freehold Reform Act 2024 (discussed previously) will replace the assumptions in s.9(1) to s.9(2) of the 1967 Act.

RATEABLE VALUE OF £500 OR MORE (£1,000 IN GREATER LONDON)

For properties where the rateable value is above the limits set out, the valuation tends to be higher as there is no need to assume in this calculation that the tenancy has been extended in the same way but that the tenant, upon the expiry of the term, will remain in occupation under Part 1 of the Landlord and Tenant Act 1954 or Schedule 10 of the Housing and Local Government Act 1989.

The 1954 Act essentially applies to tenancies granted prior to 1 March 1990, provided that the tenancy is a tenancy of a separate dwelling in which the tenant resides. When the term of the long lease expires, the tenancy continues until the tenant terminates it or the landlord proposes a Rent Act 1977 statutory tenancy on terms which the tenant accepts or the court determines or the court grants an order for possession, the landlord having established statutory grounds for possession.

The 1989 Act essentially applies to tenancies granted on or after 1 March 1990, provided that the tenancy is a tenancy of a separate dwelling in which the tenant, who is an individual, resides as his or her only or principal home. When the term of the long lease expires, the tenancy continues until the tenant terminates it, the landlord proposes an assured tenancy under the Housing Act 1988 or the landlord establishes a statutory ground for possession and an order for possession is granted by the court.

Where a long lease at a low rent expires on or after 15 January 1999, the provisions of the 1989 Act, and not those of the 1954 Act, will apply to it, even if the lease was granted prior to 1 March 1990.

Leasehold Reforms

The law on long residential leaseholds is undergoing a period of significant reform, so it is important to keep up to date with the changes in this area.

The Leasehold Reform (Ground Rent) Act 2022, which came into force on 30 June 2022, no longer allows ground rents to be charged for most new long leasehold residential properties in England and Wales. The provisions apply to a lease granted on or after the commencement date with a term that exceeds 21 years, where it was granted for a premium (purchase price) and where it is not an excepted lease as defined by the Act (s.1).

The Leasehold and Freehold Reform Act 2024 aims to reduce the cost of lease extensions and freehold purchases by removing any requirement to pay marriage value (the increase in property value due to the lease extension: s.37) and potentially eliminating the need for leaseholders to pay their freeholder's professional fees. Leaseholders should be able to extend their leases by 990 years (s.33) with a ground rent reduced to zero. The provisions also change the qualifying criteria to give leaseholders the right to extend their lease and buy their freehold (s.27) and take over the management of their buildings (s.49). The Act will also ban the grant or assignment of certain long residential leases (s.1). While the act has received Royal Assent (see Chapter 1), most of the provisions are not yet in force.

Further reform appears to be on the horizon. In the King's Speech 2024, the government stated its intention to publish a draft Leasehold and Commonhold Reform Bill to enact remaining Law Commission recommendations, to reinvigorate commonhold tenure, regulate ground rents for existing leaseholders, remove the threat of forfeiture and strengthen the rights of freehold homeowners.

Residential tenancies of flats: Landlord and Tenant Act 1987

Tenants' right to purchase interest of landlord where landlord is in breach of repairing covenants

Part III of the Act applies to qualifying residential tenants of flats under leases for terms exceeding 21 years. Its effect is that a majority of the qualifying tenants are entitled to compulsorily purchase the landlord's interest in premises to which Part III applies, if:

- a landlord fails to remedy a breach of his repairing or management covenants, within a reasonable period of time after receiving notice from the tenants and is unlikely to do so or
- a manager has been in place for three years (s.29).

The landlord's interest is transferred to a nominated person on terms either agreed by the parties or determined by a rent assessment committee.

Tenants' right to purchase interest of which landlord intends to dispose

Part I of the Act applies to qualifying residential tenants of flats. If the landlord decides to dispose of an interest in premises to which Part I applies, qualifying tenants are given the right to purchase the interest. Part I applies to disposals such as the sale or lease of the landlord's freehold reversion or, if they are a tenant, the assignment of their lease. It does not apply to disposals by mortgage, compulsory purchase, to family, charitable gifts, where required by an option or under a will (s.4).

Where the landlord proposes to dispose of an interest in the premises, they must notify the qualifying tenants of their intention, offering the interest to them (s.5). Failure to serve notice without reasonable excuse is now a criminal offence (Housing Act 1996). The tenants must be given time to accept the offer to purchase (s.6), which requires a majority of the qualifying tenants.

If the tenants accept the offer to purchase or the landlord accepts their counteroffer, they must be given another two months to nominate a person to take the landlord's interest. If they fail to nominate such a person within the two-month period, the landlord has 12 months to dispose of the premises for consideration not less than that specified in the notice. If the tenants do not accept the offer within the specified period, the landlord has 12 months to dispose of the premises, though the consideration must not be less than that specified in the notice.

Tenants' right to apply to the court for variation of the terms of the lease

Part IV of the Act permits a residential tenant (or tenants) of a flat (or flats) under a lease (or leases) for a term exceeding 21 years to apply to the court for variation of the terms of the lease(s). The court may order variation if grounds are established (s.35). Broadly, variation may be ordered where the terms of the lease are inadequate with regard to matters such as repair and maintenance of the building, the flat or installations, provision of insurance or computation of service charges.

Rights of tenants of flats with long leases to purchase the freehold or an extended lease (Leasehold Reform, Housing and Urban Development Act 1993)

Qualifying tenants of flats under leases exceeding 21 years may be entitled to purchase the freehold of the premises containing the flats if several conditions are satisfied and the tenants follow the detailed procedures in the 1993 Act, as amended by the Commonhold and Leasehold Reform Act 2002. The Act allows occupiers to collectively acquire the freehold of a block of flats with other leaseholders or to be granted a new lease of an individual flat. Conditions include that the premises must be self-contained, they must contain at least two flats occupied by qualifying tenants and the qualifying tenants must comprise at least two-thirds of the total number of flats.

The procedural steps that the tenants must follow include the nomination of a person to conduct the proceedings on their behalf. This 'person' is very often a company being a right to enfranchise company ('RTE company'), which is a private company limited by guarantee, its memorandum of association stating that its object is to exercise the right of collective enfranchisement and the service of notice by the RTE company upon the landlord.

The landlord may, by serving a counter-notice, dispute the tenants' right to purchase. In certain circumstances, they may be entitled to defeat the tenants' attempt to purchase by asserting their intention to redevelop.

The result of the process, if successful, is that the freehold is transferred to the RTE company, the landlord receiving an open market price, determined by reference to statutory provisions, i.e. by s.32 and Schedule 6 to the Act. The price will be the sum of the value of the freeholder's interest, the freeholder's share of any marriage value and an amount of compensation to the freeholder in lieu of enfranchisement.

Alternatively, a qualifying tenant of a flat who has a lease exceeding 21 years may be entitled to an extended lease. Again, certain conditions must be satisfied and procedures followed (e.g. notice must be served on the landlord). Again, in certain circumstances, the landlord may be entitled to defeat the tenant's attempt to purchase by asserting his intention to redevelop. The end result of the process, if successful, is that the tenant obtains a fixed term tenancy for a term of 90 years, the new lease commencing when the existing lease expires. The tenant pays a premium calculated in accordance with statutory provisions.

Section 36 of the Leasehold and Freehold Reform Act 2024 (again, not yet in force) will replace the method of valuation in collective enfranchisement and lease extension. The new method of calculation will be s.37 (discussed previously).

The right to manage: Commonhold and Leasehold Reform Act 2002

The Commonhold and Leasehold Reform Act 2002 gives tenants of flats the right to take over management of the building in which their flat is situated, without having to prove any fault on the part of the landlord. No compensation is payable to the landlord on exercise of this right. The right may be exercised if the premises consist of a self-contained building which contains two or more flats held by qualifying tenants and qualifying tenants hold no less than two-thirds of the total number of flats (s.72). A qualifying tenant is a tenant under a lease the term of which exceeds 21 years.

The procedural steps which the tenants must follow include the creation of a Right Manage Company or an 'RTM company', which is a private company limited by guarantee, and its memorandum of association must state that its object is to exercise the right to manage the premises. There are also a number of requirements regarding notice (ss.78 and 79).

A number of changes have come into effect following the Leasehold and Freehold Reform Act 2024. From 3 March 2025, buildings of up to 50% non-residential (i.e. commercial) use now qualify for right to manage claims (s.49), a landlord's costs in uncontested claims are generally non-recoverable (s.50) and right to manage applications must go to the First Tier Tribunal before they can progress to the High Court (s.52).

Council and other public sector tenants: Housing Act 1985

The Housing Act 1980, the provisions of which are now contained in Parts 5 and 6 of the Housing Act 1985, introduced the right to buy for tenants of a public sector landlord and also conferred on public sector tenants a security of tenure regime similar in nature to that enjoyed by tenants in the private sector. The Housing Act 1985 applies to a tenancy of a house let as a separate dwelling (s.79), provided that the tenant occupies the dwelling-house as his or her only or principal home (s.81) and the landlord is a local authority or other specified public body (s.80). It also applies to a licence to occupy a dwelling-house (s.79), other than a temporary licence given to an occupier who entered the premises as a trespasser.

Tenancies to which the Act applies are known as *secure tenancies*.

The Housing and Planning Act 2016 made a number of changes to the secure tenancy regime governed by the Housing Act 1985 in England. From its enactment all new secure tenancies, issued post the 2016 Act, will be fixed term tenancies for a term of between two and ten years and will be termed as 'new English secure tenancies'. Towards the end of the fixed term, landlords are required to carry out a review and must decide whether:

- to offer to grant a new secure tenancy,
- seek possession of the property but offer an alternative dwelling under a secure tenancy or
- seek possession of the property without offering another dwelling-house.

This review must be carried out while the term has between six and nine months to run and the tenant must be notified at least six months before the expiry of the fixed term. The tenant has a right to review that decision, which must be made within 21 days of the tenant being informed of the outcome of the review.

Types of tenancy to which the Act does not apply

- Tenancies for terms exceeding 21 years
- Where the tenant occupies the premises for the purposes of employment, where the local authority (or another specified public body) is the employer
- Where the house is used for temporary accommodation, because it is on land acquired for the purposes of development
- Where the house is let to the tenant as temporary accommodation whilst work is carried out on a home, if the tenant is not a secure tenant of that home
- Tenants of agricultural holdings
- On-licensed premises
- Certain student lettings
- Provision of housing for homeless persons
- Provision of housing for persons moving into the district in order to take up employment
- Tenancies where Part II of the Landlord and Tenant Act 1954 applies

Loss of security

A tenancy ceases to be secure if the tenant gives up possession of the house or sublets the entire house (s.93) or (in the case of a fixed term tenant) if he or she dies (s.90). Secure tenancies cannot normally be assigned (s.91).

Succession

The Housing Act 1985 was amended by the Localism Act 2011 to provide slightly different situations in England and Wales. In England there is no automatic right of succession for a tenant's family member (other than the spouse or civil partner). In Wales, when a secure periodic tenant dies, a spouse or a member of the tenant's family who had resided with the tenant for the 12 months before death and had occupied the house as a principal home may succeed to the tenancy (s.89). Only one succession is allowed.

Possession

Like the provisions of the Rent Act 1977 and the Housing Act 1988, the Housing Act 1985 precludes any landlord from regaining possession through the normal common law rules. A secure fixed term tenancy may end by expiration of the term or by court order in certain circumstances (s.86). For pre-2016 Act secure tenancies, once the fixed term ends, a statutory periodic tenancy will automatically arise (s.86) unless the tenant has been granted another tenancy of the dwelling-house which commences as soon as the original tenancy ends. The period of the periodic tenancy will be determined by reference to how the rent was paid under the fixed term tenancy. The terms will be a modified form of those of the original tenancy.

A secure periodic tenancy may only be terminated by court order. For post-2016 Act secure tenancies, the 'new English secure tenancies', the situation at the end of the fixed term has been outlined previously.

For secure periodic tenancies, the landlord must first serve a notice on the tenant informing them of their intention to seek possession of the property (s.83). This notice must be in the prescribed form and must specify the grounds upon which the landlord is relying for bringing the tenancy to an end. The court can only make an order if grounds for possession are established by the landlord (s.84). The grounds relied upon must be contained in a notice which the landlord serves on the tenant prior to bringing proceedings. The date specified by the notice for commencement of possession proceedings cannot be earlier than the first possible date of common law termination. The grounds are contained in Schedule 2.

On **grounds 1–8**, the court will only make an order for possession if it is reasonable to do so, the court possesses the power to adjourn proceedings, postpone or suspend an order or impose conditions in relation to rent. The grounds are as follows:

Ground 1 – The tenant has failed to pay rent or is in breach of the covenant.

Ground 2 – The tenant, or persons residing with or visiting him or her, have caused or been guilty of conduct likely to cause a nuisance or annoyance to persons residing, visiting or engaged in a lawful activity in the locality, have been convicted of using or allowing the use of the house for illegal or immoral purposes or of an indictable offence committed in its locality.

Ground 2ZA – Inserted by the Anti-Social Behaviour, Crime and Policing Act 2014 and covers tenants committing offences connected with riot.

Ground 2A – The tenant or a spouse or partner has left because of the violence of or threats of violence by his/her spouse or partner towards him/her or a member of his/her family and is unlikely to return.

Ground 3 – The house has deteriorated in consequence of acts of waste by the tenant or persons residing with him or her. If these are the acts of a lodger or subtenant, the tenant must have failed to take reasonable steps to remove this person.

Ground 4 – Furniture provided by the landlord has deteriorated in consequence of ill-treatment by the tenant or persons residing with him or her. If these are the acts of a lodger or subtenant, the tenant must have failed to take reasonable steps to remove this person.

Ground 5 – The tenant was granted the tenancy in consequence of a false statement made by the tenant intentionally or recklessly.

Ground 6 – The tenant acquired the tenancy by exchange and either paid or received a premium.

Ground 7 – The building which contains the house is mainly used for non-residential purposes, the tenancy was granted because the tenant is employed by the landlord and the nature of the tenant's behaviour or that of a person who resides with him or her has been such that, taking into consideration the use of the rest of the building, it is not proper for the tenant to remain in occupation.

Ground 8 – The tenant was a secure tenant of a different house and accepted the present house on the basis that he or she would leave when work on the original house was complete, that work now being complete.

Regarding **grounds 9–11**, suitable alternative accommodation must be available at the time when the order is to take effect. The grounds are as follows:

Ground 9 – The occupier is liable for an offence in respect of the overcrowding of the house.

Ground 10 – The landlord intends to demolish, reconstruct or carry out work on the building and cannot reasonably do so without possession.

Ground 10A – The dwelling-house is in an area that is the subject of a redevelopment scheme and the landlord intends to dispose of the dwelling in accordance with that scheme.

Ground 11 – The landlord is a charity and the tenant's continued occupation of the house would conflict with its objectives.

Regarding **grounds 12–16**, the court will only make an order for possession if it is reasonable to do so. The court has the power to adjourn proceedings, postpone or suspend the order or impose conditions in relation to rent. Further, suitable alternative accommodation must be available at the time when the order is to take effect. The grounds are as follows:

Ground 12 – The building which contains the house is mainly used for non-residential purposes, the tenancy was granted because the tenant is employed by the landlord, the employment has ended and the landlord needs the house for a new employee.

Ground 13 – The house is adapted for the disabled, no disabled person is living there and the landlord needs it for a disabled person.

Ground 14 – The landlord is a housing association or trust which only lets houses to people who are difficult to house, no such person is living in the house and the landlord needs it for such a person.

Ground 15 – The house is part of a group for people with special needs, there being special facilities in the area for such persons, no such person is living in the house and the landlord needs it for such a person.

Ground 15A – The house is more extensive than the tenant needs (England).

Ground 16 – The house is more extensive than the tenant needs (Wales).

The Housing and Planning Act 2016 inserts a new ground for possession relating to the new English secure tenancy. If, upon review at the end of the tenancy, the landlord wished not to renew the tenancy then the court will make an order for possession when it is satisfied that that the landlord has satisfied all of the procedural requirements, that the tenancy under review has ended, that proceedings were commenced before the end of three months beginning with the date on which the tenancy ended and the only fixed term still in existence is a new secure tenancy. Naturally if the court thinks that the landlord was wrong in law to refuse a new tenancy, then it can refuse to grant an order for possession.

Tenants' right to buy

In England, a person who has been a secure tenant for three years (reduced from five years by the Deregulation Act 2015, still five years in Wales) or more, may be entitled to purchase the house. The Housing Act 2004 sets out that any tenant exercising this right to purchase must retain ownership of the property for at least five years to avoid having to repay some of the discount. If they wish to sell within ten years, then they must give the original social landlord the right of first refusal to purchase.

The right to buy extends to the right to buy the freehold interest in the property or to be granted a long lease of the dwelling-house. This will depend upon the nature of the property and the landlord's interest. Usually, the right to buy extends only to the secure tenant or a person closely connected to that secure tenant. If the secure tenancy is a joint tenancy, then naturally the right belongs jointly to all of the tenants. Any tenant wishing to exercise this right must serve the appropriate notice on the landlord in the prescribed form and then the landlord has a period of four weeks to reply, accepting the right to buy or setting out the reasons why they do not accept the right to buy. Again, the landlord's notice must be in the prescribed form.

The price will be the value of the dwelling-house, essentially its market value, less the discount. The discount is dependent upon the qualifying period, which is the amount of time the tenant has been a public sector tenant. The discount will be 32% plus 1% for each complete year by which the qualifying period exceeds the minimum qualifying period, which in England is three years and in Wales five years, up to a maximum of 60% discount for a house. The duty to complete the sale of the house can be suspended if the tenant fails to pay any rent owed for a period of more than four weeks and cannot be reinstated until all payments are brought up to date.

The introductory tenancy (Housing Act 1996)

Instead of granting secure tenancies to new tenants, local authorities may now elect to grant introductory tenancies. This effectively provides a new tenant with a property on a trial basis. Essentially, an introductory tenancy is a periodic tenancy which will become a secure tenancy after a trial period of one year unless one of a number of specified events occurs within the year (e.g. if the local authority commences possession proceedings within the year and obtains a court order for possession, whether or not possession is obtained within the year). The trial period can be extended by a further six months by the landlord, but the landlord must explain that decision to the tenant. In order to obtain possession, it is not necessary to prove statutory grounds for possession, but the landlord must notify the tenant of the reasons for seeking possession. An order from the court must be obtained to remove a tenant from an introductory tenancy and the court must be satisfied that the tenant has had due notice of the reasons that they are being asked to vacate the property.

Renting Homes (Wales) Act 2016

This act came into effect on 1 December 2022 and introduced the term Occupation Contracts into residential lettings in Wales and tenants are now referred to as Contract-holders. Contracts are subdivided into secure contracts (usually with a community landlord) and standard contracts (usually with a private landlord). Standard contracts are then further divided into fixed term standard contracts and periodic standard contracts, which differ in relation to their variation, transfer and termination. Landlords must now provide six months' notice to end a contract rather than two months previously and that notice cannot be issued within the first six months of the contract.

Landlords are also able to ask for possession on the grounds of breach of contract, serious rent arrears and if a contract holder gives notice to leave but then does not do so. Under periodic standard contracts, serious rent arrears are defined as:

- Where the rental period is a week, a fortnight or four weeks, if at least eight weeks is unpaid.
- Where the rental period is a month, if at least two months' rent is unpaid.

- Where the rental period is a quarter, if at least one quarter's rent is more than three months in arrears.
- Where the rental period is a year, if at least 25% of the rent is more than three months in arrears.

A landlord under an occupation contract may make a possession claim on one or more of the estate management grounds as set out in Schedule 8 of the Act and these include:

- **Ground A** – That the landlord intends to demolish or reconstruct the building.
- **Ground B** – The property is needed as part of a redevelopment scheme.
- **Ground C** – The landlord is a charity and the ongoing occupation would conflict with the objectives of the charity.
- **Ground D** – That the dwelling is one which is suitable for disabled people.
- **Ground E** – That the landlord is a housing association, and they need the property to accommodate people who are difficult to house.
- **Ground F** – That the landlord wishes to make the property available to groups of people with special needs.
- **Ground G** – Around those succeeding to the occupation contract and the dwelling is now being more extensive than reasonably required by the contract holder.
- **Ground H** – Join contract-holders.
- **Ground I** – Other estate management reasons.

Landlords under a secure contract, a periodic contract or fixed term standard contract for less than seven years must ensure that the dwelling is fit for human habitation both at the commencement of the contract and for the duration of the contract. The landlord must also keep in repair the structure and exterior of the dwelling and the service installations in the property. Contract holders in Wales now have enhanced rights to pass on their home upon death to a spouse, adult child or carer who would inherit the occupational contract.

Tenancies held under the Rent Act 1977 will remain unaffected.

Homes in Multiple Occupation (HMOs)

There is a requirement that homes in multiple occupation (five or more people forming two or more separate households, per the Housing Act 2004, as amended), are licensed. New regulations, operative from 1 October 2018, altered and extended properties coming within the definition (Licensing of Houses in Multiple Occupation (Mandatory Conditions of Licenses) (England) Regulations 2018) and increased minimum room sizes. Breach of licensing conditions can result in a civil penalty under the Housing and Planning Act 2016 and having control or managing a qualifying property without a license in an offence under the 2004 Act.

Where a landlord is operating an HMO without a license, tenants may be able to reclaim rent or housing benefit through a Rent Repayment Order (RRO), available through the First-tier Tribunal (Property Chamber).

Discussion questions

1 What are the key differences between a lease and a licence in the context of renting property, and what are the practical implications of this distinction for both landlord and tenant?

2. How do the leasehold reform provisions, some of which are yet to be implemented, aim to change the balance of power between landlords and tenants, specifically regarding 'no-fault' evictions (Section 21 notices), and what are the potential benefits and drawbacks of these changes?
3. Discuss the implied obligations owed by a landlord to a tenant, and consider how these obligations have evolved through case law.
4. What are the legal formalities required for creating a valid lease and how might a lease that fails to meet these formalities still be enforceable?
5. What are the rules surrounding a landlord's right to access a property, and what constitutes reasonable notice for a landlord to enter a rented property for inspections, repairs or viewings?
6. How does the Landlord and Tenant Act 1954 impact commercial leases, particularly regarding the renewal of tenancies? What are the grounds on which a landlord may oppose the grant of a renewal tenancy, and do these grounds strike a fair balance between landlord and tenant interests?

13 Planning law

Emma Pierce-Jenkins

If people are to enjoy and maintain a reasonable quality of life in the place where they live, then issues relating to land use planning should obviously concern them. The very word 'planning' suggests change and the legislation and policies developed should aim to maintain and improve the environment as an attractive place in which to live and work, operating in the wider public interest.

Central and local planning responsibilities

Any coherent system of planning control is necessarily dependent on a proper administrative structure. The local government structure was established by the 1890s and the first Planning Act followed in 1909. The emphasis in that early Act was directed towards planning being used as a tool to secure better sanitary conditions, but it introduced for the first time the important issues of 'amenity and convenience'. Early planning Acts were based on the idea of local councils preparing planning schemes to control land use in suburban areas. General control of land use was introduced in 1947, creating the modern system of planning law. The main principles are now contained in the Town and Country Planning Act 1990 (the 1990 Act), the Planning and Compulsory Purchase Act 2004 (the 2004 Act), the Planning Act 2008 (the 2008 Act) and the Localism Act 2011. However, since 2011 there have been a considerable number of less significant Acts which have led to a number of more minor changes (for example, the Infrastructure Act 2015, the Housing and Planning Act 2016, the Neighbourhood Planning Act 2017 and the Levelling Up and Regeneration Act 2023). The details followed, usually within statutory orders, two of the most important of which are the General Development Management Procedure Order 2015 (as amended) and the General Permitted Development Order 2015 (as amended) both of which allow Central Government to control the Local Authority implementation of planning law. It is therefore the Secretary of State for the Ministry of Housing, Communities and Local Government, who has power over local planning authorities (LPA) – county councils, district councils, metropolitan authorities and other bodies vested with planning responsibilities, such as National Park Authorities.

What is development?

The key concept in the planning system is 'development'. It is defined by the 1990 Act, s.55 as: 'The carrying out of building, engineering, mining or other operations in, on, over or under the land, or the making of any material change in the use of any building or other

DOI: 10.1201/9781003657101-13

land'. If an activity comes within this definition, then, as a broad general rule, planning permission is required.

The definition of the term 'development' has two aspects: 'operations' and 'material changes of use'. An operation results in changes to the physical form of the land, which will have some degree of permanence. Material change of use consists of the activities done in or on the land which do not change its physical form.

Building operations

'Building' is defined to include structure or erection, but does not include the plant and machinery within the building. 'Building operations' are defined to include 'demolition of buildings, rebuilding, structural alterations of, or additions to, buildings and other operations normally undertaken by a person carrying on business as a builder'. These definitions, per s.55, at first sight seem to be virtually all-embracing, but s.55 of the Act does say that development does *not* include: 'The carrying out of work for the maintenance, improvement or other alteration of any building, being works which affect only the interior of the building or which do not materially affect the external appearance of the building'.

Engineering operations are not specifically defined in the 1990 Act except to the extent that they include the formation or laying out of means of access to a highway, whether for vehicles or pedestrians. The definition, however, specifically excludes works by a highway authority on land within the boundaries of a road for the maintenance or improvement of the road. Similarly, works by local authorities or statutory undertakers for the purpose of inspecting, repairing or renewing sewers, pipes, cables and other apparatus are excluded. Mining operations include matters such as the removal of materials from mineral-working deposits and the extraction of minerals from disused railway embankments. Other operations include matters such as installing metal grilles over the windows and doors of a shop.

Change of use

Only those changes of use which are 'material' are classed as development. There is no definition of 'material change of use' in the 1990 Act and, inevitably, what is material is a question of fact and degree. The 1990 Act aims to give some assistance by laying down certain specific instances of material change of use. These are:

1 Where a single dwelling-house is converted into two or more separate dwellings.
2 Where there is to be a display of advertisements on any external part of a building not previously used for that purpose.
3 Where refuse or waste materials are to be deposited, including extending an existing tip or raising its height above the height of the adjoining land.

Conversely, certain changes of use are specified not to involve development. These are:

1 Using buildings or land within the curtilage of a dwelling-house for a purpose incidental to the enjoyment of the house.
2 Using land for purposes of agriculture or forestry.
3 Changing between uses within the same 'use class'.

In 2020 there was a fundamental change to The Town and Country Planning (Use Classes) Order 1987 (as amended) and a streamlining of the different use classes. This included a new broader Class E. Class E now covers a wide range of commercial, business and service uses, ending the distinction between retail (previously Class A1), offices (Class B1a), food and drink establishments (Class A3), etc. This change immediately opened up significantly greater opportunities to change, for example, town centre uses, without the need for planning permission.

Class B2 and B3 remain, as General Industry and Storage and Distribution, respectively, and Class C, remains focused on various residential uses. Not all uses are specified in these Classes and if not included in Class E, B, C or F (e.g. use as a theatre, taxi business, scrapyard or for the sale of fuel for motor vehicles), they are considered *sui generis* (effectively, in a class of their own). As well as allowing land and buildings to be used for different purposes within the same use classes without constituting development, the GPDO also sets out a series of 'Permitted' changes of use that are not controlled by the Local Authority, as they do not constitute 'development'. The main expansion is the range of uses that can change to residential, without being considered as 'development', but also now a range of permitted changes are allowed where the use is *sui generis*, where previously this was not the case.

If the established use or the use for which permission was originally granted is not listed in the Use Classes Order, then its rules cannot assist and a normal planning application will be necessary. It is important to check that planning permission for the original use was not subject to conditions which override the provisions of the Use Classes Order. It is important to note that if the change of use requires new buildings or alterations or extensions to existing buildings, planning permission is still likely to be required because the building work will amount to a 'building operation'. This may then fall under 'permitted development rights'.

Where development is proposed, it will require planning permission. This can be gained through either 'permitted development rights' or the submission of a planning application, usually to the LPA.

Permission under the General Permitted Development Order 2015 (as amended)

To reduce the number of applications for planning permission made to Local Planning Authorities, to reduce applicant costs and to speed up the planning system the Secretary of State grants permission for a whole range of, generally, small-scale development and changes of use. This is covered by the General Permitted Development Order (GPDO) made by the Secretary of State, where permission is automatically given for all classes of development set out in the GPDO. Introduced in 1995, the order was subsequently amended numerous times as the government expanded permitted development rights, particularly in the period 2012–2014, but was usefully consolidated in 2015. It has, however, been subject to further expansion since 2015. Schedule 2 of the Order contains 20 parts, each containing several classes of permitted development. Some commonly encountered examples of development which are permitted by the GPDO include, for example:

1 Development within the curtilage of a dwelling-house (Schedule 2 Part 1 Class A). Where it is proposed to extend a dwelling-house, provided that the extension (and

any previous extensions, sheds or outbuildings) does not exceed 50% of the curtilage, excluding the footprint of the original house, such development is permitted. There are other conditions to be satisfied, e.g. that the height of the building when extended must not exceed the height of the original dwelling-house, the extension must not front on to a highway and must not extend more than a certain distance beyond the rear of the original house.

2. Changes of Use are covered in Part 3, which allow for a range of changes from one use class to another, including a number of opportunities to change the use of buildings to residential, e.g. office to residential and agricultural to residential (linking back to the Use Class Order 1987 but also allowing building works to be undertaken).
3. Construction of porches, garages and the installation of satellite antennae, subject to size, location and height limitations. It should be noted that the paving over of a front garden to provide off-street parking (exceeding 5 m^2) is now not covered by the GPDO unless the surface is made of porous materials or provision is made to direct run-off water from the hard surface to a permeable or porous area.
4. The installation of certain types of domestic microgeneration equipment, such as solar panels, wind turbines and ground and air source heat pumps.
5. Minor operations, e.g. erecting, constructing, maintaining, improving or altering gates, fences or walls.
6. Temporary buildings. Where building operations for which planning permission has been granted are in progress, this class allows temporary buildings needed in connection with the building work, such as site huts or workmen's lavatories, to be erected. But they must be removed at the end of the building operations.
7. Temporary uses. Land can be used for most temporary purposes for up to 28 days each year. This rule allows land to be used, for example, for an occasional agricultural show, car boot sale, point-to-point or as a site for a summer fête (provided it is not within the curtilage of a dwelling). The permission also covers the erection of movable structures (e.g. marquees) for these purposes, temporarily erected and easily moved. If the purpose is to hold a market or motor sports, temporary use is limited to 14 days each year.
8. Building or engineering operations required for agricultural purposes may be carried out on agricultural land, e.g., the erection of barns and steadings. This permission is restricted in terms of size and height limitations and distances from highways, but can be as large as 1,500 m^2 (1,000 m^2 if housing livestock).
9. Additions to industrial buildings and warehouses, limited in size and height.

It should be noted that permitted development under certain classes of the GPDO can only operate once. For example, whilst extending a house by up to 50% of the curtilage may be authorised by the GPDO, this does not authorise a further extension of the same house by another 50% of the remaining curtilage at a later date.

Even if work looks as though it is covered by the GPDO, the order includes extra restrictions in certain areas such as National Parks, National Landscapes (Areas of Outstanding Natural Beauty), Conservation Areas and World Heritage sites. In addition, the LPA may have made an 'Article 4 Direction' removing certain permitted development rights to allow the LPA to determine whether development is appropriate in the interests of good planning in a particular area. There is likely to be an 'Article 4 Direction' if the place is a conservation area or if a particular street has a special character or appearance, so it can prevent gardens from being changed to off-street parking and traditional windows from being ripped out and replaced. Landowners in such areas should have been served with a copy of the direction. When using permitted development rights, a condition imposed upon the grant of planning permission may also restrict development which would otherwise be permitted under the GPDO. Sometimes, therefore, a standard planning application may offer greater flexibility even if it does come with a higher fee.

In addition, since 2006, local planning authorities can promote 'local development orders' for their area, giving additional permitted development rights over and above those set out in the GPDO for certain types of development in certain specified areas. Although not extensively used, the Localism Act 2011 extended the use of local development orders to allow parish councils or a formally constituted neighbourhood forum to draft Neighbourhood Development Orders. These will also permit certain uses (subject to certain conditions) in some or all of the area to reduce the need for a standard planning application.

If a development proposal is not covered by the GPDO or a Local or Neighbourhood Development Order, then it will usually be necessary to apply to the LPA for planning permission.

Restrictions upon development imposed by covenants relating to freehold land or in leases

It should be noted that planning permission can do nothing to help in those cases where the type of development is prevented by a restrictive covenant (see Chapter 11) or where a tenant is precluded from carrying out the development in question by the express terms of the lease (see Chapter 12).

Making a planning application

Most applications for planning permission are made to the LPA and may be made by anyone, whether or not he or she own the land or have any interest in it. The actual owner does not need to give consent, but the applicant has to give notice of the application to owners or tenants of the relevant land (but not to tenants whose leases have less than seven years to run). The necessary forms (which are standard across the country) can be supplied by the LPA, but applications are now nearly all made online via the Planning Portal (www.planningportal.co.uk). Unless the application is very straightforward, it may be sensible to appoint a surveyor, architect, planning consultant or solicitor to make the application on your behalf, but this is not a requirement.

Since 1980, fees have been payable in respect of planning applications, and the level of these can be varied from time to time. The fee level depends very much on the type and scale of development and the type of planning application and has traditionally been set nationally. However, the Planning and Infrastructure Bill 2024 proposes to allow Local Planning Authorities to set their own fee based on the costs of the service. The application will not be considered until the fee is paid. If there is any doubt as to whether planning permission is required for a development, an application can be made to the LPA for a certificate of lawfulness of proposed use or development.

A full application will require detailed site plans and elevation drawings, as well as a range of supporting documents such as ecological assessments, drainage assessments, highways reports, etc. (these very much depend upon the site and the proposal), so that the site can be adequately identified and the development proposals fully understood. Details of application requirements can be found in the LPA's Validation Checklist or can be discussed as part of pre-application advice sought from the LPA. If granted, consent will usually be subject to a number of planning conditions, usually including that development starts within three years of the consent. Engaging professional assistance may be costly, and where there might be some doubt as to whether the proposal accords with local planning policy or there are concerns over issues that may be 'material' to the decision, a developer may simply want to discover the attitude of the planning authority in principle to what is proposed. In such a case, an outline planning application is usually more appropriate. Where outline consent is granted, a developer may still have to commission detailed plans, but can then go ahead with more confidence in the business venture, armed with the outline permission. The granting of outline consent must, however, be followed by a second application for 'approval of reserved matters' before any development can be undertaken. This requires a further application and further details to be provided on issues relating to all matters reserved (matters include access, scale, landscaping, appearance and layout), usually within three years of the outline consent. Similarly, for smaller-scale housing-led developments (no more than nine dwellings unless designated in the Brownfield Sites register), it is possible to use 'Permission in Principle', which is also a two-stage process. The initial application only considers location, land use and the amount of development, establishing the principle of development, and the second stage involves a 'technical details consent' when the details of the proposal are considered. It is the reserved matters and technical details consent stages that require applicants to produce significant supporting and technical information to support their proposals.

Once any planning application has been submitted, a copy, together with any supporting documents, will be publicised in the planning register, which is a public document available for inspection by anyone. The Development Management Procedure Order 2015 requires that all planning applications must be publicised by the LPA, usually by a site notice and/or by serving notice on the owners or occupiers of adjoining land. In some instances, a newspaper advertisement and a site notice may be required, for example, where an 'environmental statement' accompanies an application or where development involves a departure from the development plans or affects a public right of way. Where there is a major development, such as mineral extraction, a newspaper advertisement and either a site notice or the serving of written notice upon all adjoining owners and occupiers is required. The planning authority may also be required to notify specific bodies of specific types of development (e.g. the Secretary of State or a parish council). There are also specific provisions relating to the publication of certain types of development (e.g. development affecting the character or appearance of a conservation area or affecting the setting of a listed building).

The planning authority may also be required to consult particular bodies in certain circumstances (e.g. the local highway authority, where development involves the creation or alteration of access to the highway). Where a development would involve a radical departure from the local plan, the planning authority must send a copy of the application and various other documentation to the Secretary of State. The planning authority must then wait for 21 days (14 for permission in principle) before granting planning permission, though the planning authority will have to refer the application directly to the Secretary of

State if he or she uses the power to 'call in' the application. This is likely to happen only in the case of large-scale controversial development proposals.

Making Decisions on Planning Applications

Once the application, its supporting information and necessary consultation are complete, the planning authority must now make a decision on the application. This decision is usually taken by a professional planning officer working for the LPA (more than 90% of decisions) under the Council's scheme of delegation but where an application for planning permission generates a lot of public interest, the decision proposed is a departure from the development plan or the scheme is otherwise controversial, it is likely the decision will be made by a committee of elected councillors who will ordinarily meet every three to four weeks. This committee is supported by the planning officer, who will make a recommendation and is open to the public to attend and, with prior notice, address the committee. The decision will take account of the views of organisations it has consulted and comments it has received. However, the decision should be made in accordance with the development plan unless material considerations indicate otherwise. Examples of material considerations are provided in the planning practice guidance but can include decisions from previous planning appeals, national planning policy, the demands of the proposed development upon roads and other services and its impact upon adjacent land, etc., *H J Banks & Co Ltd v Secretary of State for Housing, Communities and Local Government* (2018) Admin, confirmed carbon dioxide emissions are a material consideration in planning decision-making. It is also important to recognise that the weight to be given to such material considerations lies with the decision maker.

A decision by the planning authority should be taken within eight weeks for a minor development, 13 weeks for a major development (ten or more houses or 1,000 square metres of commercial floor space) or within 16 weeks if an environmental impact assessment is required. For permission in principle for minor development, the period is just five weeks. That decision may be:

- Unconditional permission or approval
- Conditional permission
- Refusal of permission
- Refusal to make a decision

In the vast majority of cases, permission is conditional (not permission in principle), with more than 85% of decisions being to approve. Conditions should be used to help overcome concerns that may have led to refusal of an application; however, in accordance with the requirements of the NPPF (para 55), they must be necessary, relevant to planning, relevant to the development to be permitted, enforceable, precise and reasonable in all other respects. For instance, permission may be given for building operations to take place only during a limited period or subject to work being commenced within a particular period. Equally, conditions may relate to the materials to be used, the trees to be protected or the landscaping to be included. Conditions that place requirements on developers prior to commencing work should be avoided wherever possible and, where required, should be agreed with the applicant before consent is granted.

A planning authority may refuse to take a decision where an application is substantially the same as one which the Secretary of State has refused within the last two years or as one in relation to which the Secretary of State has dismissed an appeal within the last two years. The planning authority may only do so if, in the opinion of the authority, there has been no significant change in the development plan (in a sense material to the application) or in any other material consideration.

Once granted a planning permission usually has a lifespan of three years, within which time the development authorised by the permission must be commenced. A grant of outline planning permission or permission in principle requires a further application for the approval of 'reserved matters' or 'technical details' to be made, usually within three years of the grant of the original permission and work to commence within two years of final approval.

Larger proposals in sensitive locations are likely to require significantly more work by the applicant to assess the likely significant environmental effect, which must be taken into consideration in decision-making. Regulations in 1988 (Town and Country Planning (Assessment of Environmental Effects) Regulations 1988) introduced the concept of environmental impact assessment into the planning process, to implement a European Community Directive. Those regulations were replaced by new regulations in 1999 (Town and Country Planning (Environmental Impact Assessment) (England and Wales) Regulations 1999) and again in 2017 (Town and Country Planning (Environmental Impact Assessment) Regulations 2017). An environmental impact assessment is required in the case of developments such as oil refineries, chemical installations, nuclear power stations, railway lines for long-distance traffic and aerodromes with runways exceeding a specified length (as set out in Schedule 1 of the 2017 regulations). An assessment may be required in relation to developments for industries such as agriculture, mining, civil engineering, rubber, paper and textiles, if the development is likely to have significant environmental effects because of its nature, size or location (as set out in Schedule 2 of the 2017 regulations). The LPA can be asked to provide a 'screening opinion' if there is any doubt as to the need for an environmental impact assessment. If an Environmental Impact Assessment is required, the planning application must be accompanied by an Environmental Statement that assesses the likely significant environmental impacts and highlights ways in which to mitigate such impacts.

Planning obligations and the Community Infrastructure Levy

In addition to the use of planning conditions and where conditions are deemed insufficient, ss.106, 106A and 106B of the Town and Country Planning Act 1990 enable developers to enter into planning obligations. Such obligations are entered into voluntarily by an applicant in order to mitigate some of the impacts on local amenity associated with the development and enable consent to be granted where it would otherwise have been refused. Essentially, a planning obligation arises where a person interested in land enters into an obligation, usually with the LPA but possibly with another third party (such as a housing association), which:

- restricts the use of the land;
- requires specified operations or activities to be carried out in, on, under or over it;
- requires it to be used in a specific way or
- requires specified sums of money to be paid to the planning authority.

A planning obligation may be enforced by the planning authority, whether the planning obligation takes the form of an agreement between the developer and the local authority or whether it amounts to an undertaking by the developer which the Secretary of State has approved upon appeal following refusal of planning permission. It is possible to apply to the planning authority to have planning obligations modified or discharged (not usually within five years) and to appeal to the Secretary of State if the planning authority refuses to modify or discharge a planning obligation. It is not possible to appeal against the original planning obligation.

Essentially, a planning authority should only seek planning obligations which are both necessary to the granting of permission and relevant to the specific development. For example, it might be appropriate for the developer of a new football stadium to enter into an obligation to provide a specific number of car parking places on land close to the prospective stadium or to pay specified periodic sums of money to the planning authority in respect of the provision by them of an adequate car park.

In addition to planning obligations the Planning Act 2008 introduced the optional Community Infrastructure Levy (CIL). An LPA can set a rate of CIL through the development plan process, which will have to be paid on all developments that are granted planning permission, save for small-scale 'householder' development. The aim of the CIL is to cover the cost of new infrastructure required as a result of the development. This includes roads and other transport facilities, flood defences, schools and medical facilities, sporting and recreational facilities and open spaces. To avoid over-burdening applicants and developers where CIL is in place, there should be fewer requirements for Section 106 agreements, unless relating to the provision of such things as affordable housing or the transfer of land for open space provision.

Since April 2010 applications for what are known as 'nationally significant infrastructure projects' must be made to the relevant Secretary of State (via the Planning Inspectorate's National Infrastructure Planning Unit) for a Development Consent Order. As the name implies, these applications are for large-scale infrastructure projects such as railways, wind farms, power stations, airports and sewage treatment works (the list of NSIP was extended in 2013 to include certain business and commercial schemes, produced by the relevant government department). Decisions will be made by the relevant Secretary of State, following extensive pre-application consultation by the applicant and a detailed examination by a Planning Inspector who makes a recommendation to the Secretary of State. Details of relevant considerations and the weight to be given to them are clearly set out in a series of subject-specific National Policy Statements produced by the relevant Government Department for the infrastructure.

Appeals

If planning permission is refused or the applicant is aggrieved by any conditions imposed or the planning authority has failed to make a decision within, as appropriate, 8, 13 or 16 weeks, the applicant (and only the applicant) may appeal to the Secretary of State. In most cases, the decision is actually made on behalf of the Secretary of State by an inspector appointed by the Secretary of State. An appeal must be lodged within six months of the date of the planning authority's decision, other than for householder applications, where the period is 12 weeks. There are three methods for dealing with appeals – written representations, hearings or public inquiries. The Secretary of State will determine which method will be used, although he or she will take account of the parties' preferences.

1 *Written representations.* This procedure avoids the expense of an inquiry. An inspector considers written representations made by both parties and then makes a site visit. This

written representation procedure is used in the majority of cases. There is an expedited version of the written representations procedure for householder appeals through the Householder Appeals Service. No further representations are allowed beyond the documentation lodged with the initial application and such appeals should be determined within eight weeks.

2 *Hearings*. These take the form of informal discussions between the inspector and the parties following submission of written statements. Relevant third parties will usually be invited to attend, as well as the applicant, but the discussion will be led by the inspector.
3 *Public inquiry*. Inquiries tend to be formal and legalistic. The procedure is governed by detailed procedural rules. As with all planning appeals, the rules of natural justice must be observed so that both parties are given a fair hearing. The proceedings will be held in public and they give an opportunity to persons closely affected by the development to put their objections and cross-examine the other party's witnesses.

Following an appeal to the Secretary of State, it is possible for a person aggrieved by the decision to apply to the High Court for review of the decision within six weeks of its making. Such an application is made under s.288 of the 1990 Act. The statutory grounds of review are that an order or action was not empowered by the Act or that requirements of the Act were not complied with.

There is no third-party right of appeal and, therefore, if an applicant for planning permission does not appeal to the Secretary of State, a person aggrieved may, in some circumstances, only be able to challenge the planning authority's decision by applying to the High Court for judicial review. Such applications must be made promptly (which essentially means within three months or preferably sooner) and can only be made against an error in law in approving the application.

Development plans

The system of planning in the UK is often referred to as a 'plan-led' system. This is because the role of the development plan is central to the determination of planning applications. Under s.70 of the 1990 Act, local planning authorities must have regard to the provisions of the development plan when determining planning applications. Under s.38(6) of the 2004 Act, their determination must be made in accordance with the plan unless there are other material considerations which indicate otherwise.

The government made major changes to the hierarchy of planning policy when it abolished Regional Spatial Strategies through the Localism Act 2011 and introduced Neighbourhood Plans. This was followed very closely by the introduction of the National Planning Policy Framework (NPPF) in 2012, which replaced a plethora of national planning policy documents with one slim national policy document for England. The NPPF was first updated in July 2018, then again in February 2019, July 2021, September 2023 and December 2024, but has retained the presumption in favour of sustainable development. This policy document sets out key requirements for local planning authorities to follow when both making planning decisions and drafting local development plans. The TCPA 1990 requires the LPA to produce a development plan or Local Plan, and this plan must be in general conformity with the NPPF.

A Local Plan comprises a Written Statement and a Proposals/Policy Map. The written statement sets out the LPA's vision and objectives for the future 15–20 years, with a view to making sufficient land available to meet future development needs. The plan sets out a series of strategic policies for meeting the local community's economic, environmental and social aims in relation to the development and use of land. The focus is usually on meeting the need for more new homes and delivering economic growth. Since the Planning and Compulsory Purchase Act in 2004, the plan has not only be restricted to matters that involve the use of land and can be implemented through the planning system, but now include policies on issues that relate to the wider use and development of land to create sustainable communities, such as regeneration, economic development, education, housing, health, waste, energy, recycling, protection of the environment, transport and cultural and social issues.

The Proposals/Policy Map highlights areas proposed for future development (e.g. highlighting key settlements for growth and allocating specific sites for development) and areas that might be particularly sensitive to development (e.g. flood risk areas, protected landscapes and valuable nature conservation sites) and supports the written statement. Such Plans are drafted by the LPA but are examined by the Planning Inspectorate on behalf of the Secretary of State, in order to establish if the plan is 'sound'.

Since 2011 parishes and neighbourhoods have also been able to create smaller development plans known as Neighbourhood Plans. These are drafted by the community, usually led by the parish council, but where there is no parish council (as in many urban areas) a Neighbourhood Forum can be formed. The Neighbourhood Plan must conform with the strategic policies of the Local Plan and the requirements of the NPPF, but otherwise can include policies and proposals to meet community aspirations for their area, including the identification of land for future development and land to be protected from development. A Neighbourhood Plan is examined by an independent examiner and then subject to a local referendum. If more than 50% of the community supports the plan in the referendum, then it is made and becomes a formal part of the development plan for that area.

Since the Town and Country Planning Act 1990, as amended in 1991, the policies and proposals contained within the Development Plan are of fundamental importance when Local Planning Authorities make decisions on development proposals. This continued in section 38(6) of the Planning and Compulsory Purchase Act in 2004, requiring that 'where making any determination under the planning Acts, regard is to be had to the development plan, the determination shall be made in accordance with the plan unless material considerations indicate otherwise'. As a result of the Levelling Up and Regeneration Act 2023, this has been changed to give the same legal weight to the Government's National Development Management Policies, as to the Development Plan and to add 'strongly' before 'indicate otherwise' (but this is not yet in force at the time of writing). Future decisions on applications for planning permission will be made against policies in the National Development Management Policies, the Local Plan and the Neighbourhood Plan, but if they disagree, the more strategic policy will be given greater weight.

As these development plans are being prepared, they are subject to a number of stages of public consultation and Local Planning Authorities regularly work with landowners in their search for potential development sites. As these policies and proposals are a primary consideration in decisions on planning applications, it is essential that landowners and developers engage in these consultation processes to influence the final plan. Plans are regularly reviewed, at least every five years.

Planning is a devolved matter, and the position is slightly different in Wales. In Wales, there is a National Development Framework 2021, which forms part of the development plan, setting the context for local and community planning over a 20-year period. The Wales Spatial Plan also sets out national priorities, providing the context for the application of national and regional policies for specific sectors, such as health, education, housing and the economy and a broad geographical focus for development. In addition, national policy is set out in Planning Policy Wales (2018) and a series of subject-specific Technical Advice Notes (TANs). Local planning authorities in Wales produce Local Development Plans in the same way as in England, but currently do not have a Neighbourhood Plan level of development plan. These plans must conform with both the Wales spatial plan and planning policy Wales and have now largely replaced the previous unitary development plans.

In London, the Mayor has significant planning powers and is responsible for strategic planning through the mechanism of a Spatial Development Strategy (known as the London Plan). The London Plan addresses issues of strategic importance such as transport, economic development, housing, the built environment, major cultural and community facilities and the River Thames. The London borough councils must ensure that their local development plans are in compliance with the London Plan.

Planning enforcement

The planning rules need to provide for those cases where development is carried out without valid permission or where a developer breaches the conditions imposed on the permission. Breaches of the planning rules can come to light in a number of ways. Many are reported, often anonymously, by neighbours or persons affected by the development. Whilst others become obvious when a subsequent permission is applied for. Breach of the rules is not a criminal offence as such. The planning authority uses an enforcement notice where some action is necessary, but it is discretionary and needs to result in some public benefit. Before serving an enforcement notice, if evidence of a breach is not obvious, the planning authority may serve a planning contravention notice, requiring the developer to provide information about operations on the land, the use of the land, etc. It is also possible for a landowner to establish whether any existing use of a building or land or whether any operation carried out on the land, is lawful by applying to the LPA for a certificate of lawfulness of existing use or development.

It must be emphasised that, where possible, the planning authority will seek to resolve matters without the need to resort to enforcement procedures. For example, the development may be of a type which would clearly have been permitted if permission had been sought at the appropriate time. In such a case, the developer is encouraged to make a late 'retrospective' application, which will be considered in the ordinary way by the planning authority. Only as a last resort, when persuasion to remedy the breach has failed, will an enforcement notice be issued. Since the Levelling Up and Regeneration Act 2023, the LPA is able to take enforcement action for any breaches identified within ten years of them being substantially complete. Previously, unauthorised operational development or unauthorised change of use to a single dwelling-house became immune from action after just four years. It is still possible to use this 'four-year rule' where evidence shows that the development had been substantially complete before the change in legislation. It is for the developer to provide evidence that they are immune, i.e. evidence of the development being complete and/or continuous use for ten years. The only exception to this is where the breach has been deliberately concealed, in which case, using the rules introduced in 2012, there is, potentially, no period of immunity.

Where action is deemed appropriate, an enforcement notice must specify the alleged breach of the planning laws and the steps which must now be taken to remedy the breach, together with the time limit for taking the necessary steps. The measures to be taken can vary considerably. For example, where a building has been erected without permission, at one extreme, the step may consist of ordering it to be demolished and the land restored to its previous use, or at the other, it may simply require some tree planting or landscaping to disguise the building. Measures should be proportionate to the breach and local authorities should consider the least extreme actions first. The extent of action will usually reflect the level of harm that is caused by the development that has been undertaken but enforcement is also important to build trust in the planning system.

Once an enforcement notice is issued, a developer will have the opportunity to appeal to the Secretary of State before the notice takes effect (this is clearly stated on the notice and should be no less than 28 days from the date of issue). An appeal against an enforcement notice can be made on a number of grounds, e.g. that planning permission ought to be granted, the matters specified in the notice do not constitute a breach of planning laws or the time specified to take the necessary steps is unreasonably short. Section 119 of the Levelling Up and Regeneration Act means that, since April 2024, it is no longer possible to appeal on the first ground (planning permission ought to be granted) if planning permission has previously been sought and refused by the LPA or on appeal. Lodging an appeal has the effect of suspending the enforcement notice until the appeal has been resolved. This means that the operation or change of use can go on as before until the appeal is resolved. That could be disastrous when, by allowing the operation to continue, it would be impossible to subsequently correct matters. In such circumstances it may be appropriate to issue a stop notice or a temporary stop notice. Failure to comply with such a notice is a criminal offence. If an enforcement notice is not issued following a temporary stop notice or the enforcement notice that accompanies a stop notice is quashed on appeal on certain grounds, the LPA must compensate an owner or occupier of land for losses occasioned by the stop notice. This can include a contractor who has been held up on a site. In addition to the stop notice, the Planning and Compensation Act 1991 introduced the right for an LPA to apply for an injunction to restrain any actual or apprehended breach of planning control.

If an enforcement notice is upheld on appeal or if no appeal is made against it, then it must be complied with within the prescribed time limits. Non-compliance with an enforcement notice is a criminal offence, punishable by a fine. Moreover, in cases of non-compliance, the LPA may enter the land and carry out the necessary steps at the expense of the owner. Finally, it should be noted that, in respect of the breach of a planning condition, instead of serving an enforcement notice, the planning authority may serve a breach of condition notice. If the developer fails to comply with the condition within the specified time, he is guilty of a criminal offence. There is no appeal to the Secretary of State.

Planning and the preservation of amenity

A vital part of the planning process is the preservation and enhancement of amenity. That term is not defined by the planning legislation, but the dictionary defines it as 'pleasantness, as in situation and characteristics'. The ordinary planning rules can do much to preserve amenity by refusing permission in appropriate cases or attaching conditions to preserve or improve amenity in others. Additionally, certain specific forms of control are particularly geared to protect amenity. The LPA can serve an amenity notice (s.215 Town and Country Planning Act 1990) where they feel an untidy site is having an adverse effect on local amenity, such a notice will set out what is required to tidy the site. Where such work is not undertaken, a fine can be applied and/or the LPA can enter the land to tidy the site at the owner's expense. There is a right to appeal such a notice.

Trees

Local planning authorities may make tree preservation orders in the interests of amenity. These may be designed to protect individual trees or areas of woodland. In such cases, once an order is made, permission must be sought from the LPA to fell or lop a tree (unless the tree in question is dead, dying or dangerous, in which case emergency work can be undertaken). Interference with protected trees is a criminal offence, punishable by a fine. If such permission is granted, the LPA will often require replacement planting. Ordinary planning permission which has the potential to damage or destroy trees, may be granted subject to the preservation of trees on the site or subject to the planting of additional trees.

Before felling or lopping a tree in a conservation area (see later on for details on conservation areas), the LPA must be notified, even if the tree is not protected by a tree preservation order. The LPA can consent to the works, make a tree preservation order to protect the tree or do nothing. Under the last option the works to the tree can be carried out six weeks after the date of the notice to the LPA.

Hedgerows

Subject to exceptions, it is necessary to serve a hedgerow removal notice on the planning authority before removing a hedgerow to which the Hedgerows Regulations 1997 apply (e.g. a hedgerow in or adjacent to agricultural land with a continuous length of 20 metres or more unless within the curtilage or forming part of the curtilage of a dwelling-house). Within 42 days the planning authority may serve a hedgerow retention notice if the hedgerow is an important hedgerow (as defined by the regulations), thus preventing its removal. An appeal against such a notice may be made to the Secretary of State. Removal of a hedgerow in breach of such a notice is a criminal offence. Where hedgerow removal is part of the granting of planning consent, these regulations do not apply, but it is likely that replacement planting will be required via a planning condition.

Listed Buildings and Conservation Areas (Planning (Listed Buildings and Conservation Areas) Act 1990)

Buildings of special interest are usually referred to as 'listed buildings'. They are listed by the Secretary of State for Culture, Media and Sport after consultation with experts such as Historic England to determine if they are of special architectural or historic interest. There are three grades of listing to indicate a building's relative importance: Grade I, Grade II* and Grade II. Buildings that are listed are often proposed to the Secretary of State by the LPA. The owner of the building has no formal right to object to its listing. He or she may know nothing of a decision to list the building until notified by the LPA. The listing will also be registered in the local land charges register. The LPA may also seek to protect a building which they consider is of special architectural or historic interest and is in danger of being demolished or substantially altered by the use of a 'building preservation notice'. This gives the authority six months to initiate the procedures for listing with the Secretary of State.

If a building is listed, any work (internal or external, whatever the grade of listing) to alter or demolish it must have 'listed building consent', an application for which should be made to the LPA. Demolishing or altering a listed building without consent may amount to a criminal offence. Further, the planning authority may serve a listed building enforcement notice, requiring the restoration of the building or the alleviation of the effects of alteration. If such consent is refused or is conditional, the owner can then appeal to the Secretary of State and argue *inter alia* that the building should never have been listed. This argument is significant because the owner had no right initially to object to the listing. Many owners do not welcome listing of their property, not only is the planning control much more rigorous, but it may also cause considerable loss if it was hoped to sell the building to realise its development value. This was the problem facing the parties in the case of *Amalgamated Investment Co. Ltd v John Walker and Sons Ltd* (1977), where a property with development potential was agreed to be sold for £1.7 million. The building on the land was listed before the sale went through and the value of the land then slumped to £250,000! However, the hope is always that new uses can be found for old buildings, as this is often the best way to ensure their conservation.

Where listed building consent is refused or is conditional, an owner may serve a 'listed building purchase notice' on the planning authority, requiring them to purchase the listed building if he or she can show that it has become incapable of reasonably beneficial use. Planning authorities also possess limited powers to repair listed buildings at the expense of the owner and, if authorised by the Secretary of State, to acquire such buildings by compulsory purchase.

The concept of Conservation Areas stems from the Civic Amenities Act 1967. Prior to 1967 the emphasis was on the preservation of individual buildings, not areas. Local planning authorities are now required to determine those areas which are of special historical or architectural interest, where it is desirable to preserve or enhance their appearance. Effectively these are areas which are rich or potentially rich in listed buildings or of scenic beauty. Where an area is to be designated as a conservation area, the LPA must give notice in the local press and enter appropriate notices in the land charges register. Once a conservation area has been designated, a conservation area appraisal should be undertaken from time to time by the LPA, to ensure that planning policies and decisions are suitably protecting and enhancing the conservation area.

When the LPA is considering designating a conservation area, it will take account not only of individual buildings but also groupings of buildings, their relationship to each other and the quality and character of the space between buildings. The group of houses and shops in a marketplace of a small town may not be particularly worthy of note individually, but the overall impact of the area may be sufficiently pleasing to warrant creating a conservation area. New development within a conservation area is not forbidden, but in determining whether to grant planning permission, the planning authority must pay special attention to the desirability of preserving or enhancing the character or appearance of the area. This, in turn, is likely to result in higher standards of design and materials, as well as careful consideration of the siting of any new buildings. Further, demolition in a conservation area may require conservation area consent and fewer or more stringent permitted development rights exist.

Advertisements

The Secretary of State can, by regulations (Town and Country Planning (Control of Advertisements) (England) Regulations 2007), control the display of advertisements so far as it appears to be expedient in the interests of amenity or public safety. This is generally designed to control the display of outdoor advertising, which includes any model, sign, placard, board, notice, device or representation, whether illuminated or not. Various classes of advertisements are excluded from the regulations, however (e.g. traffic signs). The power of the planning authority relates to issues of amenity (e.g. historic and architectural considerations) and public safety (e.g. road safety); it could not therefore refuse permission on grounds of social desirability.

Under Regulation 1(3) of the Regulations, certain advertisements are exempted from control. These include advertisements on vehicles and advertisements incorporated into the fabric of a building or displayed inside a building.

Under Regulation 6, deemed consent is granted for such advertisements as the functional advertisements of local authorities and statutory undertakers, nameplates for businesses or companies and temporary advertisements, including signs relating to the sale or letting of premises. If an advertisement is not exempt from control under Regulation 3 or there is no deemed consent under Regulation 6, planning permission is required from the planning authority, though there is a right of appeal to the Secretary of State. Displaying an advertisement in breach of the Regulations is a criminal offence. Moreover, the LPA may require the removal of such an advertisement or may remove such an advertisement.

Financial problems – compensation and betterment

When a planning decision has been made (e.g. to create a new motorway or build a new airport runway), there will be owners of property adjacent to the development whose property needs to be acquired by compulsory purchase for the purposes of the development. For some owners, the value of their property will be significantly reduced by the development, while for others, the value of the property will be increased by the planning decision.

Local authorities and development corporations have the power to acquire land compulsorily for planning purposes. In such a case the appropriate authority or corporation will have initiated the necessary legal steps to acquire the land. There are two situations, however, where a landowner can initiate proceedings to force the local authority to purchase the land:

1 *Purchase notices.* Where planning permission has been refused or made subject to onerous conditions, an owner may argue that the land is incapable of reasonably beneficial use. He or she may then serve a purchase notice on the local authority, requiring it to buy the property.
2 *Blight notices.* Where an owner is unable to sell the land at a fair market price because the prospect that it may be compulsorily purchased (as revealed by a development plan) has rendered it virtually unsaleable, he or she may serve a blight notice on the appropriate public authority requiring it to buy the land.

Further, where a revocation or modification of planning permission results in wasted expenditure, loss or damage, compensation may be payable.

Discussion questions

1. **The Role of the Local Plan:** How does a local authority's Local Plan influence decisions on planning applications, and to what extent should it be strictly adhered to versus allowing flexibility for unique cases?
2. **Permitted Development Rights:** What are the advantages and disadvantages of permitted development rights, and how do they balance private property rights with broader public planning interests?
3. **Material Considerations:** What factors are considered 'material' in planning decisions, and how should local authorities weigh conflicting material considerations, such as economic benefits versus environmental impacts?
4. **Planning Appeals:** How effective is the appeal process in ensuring fairness and accountability in planning decisions, and what challenges exist for both applicants and local authorities?
5. **Enforcement of Planning Controls:** What are the key powers available to local authorities to enforce planning law, and how can enforcement action impact both developers and communities?
6. **Public Participation:** How important is public consultation in the planning process, and what mechanisms could improve community involvement in shaping local development?

14 Highways

Mark Allison

Even a short journey by road is enough to demonstrate the complexity of highway law. One might encounter roadworks, diversions for repairs or signage indicating the construction of a new route. Overhanging branches may present a hazard; road surfaces may be worn or uneven; construction materials might spill into the carriageway; demonstrators could obstruct traffic; animals may stray onto the road; or snow and ice might make conditions treacherous. Such incidents can lead to inconvenience, but also injury, loss or delay. Who bears responsibility? Against whom can a complaint be made? And when might a claim for damages succeed? These are the kinds of questions explored within the legal framework governing highways.

Definition

The first question to be posed is, what is a highway? Simply, the answer is any portion of land over which the public has the right to pass and repass. While many might imagine a highway as a busy urban road full of motor traffic, the legal definition is significantly broader. Some highways have existed for over a thousand years, and the law governing them has evolved over nearly as long; a fact which helps explain its complexity.

Most of the relevant statutory law on the point is now contained in the Highways Act 1980 (the 1980 Act). The main aim of the Act was to bring together many rules previously contained in Highways Acts passed between 1959 and 1971, and in the Private Street Works Act 1961, but account was also taken of recommendations made by the Law Commission. The enormity of the subject can be grasped when it is seen that the 1980 Act runs to 345 sections and 16 supporting Schedules.

To clarify terminology, highways are subdivided into various categories:

1 *Footpath* – a highway over which the public has a right of way on foot only.
2 *Footway* – a path forming part of a highway that is also a carriageway (see later on), where the public has a right of way on foot only. A footway is what is commonly referred to as the pavement.
3 *Carriageway* – a highway over which the public has a right of way for the passage of vehicles, i.e. what is commonly called a road.
4 *Public bridleway* – a highway over which the public has a right of way on foot or on horseback.
5 *Restricted byway* – a highway over which the public has a right of way on foot, horseback or in non-mechanically propelled vehicles (i.e. horse-drawn carriages and bicycles).

DOI: 10.1201/9781003657101-14

6 *Byways open to all traffic (BOAT)* – a highway over which the public has a right of way for all types of traffic, though it is predominantly used in the manner of a footpath or bridleway. BOATs are often used by 4×4 vehicles and trail bikes and are sometimes referred to as 'green lanes'.
7 *Cycle track* – a way constituting or comprised in a highway over which the public has a right of way on pedal cycles (not being motor vehicles) with or without a right of way on foot.
8 *Walkways* – a footpath created in pursuance of an agreement between a local highway authority or district council and a building owner to provide ways over, through or under the building 'for the dedication . . . of those ways as footpaths subject to such limitations and conditions, if any . . . as may be specified in the agreement'. (This right of way may apply to shopping centres. The reference to 'limitations and conditions' allows for the centres to be closed at night without infringing any right of way.)
9 *Toll roads* – toll roads have not existed in this country since the abolition of turnpikes in the late nineteenth century. The first modern toll road, the Birmingham Northern Relief Road (M6 Toll Road), opened in December 2003.

Certain obvious questions arise in connection with highways:

- How does a highway come into existence?
- Who owns the highway?
- Who is responsible for the upkeep of the highway?

Acquisition

There are two ways in which a highway can be created: by dedication or by statute.

Dedication

A highway can come into existence if the owner of the land concerned dedicates the right to cross it to the general public, which right is then accepted by the public actually using it. Dedication sounds like a very formal act, but in fact it can occur impliedly. Several points about dedication should be noted:

1 The owner of the land must intend to dedicate it as a highway. This intention is shown in the case where a formal dedication is made, e.g. if a formal document is used. Often, the intention will be implied from the circumstances, e.g. people are passing and repassing over the land, and the owner takes none of the steps available to stop the creation of a highway (see what follows).
2 The highway must be dedicated to the public generally.
3 The person dedicating the land as a highway must be granting a right for all time. Effectively, therefore, only an outright owner of land can dedicate.
4 The right to pass and repass must be accepted by the public actually using the highway.

Some assistance in proving the existence of a highway created by dedication is given by statute. Section 31 of the 1980 Act provides that where a way has been enjoyed by the public 'as of right' (i.e. without force, secrecy or permission), without interruption for 20 years (this does not apply to use by mechanically propelled vehicles), then it is deemed to be dedicated as a highway unless the landowner can demonstrate a lack of intention to dedicate. This is a rebuttable presumption.

To rebut this presumption the owner could evidence that notices had been displayed to that effect, e.g. 'No public right of way' or 'This land is not dedicated as a public highway'. Alternatively, the owner can, under s.31(6) of the Act, lodge a map of the land with the local authority showing the ways (if any) which he or she admits as being dedicated as public highways and can thereafter make a statutory declaration at any time within ten years of lodging the map (or a previous declaration) that no other ways over the land have been dedicated.

The rule created by s.31 is important because it places responsibility on the owner to show that there is no highway, rather than on the public to show that there is. Even if a 20-year period has not elapsed, it would still be open to a member of the public to try to prove the existence of a highway. It would then be necessary to prove that the owner intended to dedicate it. This might be possible, for example, by showing that the owner had allowed repairs to be undertaken by the highway authority. Merely because a highway is acknowledged to exist by the landowner, it does not necessarily follow that it is maintainable at public expense (see later on).

The most common method of dedicating a new highway is under s.38 of the 1980 Act. A developer, usually of a new housing estate or industrial/retail park, will enter into a 's.38 Agreement' with the local highway authority to build the roads in the development to the standard specified by the authority. After a short period, usually one year, the highway authority agrees to adopt the roads as highway maintainable at the public expense (see what follows for details).

Creation of highways by Act of Parliament

The main rules are now consolidated in the 1980 Act, s.24. Under its provisions, highway authorities may construct new highways, including particular types such as motorways, where access and use are restricted to certain classes of traffic.

It is also worth noting that strategic road projects may now proceed under the Planning Act 2008 as Nationally Significant Infrastructure Projects (NSIPs). In such cases, development consent is granted through a Development Consent Order (DCO) rather than a traditional Act of Parliament. These powers are typically used for major infrastructure such as new motorways, trunk roads or significant improvements to the strategic road network and are usually promoted by National Highways (formerly Highways England).

Ownership

Although the highway authority, when building new roads, may have acquired the land on which the highway is built, it is quite usual for the land under a highway to remain in the ownership of a private person. Often the landowners on either side of a highway own the land to the centre line of the road. Naturally, the use to which they can put such land is very limited, being subject to the public's right of passage and the statutory powers of highway authorities and utility undertakers to carry out works under the Highways Act 1980 and New Roads and Street Works Act 1991, respectively. At common law, owners did have rights, such as to tunnel under the land, but many of these have been eroded by legislative controls. Ownership of a highway may still be important, however, particularly where it is not maintainable at public expense (see later on) or in the context of access rights, development restrictions or compensation where a highway is stopped up.

Upkeep of the highway

The position with regard to repair and maintenance of highways is more easily understood if a brief mention is made of the historical development of the rules. At common law (before the introduction of any statutory changes), the duty and expense of repair rested with parishes (*Steel v Prickett* (1819) KB), of which there were originally more than 10,000. The drawbacks of such a fragmented system were obvious, especially once methods of communication and transport were quicker. Changes made by statutes gradually produced the result that 'highway authorities' were created with responsibility for repair and maintenance. Today the position is that highway authorities are either:

1 The Secretary of State for Transport or
2 County councils, London borough councils and unitary authorities.

The responsibility of the Secretary of State is for trunk roads, i.e. the principal roads constituting the national system of routes for through traffic in Great Britain. These roads are now managed by National Highways, formerly Highways England. However, the work of repair and maintenance is commonly delegated to the councils acting as agents. The councils are the highway authorities responsible for all other roads maintainable at public expense. In turn, repair work is commonly delegated by a county council to district councils.

Section 36 of the 1980 Act deals with the question of which highways are maintainable at public expense. Effectively, these are:

1 Those highways in existence before 31 August 1835.
2 Highways created since 1 September 1835 which have been adopted by the highway authority. Adoption can occur in a number of ways, e.g. by the dedicator serving notice on the highway authority which is accepted within the terms of s.37 or by agreement (s.38). Most important, however, are the adoption procedures with regard to private streets provided for by Part XI of the 1980 Act.

By s.41 the local highway authority is under a duty to maintain the highway. This means, according to *Rider v Rider* (1973) CA, that an authority must

> reasonably maintain and repair the highway so that it is free of dangers to all users who use that highway in the way normally to be expected of them – taking account of the traffic reasonably to be expected on the particular highway.

The duty is not one to ensure that a highway has to be free from surface water, snow and ice at all times or take preventative steps to stop ice from forming or snow from settling. Rather, it is limited to the taking of reasonable steps such as preventive or clearance measures which are sufficient to keep the surface reasonably safe (*Goodes v East Sussex County Council* (2000) HL). However, it is to be noted that if surface water or ice results from a design defect or inadequate drainage, then there may be a breach of duty.

Recent case law confirms that whether the authority has breached s.41 will depend on the adequacy of its inspection and maintenance regime, and the reasonableness of its response to known or foreseeable risks. In particular, the courts have reinforced that claims are highly fact-sensitive (e.g. *Brewster v Oxfordshire County Council* (2020) QBD; *Rogerson v Bolsover District Council* (2019) CA).

A highway includes any verge, and a difference in height between the carriageway and verge could, in the event of an accident, lead to a finding that the highway authority had breached their duty of care (*West Sussex County Council v Russell* (2010) CA).

There is also a duty to ensure that trees do not pose a danger or obstruction, requiring a system of inspection by either private landowners or local authorities, depending on the position of the tree. In 2002 Birmingham City Council was convicted of Health and Safety at Work, etc. Act 1974 breaches for failures in tree maintenance when three people were injured by a diseased ash tree. In more recent authority, the Court of Appeal reaffirmed that liability may arise where there has been a failure to maintain a reasonable system of tree inspection and risk management (*Witley Parish Council v Cavanagh* (2018) CA).

Standard of the highway

When considering the state of repair of a highway, all highway authorities should bear in mind the words of Lord Atkinson in *Sharpness New Docks & Gloucester and Birmingham Navigation Co. v Attorney General* (1915) in the House of Lords,

> it is the duty of the road authorities to keep their public highways in a state fit to accommodate the ordinary traffic which passes, or may be expected to pass, along them. As the ordinary traffic expands or changes in character, so must the nature of the maintenance and repair of the highway alter to suit the change.

In fact, most highway authorities will wish to improve highways under their care in so far as their budget allows them to do so. The 1980 Act makes extensive provision for such improvements. Part V of the Act contains rules, e.g. with regard to dual carriageways and roundabouts, cycle tracks, footways, guardrails, refuges, subways and footbridges, levelling of highways, improvement of corners, fencing and lighting of the highway and roadside planting.

Where a person is affected by a failure to repair the highway, he or she may wish to exercise the enforcement procedure provided by s.56 of the 1980 Act. To use this, a notice is served on the highway authority (or other party alleged to be responsible for maintenance) requiring the repair to be done. If the highway authority disputes liability, the case is referred to the Crown Court, which may order the highway to be put in proper repair within a specified period. In assessing whether a highway meets the requisite standard of repair, the court will consider the character and ordinary use of the highway (*Kind v Newcastle upon Tyne City Council* (2001) Admin, illustrates the application of these principles). If the authority admits liability but disputes the extent of the repairs necessary, the matter is settled by the magistrates' court.

The failure to repair the highway may have led to injury or damage being caused. Until 1961, it was very difficult to bring a civil action for damages against a highway authority, as the rule was that the authority owed no liability for injury caused by failure to repair, only for injury caused by negligent carrying out of the repairs. That rule is now changed; the present law is that the highway authority can be fully liable for injury caused by its failure to maintain or repair the highway. However, by way of defence, the authority may plead that it took such care as was reasonably required to secure that the highway was not dangerous to traffic (s.58). Naturally, this will involve factors such as the type of highway involved, the usual nature of the traffic using it and the state of knowledge of the highway authority.

Recent judicial decisions have further clarified the operation of the s.58 defence. In *Rogerson v Bolsover District Council* (2019) the Court of Appeal emphasised that a highway authority must show it had in place a reasonable system of inspection and maintenance tailored to the nature and use of the highway in question. Similarly, in *Brewster v Oxfordshire County Council* (2020), the High Court held that an authority's failure to detect and respond to a reported defect within a reasonable time could undermine reliance on s.58. In both cases, the courts stressed that what is 'reasonably required' turns on factors such as the character of the road, the volume and type of traffic and the foreseeability of risk. Each claim is fact-specific, and success for the claimant or defendant will depend heavily on evidence of the authority's inspection regime, its actual or constructive knowledge of the defect and the promptness of remedial action.

People frequently seek to bring actions under these rules when they have been injured as pedestrians by a fall on an uneven or broken pavement. As the Court of Appeal held in the leading 'tripping' case of *Winterhalder v Leeds County Council* (2000) CA:

> It is quite impossible and very misleading to state that a depression or gap or a dip of any particular dimension forms a danger to pedestrians. Any case such as this is fact sensitive and will be resolved on the basis of its own facts.

An interesting example of the application of s.56 can be seen in *Wentworth v Wiltshire County Council* (1993) CA. A farmer had a long-running dispute with the council about the status of the road to his farm, which he argued was a highway maintainable at public expense. The council did not agree. After the dispute had raged for a considerable number of years, matters came to a head when the Milk Marketing Board withdrew the tanker from his farm because of the state of the road. On application to the Crown Court, the farmer was successful in obtaining a s.56 order, under which the council was given nine months to effect the repairs. Unfortunately, as a result of the withdrawal of the milk tanker, the farmer had to abandon his dairy business. He now seeks financial compensation. The High Court refused to make such an order, as he had a remedy under s.56, which he ought to have pursued earlier, and his losses in relation to his dairy business were economic loss.

Heavy traffic

The damage caused to highways is often due to excessively heavy traffic. It seems very unjust that the public should bear the cost of repair and maintenance when the damage is perhaps caused by the vehicles of one person, company or organisation. Rules to redress this injustice are contained in s.59 of the 1980 Act. This provision allows a highway authority to recover 'excess expenses' for maintaining the highway from any person who causes extraordinary traffic or traffic of excessive weight to pass along it. The excess expenses are those incurred over and above the average expenses for maintaining that or a similar highway, which have resulted from damage arising from the extraordinary traffic.

Where a person knows that he will be operating traffic likely to cause such damage (for example, in the course of major construction or mineral extraction), he can agree in advance on a payment to the highway authority. The advantage of this is that the payment will be a fixed amount, whereas if the operator waits for recovery procedures (proceedings may be taken in a county court) to be used by the highway authority, the 'excess expenses' are an unknown quantity.

The terms 'excessive weight' and 'extraordinary traffic' are not defined in the Act, but have been considered in case law, which suggests that frequency, volume and type of vehicles involved, as compared with normal use of the highway, are all relevant factors. For instance, regular use by heavy goods vehicles in an otherwise quiet rural lane may constitute extraordinary traffic. Case law indicates that this is a fact-sensitive inquiry and what qualifies as 'extraordinary' will depend on the nature of the road and its expected use.

Given the weight and frequency of construction traffic, this section may be particularly relevant to developers and contractors, who are advised to address such risks with the highway authority at the outset of works.

Streets and street works

Although the word 'street' may conjure up a mental picture of a road with buildings on either side of it, the definition provided by the 1980 Act is more vague. 'Street' is given the same meaning for both the 1980 Act and the New Roads and Street Works Act 1991 (the 1991 Act) and includes, irrespective of whether it is a thoroughfare, any highway, road, lane, footway, alley or passage, square or court and any land laid out as a way whether it is for the time being formed as a way or not. The 1991 Act goes on to define a street as including one that is not a highway maintainable at public expense.

In order for a private street to be 'adopted', i.e. to become maintainable at public expense, street works as specified by the 1980 Act must be carried out, and the adoption procedure under s.228 applied. Under this section, the street works authority may declare the street adopted once the works are completed, though objections may be raised by a majority of frontagers. Where objections are made, the authority must apply to the magistrates' court for an order authorising the adoption. These rules are important to landowners who have undertaken the necessary work and now seek to transfer responsibility for repair and maintenance. This is particularly significant for estate developers.

Street works carried out under the 1991 Act are now subject to increased coordination requirements, supported by the Department for Transport's Street Manager digital platform, introduced in 2020. This system enables better management and notification of works, particularly between local authorities and statutory undertakers, in line with the duty to coordinate imposed by the Act.

The Private Street Works Code

This Code (ss.205–218 of the 1980 Act) applies to a private street, meaning a street that is not a highway maintainable at the public expense. It allows Street Works Authorities to carry out work on such streets and recover the expense for so doing. Street Works Authorities (generally, outside London, these are the county councils or unitary authorities) and inside London the relevant London borough council have the power to execute street works in private streets (s.205). Where this power is used, it is generally referred to as the Private Street Works Code.

This involves the council preparing:

- A specification of the street works.
- An estimate of the probable expense.
- A provisional apportionment of those probable expenses between the premises liable to be charged – this means premises fronting the street. The definition of 'fronting' in the 1980 Act includes premises that adjoin the street and therefore premises with side or rear frontages that have no access to the street are liable to be charged.

Once the Street Works Authority approves these details, it must:

1. Publish notices containing relevant details in local newspapers.
2. Post notices in the street affected.
3. Serve individual notices on the owners of the premises liable to be charged for the work, stating the sum provisionally apportioned to their premises. Generally, this will be decided according to the length of frontage of the premises to the street, but s.207 provides for the provisional sum calculated to be increased or decreased, depending on factors such as the degree of benefit to be derived by the owners of the premises and any work already carried out by the owners.

Once all the required notices have been given, the owners of the premises affected may wish to object. Section 208 sets out possible grounds of objection. These include arguments that:

- The street is not a private street.
- The estimated expenses are excessive.
- The proposed works are unreasonable or insufficient.
- The provisional apportionment is unacceptable.

Hearings of these objections take place before the magistrates' court, which has the authority to quash the proposals in whole or in part or amend them.

Once any objections are resolved and the work is completed, the Street Works Authority can then recover the costs in the agreed proportions from the owners, who may be given up to 30 years to pay. The actual amount due becomes a charge on the premises. This means that the agreed cost passes as a liability to any new owner of the premises.

Once the street works have been executed, it is then possible to 'adopt' the street and make it a highway maintainable at public expense under s.228. The Street Works Authority does this by displaying a notice in the street declaring it to be a highway maintainable at public expense. If no objection is made within a month, the street becomes such a highway. But a majority of owners in the street may object and in that case, the Street Works Authority must apply to the magistrates' court for an order overruling their objections, before the street can be turned into a highway maintainable at public expense.

Advance Payments Code

The standards imposed by Street Works Authorities in the making up of streets mean that this work can prove very costly. The authority operating under the Private Street Works Code is responsible for recovering the sums due after completion of the work and money may be outstanding for long periods. The drawback of this system can be avoided in cases where the Advance Payments Code applies.

The purpose of this code (ss.219–225 of the 1980 Act) is to secure payment of the expenses of executing street works in private streets adjacent to new buildings. Section 219 provides that where (a) it is proposed to erect a building for which plans must be deposited with the local authority and (b) the building fronts on to a private street where the Street Works Authority could execute street works under the Private Street Works Code, then no building can take place until the owner of the land has paid or secured the payment of the sums required to the Street Works Authority under the Advance Payments Code. The sanction for non-compliance is a fine.

However, no offence will be committed if the Street Works Authority does not notify the developer of the sum they are required to pay within six weeks of the plans being deposited. Section 219, however, is subject to a number of important exceptions, e.g. if the building to be erected is in the grounds of an existing building or where an agreement under s.38 has been made with the Street Works Authority, whereby the person undertaking street works at his own expense will, on completion of the works, dedicate the street as a highway.

If the Advance Payments Code does apply, then the Act provides the machinery for establishing the amount to be deposited – this is 'such sum as, in the opinion of the Street Works Authority, would be recoverable under the Private Street Works Code if the Street Works Authority were to carry out the work'. Provision is made for an appeal to the Secretary of State for Transport against the amount proposed. Once an advance payment has been made, it opens the way for the 'adoption' procedure under s.229. As long as one frontager (the owner of land abutting a street) to the street has paid under the Code then a majority of the frontagers (either in terms of numbers of owners or length of frontage owned) may request the Street Works Authority to secure the carrying out of appropriate street works and then declare the street to be a highway maintainable at public expense.

Two final points with regard to street works should be noted:

1 Where a private street is in need of repair to obviate danger to traffic, the Street Works Authority can require owners fronting onto the street to undertake specified repairs within a given time. If the owners fail to carry out the repairs, the Street Works Authority may proceed with the repairs and recover the cost proportionately from the owners. Where a notice is served requiring urgent repairs, it may be in the interests of the owners to ask the authority to undertake the work of properly making up the street under the Private Street Works Code, as it will, thereafter, be a highway maintainable at public expense (s.230).
2 Section 236 gives power to a Street Works Authority to resolve to bear all or any part of the costs of works under the Private Street Works Code. This then discharges or reduces the liability of the owner.

New Roads and Street Works Act 1991

Various utilities may need to dig up a street to install or maintain apparatus beneath the surface, e.g. gas, electricity, water and telecommunications. Authority to carry out work of this type is given by Parliament to utility companies, who are known as 'undertakers'. The process of doing such work is regulated by the New Roads and Street Works Act

1991. The Act applies, subject to certain restrictions, to both highways maintainable at the public expense and private roads.

By s.51 it is a criminal offence, punishable by a fine, for a person, e.g. a company, other than the street authority (usually the highway authority), amongst other things, to place apparatus in or break up or open up a street otherwise than in pursuance of a statutory right or a street works licence. An undertaker who proposes to carry out street works must give not less than seven days' notice (three months in the case of major works) to the street authority and to any other company having apparatus likely to be affected by the works. The requirement of notice does not apply to emergency works, which are defined in the Act as works whose execution is required to put an end to circumstances that are likely to cause danger to persons or property.

Various obligations are placed upon an undertaker:

- s.66 provides that street works shall be carried on and completed 'with all such dispatch as is reasonably practicable';
- regulations made under s.74A provide for an undertaker executing street works in a highway maintainable at the public expense to pay a charge, calculated by reference to the duration of the works, to the local highway authority;
- s.70 provides that it is the duty of the undertaker to reinstate the street once works are completed and to meet the relevant reinstatement performance standards and
- s.59 places a duty on the street authority to coordinate street works and a duty on undertakers to cooperate with that coordination.

To lessen the disruption caused by street works, s.58 provides that where 'it is proposed to carry out substantial road works in a highway, the street authority may by notice . . . restrict the execution of street works during the 12 months following the completion of those works'. This notice allows for coordination of street works, e.g. if a highway authority digs up a street, other undertakers may take advantage of the street being opened up to effect any necessary repairs to their apparatus.

In England, many of these obligations and notifications are now administered through the Street Manager digital platform (introduced in 2020 by the Department for Transport), which provides a real-time system for recording and managing street and road works under the Act.

Closing and obstruction of highways

It may be necessary, from time to time, to close or divert a highway, e.g. while work is in progress on it. Alternatively, the need for the highway may have disappeared and the authority may wish to close it permanently. The procedure for 'stopping up' is contained in the 1980 Act, s.116. Generally, this involves an application by the highway authority to the magistrates' court, following the steps set out in the Magistrates' Courts (Highways) Rules 1959. (Where the closure is sought by a person other than the highway authority, he or she must ask the highway authority to take the necessary steps. There is no appeal if the highway authority refuses to activate the procedure, but it must not unreasonably withhold consent.) The magistrates can make an appropriate order once the highway authority has followed through the necessary steps, which include notifying adjoining owners and occupiers, advertising the proposed closure or diversion in the press and fixing notices at each end of the highway.

It is also possible for a council to stop up or divert a footpath, bridleway or restricted bridleway if it is considered that it is expedient to do so. This can be either on the grounds that it is not needed for public use (stopping up) or that the path should be diverted in the interests of either the owner of the land or the public (ss.118 and 119). Both possibilities are, however, subject to strict controls that make the relevant stopping up or diversion orders difficult to obtain.

Under s.247 of the Town and Country Planning Act 1990, the Secretary of State has the power to stop up or divert a highway if he or she is satisfied that it is necessary to do so in order that a development for which planning permission has been granted can be carried out. A local planning authority has similar powers in respect of footpaths, bridleways and restricted bridleways (ss.257 and 258).

Any obstruction of, or interference with, the highway, unless authorised in some way, may constitute a civil and/or criminal offence. The Highways Act 1980, Part IX, contains numerous offences (e.g. in relation to straying animals and unlawful deposits) and also creates various duties and powers.

The highway authority is under a general duty to assert and protect the rights of the public to the use and enjoyment of any highway and to prevent as far as possible the obstruction of the highway. It may enforce these rules in legal proceedings. Some specific examples of interference with the highway that are relevant to construction work should be noted:

1 Section 133. If excavation or other work on land adjoining a street causes damage to a footway (pavement), the highway authority can recover the cost of repair from the landowner or from the person causing the damage.
2 Section 131. It is an offence, without lawful authority, to make an excavation on the highway or to deposit anything on a highway that will cause it damage. The penalty imposed in respect of such offences is a fine.
3 Section 137. It is an offence to willfully obstruct free passage along the highway.
4 Section 139. Builders' skips may not be deposited on a highway without permission from the highway authority. When granting such a permission, the authority may specify the size of the skip, conditions as to its siting, lighting and guarding, means for making it visible to oncoming traffic, disposal of its contents and its ultimate removal. The owner must make sure that the skip is:

   a properly lit;
   b clearly marked with his name and address;
   c removed as soon as practicable when full and
   d complying in all respects with any conditions attached to the permission.

Breaches of these rules are punishable by fines. The section specifically states that 'nothing in this section is to be taken as authorising the creation of a nuisance'. Where the skip is deposited on a highway, even when permission has been granted, s.140 provides that the highway authority or a police constable can require it to be removed or repositioned, or may themselves remove or reposition it and recover from the owner the expense of doing so.

5 Section 168 deals with building operations affecting public safety. If building operations in or near a street cause an accident which gives rise to a risk of injury to persons in the street, then the owner of the land on which the work is taking place or

any other person liable for causing such a risk (e.g. the building contractors) is liable to a fine.

6 Section 169 provides that where scaffolding will obstruct the highway, it can only be erected if a licence is obtained from the highway authority. There is a right of appeal against the refusal of a licence or against any conditions attached to it. If a licence is granted, then the scaffolding must be adequately lit during darkness and any necessary traffic signs must be erected and maintained.
7 Section 170 controls the mixing of mortar and cement on the highway in such a way that it may stick to the highway or enter drains and sewers and solidify. The section permits mixing in a receptacle or on a plate, but any breach of the rules is an offence carrying a fine.
8 Section 171 controls deposits of building materials and the making of excavations in streets. Such activities can be carried on with the consent of the highway authority, which may impose conditions, particularly if there is a danger of damage or obstruction to the apparatus of statutory undertakers (e.g. the gas, electricity and water boards). Appeal against refusal of permission, or the conditions imposed, is to the magistrates' court. The highway authority may levy a charge in the case of depositing building materials, erecting scaffolding under s.169 or mixing mortar and cement on the highway under s.170 if the works are not completed in a reasonable period.
9 Section 172 regulates when building works must be close boarded by a hoarding or fence. This section operates whenever a building is to be erected or taken down in a street, unless the local authority dispenses with the requirement. Where a hoarding is required, conditions are likely to be imposed regarding the creation and lighting of a footway and the maintenance of the hoarding. Under s.173, it is an offence to use such a hoarding that is not securely fixed to the satisfaction of the local authority.

It should be noted that some of the foregoing sections apply to highways, whereas others apply to streets. Many create statutory offences punishable by a fine. But the behaviour constituting the statutory offence may also cause damage or injury to an individual, who may be able to bring a civil claim for damages in the law of tort. A number of torts may be relevant, e.g. a person falling into an unfenced and unlit excavation on the highway may succeed in negligence, whereas a person injured by the collapse of an unsafe hoarding around building work in a street may succeed in negligence or public nuisance. An accumulation of rubbish in a builder's skip may constitute a nuisance. For the relevant rules in tort, see Chapter 9.

The tort of greatest significance in relation to activities on or near the highway is public nuisance. Claimants have succeeded in cases where crowds blocked the highway, where smoke obscured visibility in the highway, where a dangerous lamp overhanging the highway fell onto it, where there was unlit scaffolding, where the danger consisted of broken cellar flaps and where the danger arose from a heap of slates piled at the side of the road. In *National Highways Ltd v Person Unknown* (2023), the Court of Appeal reaffirmed the continuing relevance of public nuisance in cases involving dangers arising from activity on or near the highway.

Building and improvement lines

Part V of the 1980 Act contains a number of rules, powers and duties with regard to the improvement of highways. It may be obvious to the highway authority that, because of changing or increasing traffic volume, it will eventually become necessary to widen a particular highway. The need may not yet be urgent or funding may not be available, but it would be absurd if, in the meantime, the local authority continued to allow building work to take place, thereby making ultimate widening of the highway more difficult and costly.

To prevent this situation occurring, s.73 of the Act provides that the highway authority may prescribe an improvement line for the street in question. This is the line to which it is ultimately intended to widen the street. Naturally, such a decision can affect the value of adjacent properties and the development which can be undertaken in relation to them. It would not be surprising, therefore, if the owners wished to object to the prescription of the improvement line. This can be done by any aggrieved person appealing to the Crown Court. If the appeal fails, it should be noted that s.73(9) provides for compensation to be paid to persons whose property is thus injuriously affected.

Section 74 of the Act provides for the highway authority to prescribe a building line. Once such a building line is operative, no new building shall be erected beyond the building line without the consent of the highway authority. Rather like the rules in s.73, provision is made to compensate owners whose property is injuriously affected, but unlike s.73, there is no provision in s.74 for appealing against the imposition of the building line. Schedule 9 merely provides that the highway authority must consider objections received. Any challenge would therefore be limited to judicial review on public law grounds.

Once either type of line is properly prescribed, it must be shown on a duly authenticated plan that is available for inspection by interested parties. Both ss.73 and 74 provide means whereby these lines can be removed if they are subsequently seen to be unnecessary.

Discussion questions

1 What is the legal distinction between a footpath, a bridleway and a byway open to all traffic (BOAT)? Why might this classification matter in practice?
2 To what extent does s.41 of the Highways Act 1980 impose a realistic duty on highway authorities to maintain highways in a safe condition? Refer to recent case law.
3 In what circumstances might a private landowner successfully prevent the creation of a public right of way across their land under s.31 of the Highways Act 1980?
4 Consider a new housing development with unadopted roads. What are the respective roles of the Advance Payments Code and s.38 agreements in ensuring long-term road maintenance?
5 Discuss the types of conduct by contractors that may give rise to criminal or civil liability when working on or near the highway. How do public nuisance and statutory offences interact in this context?

15 Building regulations

Jennifer Charlson

Purpose

Under s.1 of the Building Act 1984 (the Act), the Secretary of State may make regulations with respect to the design and construction of buildings, the demolition of buildings and services, fittings and equipment provided in or in connection with buildings to:

1. secure the health, safety, welfare and convenience of persons in or about buildings and of others who may be affected by buildings;
2. further the conservation of fuel and power;
3. prevent waste, undue consumption, misuse or contamination of water;
4. further the protection or enhancement of the environment;
5. facilitate sustainable development or
6. further the prevention or detection of crime.

In accordance with the Act, the Secretary of State consulted with the Building Regulations Advisory Committee and then made the Building Regulations 2010 (the Regulations). The Regulations require regulated building work to be performed to prescribe minimum standards.

Scope

Building work must be carried out so that it complies with the applicable requirements set out in Parts A to T of Schedule 1 of the Regulations. The Act gives the Secretary of State power to approve and issue documents (Approved Documents). The Approved Documents contain practical guidance on how to meet the requirements regarding structure (Part A), fire safety (Part B), site preparation and resistance to moisture (Part C), toxic substances (Part D), resistance to the passage of sound (Part E), ventilation (Part F), sanitation, hot water safety and water efficiency (Part G), drainage and waste disposal (Part H), combustion appliances and fuel storage systems (Part J), protection from falling, collision and impact (Part K), conservation of fuel and power (Part L), access to and use of buildings (Part M), overheating (Part O), electrical safety (Part P), security in dwellings (Part Q), infrastructure for electronic communication (Part R), infrastructure for charging electric vehicles (Part S) and toilet accommodation (Part T). However, failure to comply with an Approved Document does not mean that there has been a breach of the Regulations, but does provide evidence of non-compliance.

DOI: 10.1201/9781003657101-15

Registered Building Inspectors

Registered Building Inspectors (RBIs) carry out regulated building control activities, which are:

- Assessing plans
- Inspections

RBIs work for building control bodies, either as employees or contractors. Building control bodies are:

- Local authorities
- Registered building control approvers (RBCAs)

The Building (Registered Building Control Approvers etc.) (England) Regulations 2024 are applicable.

The Building Safety Regulator (BSR) regulates RBIs and building control bodies.

Process and enforcement

It is the responsibility of those commissioning and carrying out design and building work to ensure that their work complies with the Regulations. For work to which the Regulations apply, builders or developers must usually seek approval from a building control authority. The building control approval process for buildings (which are not classed as higher-risk buildings) follows three main steps:

1. Before starting the work, the applicant must notify the local authority of the work by issuing a 'building notice' or, if they choose to use an RBCA, by issuing a joint 'initial notice'. The notice must include a description of the proposed work.
2. During construction, the local authority or RBCA will inspect the work periodically to assess whether it complies with the requirements set out in the Regulations.
3. On completion of the work, the local authority will issue a 'completion certificate' if it is satisfied 'after taking all reasonable steps' that the work complies with relevant requirements. The RBCA will issue a 'final certificate' if they are satisfied that work was carried out in compliance with the Regulations.

Where building work has been carried out without approval, but in line with building regulations in place at the time of construction, a building owner may be able to obtain a 'regularisation certificate'.

Local authorities have enforcement powers to require developers to fix non-compliant work and can also pursue prosecution against people who violate building regulations. RBCAs do not have enforcement powers.

Reform

A fire which destroyed Grenfell Tower in June 2017 resulted in the death of 72 people. The Government's response included appointing Dame Judith Hackitt to lead an independent review and setting up a public inquiry to examine the circumstances leading up to and surrounding the fire at Grenfell Tower.

Building a Safer Future – Independent Review of Building Regulations and Fire Safety ('the Hackitt Review'), published in May 2018, set out principles for a new regulatory framework. The report recommended a very clear model of risk ownership with clear responsibilities for the Client, Designer, Contractor and Owner to demonstrate the delivery and maintenance of safe buildings, overseen and held to account by a new Joint Competent Authority. The review also recommended a package of regulations and guidance which is simpler to navigate but genuinely reflects the level of complexity of the building work.

The Phase 1 Report of the public inquiry, published in October 2019, focused on the events that occurred during the night of the fire. Phase 2, published in September 2024, examined those aspects of the design, construction and management of the building which were primarily responsible for the disaster. The UK government has agreed to implement all 58 Phase 1 Report recommendations. The Building Safety Act 2022 had already introduced reforms to give residents and homeowners more rights, powers and protections.

Also in October 2019, the Government launched a consultation on changes to Parts L (conservation of fuel and power) and F (ventilation) of the Regulations applicable to new dwellings to meet the Future Homes Standard due to be introduced by 2025. Reform of the Regulations, therefore, seems probable.

Discussion questions

1 Are the Building Regulations' purposes due for a review and update?
2 Evaluate the progress of the identified reforms.

16 Environmental law

Jennifer Charlson

This chapter introduces environmental law and sustainable development as applicable to the built environment. The contaminated land section is relevant to brownfield but not greenfield sites. Other environmental law concerns, which may be pertinent to sites with or without previous development, including waste management, water pollution and environmental impact assessments, are outlined.

Sustainable development

It can be argued that environmental law in a construction context aims to deliver sustainable development. The judgment in *R (Spurrier) v Secretary of State for Transport* (2019) QBD stated that: 'sustainable development' is an uncontroversial concept and referenced the United Nations 1987 Brundtland Report: 'At a very high level, the objective of sustainable development can be summarised as meeting the needs of the present without compromising the ability of future generations to meet their own needs'.

Contaminated land

This section explains the contaminated land legal regime illustrated by case law and remediation declarations. The aim of Part 2A of the Environmental Protection Act 1990 (EPA 1990) is to identify and remediate contaminated land in England which poses an unacceptable risk to human health or the environment and which is not being remediated voluntarily.

Contaminated land is defined in s.78A(2) as:

> Any land which appears to the local authority in whose area it is situated to be in such a condition, by reason of substances in, on or under land, that:
>
> a Significant harm is being caused or there is a significant possibility of such harm being caused; or
> b Pollution of controlled waters is being, or is likely to be, caused.

These are subjective criteria. It is also important to note that only some, but not all, brownfield land will fail to exceed these standards and be identified as contaminated land.

The person primarily liable for the remediation costs is the 'polluter' (Class A person), with the owner or occupier (Class B person) secondarily liable. A challenge is that the 'polluter' (Class A person) may be untraceable or insolvent, particularly if the pollution occurred many decades ago.

DOI: 10.1201/9781003657101-16

In accordance with the accompanying statutory guidance, local authorities are required to formulate and then implement an inspection strategy, but no deadline is imposed. Therefore, local authorities have made limited progress in inspecting their land, resulting in a lack of information about the extent of contaminated land in England. Nevertheless, Part 2A imposes a duty on a local authority to serve a remediation notice on an appropriate person if the authority decides the land is contaminated. A developer who purchases brownfield land is at risk of the local authority subsequently identifying them as a Class B person and serving a remediation notice. The potential liability of the recipient of a remediation notice is uncapped and failure to comply could result in a criminal conviction.

Developers have indeed been served remediation notices, as illustrated by examples concerning Circular Facilities (London) Ltd and Crest Nicholson Residential Ltd.

Circular Facilities (London) Ltd (CFL) constructed a residential estate on a site which was detected by Sevenoaks District Council (SDC) as causing significant risk of harm to the householders as a consequence of excessive levels of methane and carbon dioxide seeping from pits filled with now decomposing organic material. In 2002, SDC served a remediation notice on CFL as a Class A appropriate person. In *Circular Facilities (London) Ltd v Sevenoaks District Council* (2005) Admin, CFL appealed against the remediation notice. The case was confidentially settled out of court and the notice was revoked.

St Leonard's Court is a residential estate near St Albans, Hertfordshire. The location was previously a chemical manufacturer, which was purchased by a developer. Bromate and bromide were recognised by St Albans District Council to have leached into the soil, contaminating a water source. The Environment Agency served remediation notices on both the former chemical factory operator and the developer. In the appeal against the chemical operator's remediation notice, *R (Redland Minerals Ltd) v Secretary of State for Environment, Food and Rural Affairs* (2010) Admin, the judge concluded:

> not only that a manufacturer of chemical had caused chemical contaminants to enter soil and groundwater, but that a developer that subsequently removed buildings and hardstandings from the site during its re-development for housing, had caused chemicals to enter the groundwater due to rain falling directly on the ground and accelerating the entry of contaminants to lower levels in the ground.

The connected case was *R (Crest Nicholson Residential Ltd) v Secretary of State for Environment, Food and Rural Affairs* (2010) Admin. The judge found both parties responsible and apportioned liability between them.

The following council water-linked remediation declarations have also been made:

- Warrington Borough Council, in 2008, issued a remediation declaration for the pollution of shallow groundwater and a nearby brook (controlled waters).
- Congleton Borough Council, in 2011, decided that Malkins Bank Golf Course was contaminated land. However, the Council concurrently issued a remediation declaration stating that it would be disproportionate to remediate the area because groundwater pollution was having a limited effect on the nearby surface waters.
- Cornwall County Council, in 2011, issued a remediation declaration regarding a former gasworks site where pollutants were entering the Fal Estuary through the harbour wall.

However, no total is available for the number of remediation declarations made.

Environmental damage

The Environmental Damage (Prevention and Remediation) (England) Regulations 2015 apply to the most serious cases of environmental damage. They relate to the prevention and remediation of damage to species, habitats, sites of special scientific interest, surface or groundwater, and land.

If the regulator considers that environmental damage has occurred, it can serve a remediation notice on the responsible operator, setting out the measures that must be taken. The principal enforcement authority is the relevant local authority.

Failure to comply with a remediation notice, without reasonable excuse, is a criminal offence punishable by a fine. The regulator can carry out the remediation itself and recover costs from the relevant parties.

Waste management

Processes on brownfield sites could be interpreted as waste treatment and have been subject to judicial scrutiny. In *Kent County Council v Queenborough Rolling Mill Co Ltd* (1990) QBD, an area subject to subsidence was filled with a mixture of ballast, china, china clay and broken pottery, which as deemed to be waste when it was moved from its original location and its character was not subsequently changed either by its being sorted into different categories or by its usefulness for infill purposes. By contrast, in *Cheshire County Council v Armstrong's Transport (Wigan) Ltd* (1995) CA, building site rubble, which was crushed and returned to the original site to assist in rebuilding works, was found not to be waste because it was not discarded but merely removed, processed and returned. Therefore, construction processes may require waste treatment permits.

The Environmental Permitting (England and Wales) Regulations 2016 (EPR 2016) mandate certain operators to have environmental permits. In England and Wales, waste treatment facilities require permits from the Environment Agency or Natural Resources Wales.

In the Court of Appeal case *R (Tarmac Aggregates Ltd) v Secretary of State for Environment, Food and Rural Affairs* (2016), the proposal to use recovered material as backfill to reinstate a public footpath was considered. The Court decided that the main purpose of the operation was to ecologically improve the site. Therefore, the Court held that the use of materials was waste recovery, rather than waste disposal, and so a standard environmental permit should be issued.

Mountpace Ltd had commissioned a developer to renovate a property. The developer subcontracted the removal and disposal of waste from the site to a third-party waste contractor who illegally dumped (fly-tipped) the waste. In *Mountpace Ltd v London Borough of Haringey* (2012)Admin, the judge ruled that Mountpace Ltd was in breach of the law because of failure to exercise control over waste from its premises.

Water pollution

Regulation 12 of the EPR 2016 also states that a person cannot cause or knowingly permit a water discharge activity except under and to the extent authorised by an environmental permit. Contravention is an offence under Regulation 38.

Miller Homes Ltd was fined £100,000 in 2016 over a pollution incident concerning a housing development in Huddersfield. The developer admitted one environmental offence of an unauthorised discharge of water, containing silt and sediment, from the construction site into a nearby watercourse. In 2017, Harron Homes Ltd was also fined £120,000 for illegally polluting a watercourse from a construction site.

Taylor Wimpey was fined £480,000 in 2023 after failing to implement appropriate measures to prevent multiple pollution incidents. The construction company was found guilty of a charge relating to breaches in Environmental Permitting Regulations. Inspections carried out by officers from Natural Resources Wales revealed that the pollutant on each occasion was caused by water run-off from the site, which had been contaminated with silt.

Construction companies should therefore consider the potential environmental consequences of developments they undertake and must adhere to environmental permitting rules.

Thames Water Utilities Ltd was fined over £20 million in 2017 for a series of serious pollution incidents of the River Thames. His Honour Judge Sheridan condemned the 'disgraceful conduct' of Thames Water, the consequences of which he said were 'entirely foreseeable and preventable'.

Between 2015 and 2024, the Environment Agency successfully concluded 63 prosecutions against water and sewerage companies for pollution offences, securing fines of over £151 million.

In 2025, Ofwat fined Thames Water nearly £123 million, which included a £104.5 million penalty for breaches of rules relating to its wastewater operations. This is the largest penalty Ofwat has ever issued.

Nutrient pollution in watercourses damages freshwater habitats and disrupts natural processes, harming plants, wildlife and ecosystems. Nutrient neutrality requires that new housing developments in certain areas should not add more 'nutrient pollution' to the water catchment.

Environmental Impact Assessment

The Town and Country Planning (Environmental Impact Assessment) Regulations 2017 mandate environmental impact assessments (EIAs) for some developments. The aim of EIAs is to protect the environment by ensuring that a local planning authority, when deciding whether to grant planning permission for a project, which is likely to have significant effects on the environment, does so in the full knowledge of the likely significant effects. An EIA is a procedure which follows five main steps: screening, scoping, preparing an environmental statement, making a planning application and consultation and finally decision-making. Therefore, development may require expensive EIAs.

In *R (Bateman) v South Cambridgeshire District Council* (2011), the Court of Appeal gave important guidance on the requirement that clear reasons be given for a local authority's screening decision that an environmental impact assessment is not required. In *R (Champion) v North Norfolk District Council & another* (2015), the Supreme Court decided that initial unresolved uncertainties over the mitigation measures mandated an EIA.

Environment Act 2021

The Environment Act 2021 addresses environmental governance, waste and resource efficiency, air quality, water, and nature and biodiversity. Under the Act, legally binding environmental improvement targets were set in 2022, with an Environmental Improvement Plan for England published the following year.

Discussion questions

1. Are the most appropriate persons liable, under Part 2A of the Environmental Protection Act 1990, for the costs to remediate contaminated land?
2. Does case law encourage the re-use of waste?
3. Are construction companies sufficiently diligent in preventing water pollution from their sites?
4. Are the requirements for Environmental Impact Assessments too complex and consequently too expensive?

References

Bagehot, W (1867) *The British Constitution*. Available in Oxford World's Classics Series, pub. 2009. Oxford: Oxford University Press.

Brundtland, GH (1987) *Our Common Future*. Report of the World Commission on Environment and Development. Geneva: United Nations.

Civil Procedure Rules (CPR) www.justice.gov.uk/courts/procedure-rules/civil

Department of Trade & Industry (1998) *Fairness at Work*. Cm 3968. London: HM Stationery Office.

de Smith, A (1971) *Constitutional and Administrative Law*. London: Penguin.

Great Britain (1972) *Safety and Health at Work: Report of the Committee, 1970–72*. London: HM Stationery Office. [Chaired by Lord Robens]

Great Britain (1978) *Royal Commission on Civil Liability and Compensation for Personal Injury*. London: HM Stationery Office. Cmnd. 7054. [Chaired by Lord Pearson]

Great Britain (1990) *The Public Inquiry into the Pipe Alpha Disaster*. London: HM Stationery Office. Cm 1310. [Chaired by Lord Cullen]

Great Britain (1999) *Royal Commission on the Reform of the House of Lords*. London: The Stationery Office. Cm 4183. [Chaired by Lord Wakeham]

Great Britain (2009a) *Final Report of the Secretary of State for Work and Pensions: One Death Is Too Many. Inquiry into the Underlying Causes of Construction Fatal Accidents*. London: HM Stationery Office. Cmn 7657. [Chaired by Rita Donaghy]

Great Britain (2009b) *Responding to Human Rights Judgments*. Government Response to the Joint Committee on Human Rights. London: HM Stationery Office. Cm 7892.

Great Britain (2012) *National Planning Policy Framework*. London: HM Stationery Office.

Great Britain (2018) *Building a Safer Future – Independent Review of Building Regulations and Fire Safety*. London: HM Stationery Office. Cm 9607. [Chaired by Dame Judith Hackitt]

Great Britain (2019) *Regulation of Property Agents Working Group Final Report*. London: HM Stationery Office. [Chaired by Lord Best]

Hansard (2 July 2025) *HC Deb*. vol. 770, col. 277.

HSE (2013) *Managing and Working with Asbestos, Control of Asbestos Regulations 2012, Approved Code of Practice*, 2nd ed. London: HSE Books.

Jennings, IS (1959) *The Law and the Constitution*, 5th ed. London: Hodder & Stoughton.

Latham, M (1994) *Constructing the Team: The Final Report of the Joint Government/Industry Review of Procurement and Contractual Arrangements in the UK Construction Industry*. London: Her Majesty's Stationery Office.

Law Commission (2011) *Making Land Work: Easements, Covenants and Profits à Prendre*. Law Com No 327.

Löfstedt, R (2011) *Reclaiming Health and Safety for All: An Independent Review of Health and Safety Legislation*. Cm 8219. London: HM Stationery Office.

Natzler, D et al (ed) (2019) *Erskine May: Parliamentary Practice*, 25th ed. London: LexisNexis.

NEC (2019) *National Construction Contracts and Law Report 2018*. Newcastle-upon-Tyne: NBS.

OSCOLA (2012) *The Oxford University Standard for the Citation of Legal Authorities*. https://www.law.ox.ac.uk/sites/files/oxlaw/oscola_4th_edn_hart_2012.pdf

RIBA (2022) *RIBA Construction Contracts and Law Report 2022*. London: Royal Institute of British Architects.

Sentencing Council (2016) *Sentencing Guidelines: Corporate Manslaughter*. www.sentencingcouncil.org.uk/offences/crown-court/item/corporate-manslaughter/

Taylor, M (2017) *Good Work: The Taylor Review of Modern Working Practices*. https://assets.publishing.service.gov.uk/government/uploads/system/uploads/attachment_data/file/627671/good-work-taylor-review-modern-working-practices-rg.pdf

Welsh Government (2018) *Planning Policy Wales*, 10th ed. https://gov.wales/sites/default/files/publications/2018-12/planning-policy-wales-edition–10.pdf

Index

For Product Safety Concerns and Information please contact our EU representative GPSR@taylorandfrancis.com
Taylor & Francis Verlag GmbH, Kaufingerstraße 24, 80331 München, Germany

www.ingramcontent.com/pod-product-compliance
Ingram Content Group UK Ltd.
Pitfield, Milton Keynes, MK11 3LW, UK
UKHW050648150226
468019UK00004B/8

* 9 7 8 1 0 4 1 1 0 8 4 9 8 *